原版影印说明

1.《聚合物百科词典》(5册)是 Springer Reference *Encyclopedic Dictionary of Polymers*(2nd Edition)的影印版。为使用方便,由原版2卷改为5册:

第1册 收录 A–C 开头的词组;

第2册 收录 D–I 开头的词组;

第3册 收录 J–Q 开头的词组;

第4册 收录 R–Z 开头的词组;

第5册 为原书的附录部分及参考文献。

2. 缩写及符号、数学符号、字母对照表、元素符号等查阅说明各册均完整给出。

由 Jan W. Gooch 主编的《聚合物百科词典》是关于高分子科学与工程领域的参考书,2007年出版第一版,2011年再版。本书收录了7 500多个高分子材料方面的术语,涉及高分子材料的各个方面,如粘合剂、涂料、油墨、弹性体、塑料、纤维等,还包括生物化学和微生物学方面的术语,以及与新材料、新工艺相关的术语;并且不仅包括其物理、电子和磁学性能方面的术语,还增加了数据处理的统计和数值分析以及实验设计方面的术语。每个词条方便查找,并给出了简洁的定义,以及相互参照的相关术语。为了说明得更清晰,全书给出1 160个图、73个表。有的词条还给出方程式、化学结构等。

材料科学与工程图书工作室

联系电话 0451-86412421
　　　　　0451-86414559
邮　　箱 yh_bj@aliyun.com
　　　　　xuyaying81823@gmail.com
　　　　　zhxh6414559@aliyun.com

Springer 词典精选原版系列

聚合物百科词典

Jan W. Gooch

Encyclopedic Dictionary of Polymers

2nd Edition

VOLUME 3
J-Q

哈尔滨工业大学出版社
HARBIN INSTITUTE OF TECHNOLOGY PRESS

黑版贸审字08-2014-010号

Reprint from English language edition:
Encyclopedic Dictionary of Polymers
by Jan W.Gooch
Copyright © 2011 Springer New York
Springer New York is a part of Springer Science+Business Media
All Rights Reserved

This reprint has been authorized by Springer Science & Business Media for distribution in China Mainland only and not for export therefrom.

图书在版编目（CIP）数据

聚合物百科词典. 3, J～Q：英文/（美）古驰（Gooch, J. W.）主编. —哈尔滨：哈尔滨工业大学出版社，2014.3

（Springer词典精选原版系列）

ISBN 978-7-5603-4444-7

Ⅰ.①聚… Ⅱ.①古… Ⅲ.①聚合物–词典–英文 Ⅳ.①O63-61

中国版本图书馆CIP数据核字（2013）第292191号

责任编辑	张秀华　杨　桦　许雅莹
出版发行	哈尔滨工业大学出版社
社　　址	哈尔滨市南岗区复华四道街10号 邮编150006
传　　真	0451-86414749
网　　址	http://hitpress.hit.edu.cn
印　　刷	哈尔滨市石桥印务有限公司
开　　本	787mm×1092mm 1/16 印张 14.25
版　　次	2014年3月第1版 2014年3月第1次印刷
书　　号	ISBN 978-7-5603-4444-7
定　　价	108.00元

（如因印刷质量问题影响阅读，我社负责调换）

Acknowledgements

The editor wishes to express his gratitude to all individuals who made available their time and resources for the preparation of this book: James W. Larsen (Georgia Institute of Technology), for his innovations, scientific knowledge and computer programming expertise that were invaluable for the preparation of the Interactive Polymer Technology Programs that accompany this book; Judith Wiesman (graphics artist), for the many graphical presentations that assist the reader for interpreting the many complex entries in this publication; Kenneth Howell (Springer, New York), for his continued support for polymer science and engineering publications; and Daniel Quinones and Lydia Mueller (Springer, Heidelberg) for supporting the printed book and making available the electronic version and accompanying electronic interactive programs that are important to the scientific and engineering readers.

Preface

The second edition of Encyclopedic Dictionary of Polymers provides 40% more entries and information for the reader. A Polymers Properties section has been added to provide quick reference for thermal properties, crystallinity, density, solubility parameters, infrared and nuclear magnetic spectra. Interactive Polymer Technology is available in the electronic version, and provides templates for the user to insert values and instantly calculate unknowns for equations and hundreds of other polymer science and engineering relationships. The editor offers scientists, engineers, academia and others interested in adhesives, coatings, elastomers, inks, plastics and textiles a valuable communication tool within this book. In addition, the more recent innovations and biocompatible polymers and adhesives products have necessitated inclusion into any lexicon that addresses polymeric materials. Communication among scientific and engineering personnel has always been of critical importance, and as in any technical field, the terms and descriptions of materials and processes lag the availability of a manual or handbook that would benefit individuals working and studying in scientific and engineering disciplines. There is often a challenge when conveying an idea from one individual to another due to its complexity, and sometimes even the pronunciation of a word is different not only in different countries, but in industries. Colloquialisms and trivial terms that find their way into technical language for materials and products tend to create a communications fog, thus unacceptable in today's global markets and technical communities.

The editor wishes to make a distinction between this book and traditional dictionaries, which provide a word and definition. The present book provides for each term a complete expression, chemical structures and mathematic expression where applicable, phonetic pronunciation, etymology, translations into German, French and Spanish, and related figures if appropriate. This is a complete book of terminology never before attempted or published.

The information for each chemical entry is given as it is relevant to polymeric materials. Individual chemical species (e.g., ethanol) were taken from he *CRC Handbook of Chemistry and Physics*, 2004 Version, the Merck Index and other reference materials. The reader may refer to these references for additional physical properties and written chemical formulae. Extensive use was made of ChemDraw®, CambridgeSoft Corporation, for naming and drawing chemical structures (conversion of structure to name and vice versa) which are included with each chemical entry where possible. Special attention was given to the IUPAC name that is often given with the common name for the convenience of the reader.

The editor assembled notes over a combined career in the chemical industries and academic institutions regarding technical communication among numerous colleagues and helpful acquaintances concerning expressions and associated anomalies. Presently, multiple methods of nomenclature are employed to describe identical chemical compounds by common and IUPAC names (eg. acetone and 2-propanone) because the old systems (19th century European and trivial) methods of nomenclature exists with the modern International Union of Pure and Applied Chemistry, and the conflicts between them are not likely to relent in the near future including the weights and measures systems because some nations are reluctant to convert from English to metric and, and more recently, the International Systems of Units (SI). Conversion tables for converting other systems to the SI units are included in this book for this purpose. In addition, there are always the differences in verbal pronunciation, but the reasons not acceptable to prevent cogent communication between people sharing common interests.

In consideration of the many challenges confronting the reader who must economize time investment, the structure of this book is optimized with regard the convenience of the reader as follows:

- Comprehensive table of contents
- Abbreviations and symbols
- Mathematics signs
- English, Greek, Latin and Russian alphabets
- Pronunciation/phonetic symbols
- Main body of terms with entry term in English, French German and Italian
- Conversion factors

- Microbiology nomenclature and terminology
- References

The editor acknowledges the utilization of many international sources of information including journals, books, dictionaries, communications, and conversations with people experienced in materials, polymer science and engineering. A comprehensive reference section contains all of the sources of information used in this publication. Pronunciation, etymological, cross-reference and related information is presented in the style of the 11th Edition of the Meriam-Webster Dictionary, where known, for each term. The spelling for each term is presented in German, French, and Spanish where translation is possible. Each term in this book includes the following useful information:

- Spelling (in **bold** face) of each term and alternative spellings where more than one derivation is commonly used
- Phonetic spelling \-\ using internationally published phonetic symbols, and this is the first book that includes phonetic pronunciation information missing in technical dictionaries that allows the reader to pronounce the term
- Parts of speech in English following each phonetic spelling, eg. *n.*, *adj.*
- Cross-references in CAPITALS letters
- Also called *example* in italics
- Etymological information [-] for old and new terms that provides the reader the national origins of terms including root words, prefixes and suffixes; historical information is critical to the appreciation of a term and its true meaning
- French, German, Italian and Spanish spellings of the term { - }
- A comprehensive explanation of the term
- Mathematical expressions where applicable
- Figures and tables where applicable
- A comprehensive reference section is included for further research

References are included for individual entries where a publication(s) is directly attributable to a definition or description. Not all of the references listed in the Reference section are directly attributable to entries, but they were reviewed for information and listed for the reader's information. Published dictionaries and glossaries of materials were very helpful for collecting information in the many diverse and smaller technologies of the huge field of polymers. The editor is grateful that so much work has been done by other people interested in polymers.

The editor has attempted to utilize all relevant methods to convey the meaning of terms to the reader, because a term often requires more information than a standard entry in a textbook dictionary, so this book is dedicated to a complete expression. Terminology and correct pronunciation of technical terms is continuously evolving in scientific and industrial fields and too often undocumented or published, and therefore, not shared with others sometimes leading to misunderstandings. Engineering and scientific terms describe a material, procedure, test, theory or process, and communication between technical people must involve similar jargon or much will be lost in the translation as often has been the editor's experience. The editor has made an attempt to provide the reader who has an interested in the industries that have evolved from adhesives, coatings, inks, elastomers, plastics and textiles with the proper terminology to communicate with other parties whether or not directly involved in the industries. This publication is a single volume in the form of a desk-handbook that is hoped will be an invaluable tool for communicating in the spoken and written media.

Physics, electronic and magnetic terms because they are related to materials and processes (e.g., *ampere*).

Biomolecular materials and processes have in the recent decade overlapped with polymer science and engineering. Advancements in polymeric materials research for biomolecular and medical applications are rapidly becoming commercialized, examples include biocompatible adhesives for sutureless tissue bonding, liquid dressings for wounds and many other materials used for *in vitro* and *in vivo* medical applications. To keep pace with these advancements, the editor has included useful terms in the main body that are commonly used in the material sciences for these new industries.

A microbiology section has been included to assist the reader in becoming familiar with the proper nomenclature of bacteria, fungi, mildew, and yeasts – organisms that affect materials and processes because they are ubiquitous in our environment. Corrosion of materials by microorganisms is commonplace, and identification of a specific organism is critical to prevent its occurrence. Engineers and materials scientists will appreciate the extensive sections on different types of microorganisms together with a section dedicated to microbiology terminology that is useful for communicating in the jargon of biologists instead of referring to all organisms as "bugs."

New materials and processes, and therefore new terms, are constantly evolving with research, development and global commercialization. The editor will periodically update this publication for the convenience of the reader.

Statistics, numerical analysis other data processing and experimental design terms are addressed as individual terms and as a separate section in the appendix, but only as probability and statistics relate to polymer technology and not the broad field of this mathematical science. The interactive equations are listed in the Statistics section of the Interactive Polymer Technology program.

Interactive Polymer Technology Programs

Along with this book we are happy to provide a collection of unique and useful tools and interactive programs along with this Springer Reference. You will find short descriptions of the different functions below. Please download the software at the following website: http://extras.springer.com/2011/978-1-4419-6247-8

Please note that the file is more than 200 MB. Download the ZIP file and unzip it. It is strongly recommended to read the **ReadMe.txt** before installing. The software is started by opening the file InPolyTech.pdf and following the instructions. Detailed instructions can be found under 'Help Instructions'.

The software consists of 15 programs and tools that are briefly described in the appendix.

Abbreviations and Symbols

Abbreviations	Symbols
An	absorption (formerly extinction) (= log t_i^{-1})
A	Area
A	surface
A	Helmholtz energy ($A = U - TS$)
A	preexponential constant [in $k = A \exp(-E^{\ddagger}/RT)$]
A_2	second virial coefficient
a	exponent in the property/molecular weight relationship ($E^{\ddagger} = KM^a$); always with an index, e.g., a_η, a_s, etc.
a	linear absorption coefficient, $a = l^{-1}$
absolute	abs
acre	spell out
acre-foot	acre-ft
air horsepower	air hp
alternating-current (as adjective)	a-c
A^m	molar Helmholtz energy
American Society for Testing and Materials	ASTM
amount of a substance (mole)	n
ampere	A or amp
ampere-hour	amp-hr
amplitude, an elliptic function	am.
angle	β
angle, especially angle of rotation in optical activity	α
Angstrom unit	Å
antilogarithm	antilog
a_o	constant in the Moffit–Yang equation
Area	A
Atactic	at
atomic weight	at. wt
Association	Assn.
atmosphere	atm

Abbreviations	Symbols
average	avg
Avogadro number	N_L
avoirdupois	avdp
azimuth	az or α
barometer	bar.
barrel	bbl
Baumé	Bé
b_o	constant in the Mofit–Yang equation
board fee (feet board measure)	fbm
boiler pressure	spell out
boiling point	bp
Boltzmann constant	k
brake horsepower	bhp
brake horsepower-hour	bhp-hr
Brinell hardness number	Bhn
British Standards Institute	BSI
British thermal unit[1]	Btu or B
bushel	bu
C	heat capacity
c	specific heat capacity (formerly; specific heat); c_p = specific isobaric heat capacity, c_v = specific isochore heat capacity
c	"weight" concentration (= weight of solute divided by volume of solvent); IUPAC suggests the symbol ρ for this quantity, which could lead to confusion with the same IUPAC symbol for density
c	speed of light in a vacuum
c	speed of sound
calorie	cal
candle	c
candle-hour	c-hr
candlepower	cp
ceiling temperature of polymerization, °C	T_c

Abbreviations	Symbols
cent	c or ¢
center to center	c to c
centigram	cg
centiliter	cl
centimeter or centimeter	cm
centimeter-gram-second (system)	cgs
centipoise	cP
centistokes	cSt
characteristic temperature	Θ
chemical	chem.
chemical potential	μ
chemical shift	δ
chemically pure	cp
circa, about, approximate	ca.
circular	cir
circular mils	cir mils
cis-tactic	ct
C^m	molar heat capacity
coefficient	coef
cologarithm	colog
compare	cf.
concentrate	conc
conductivity	cond, λ
constant	const
continental housepower	cont hp
cord	cd
cosecant	csc
cosine	cos
cosine of the amplitude, an elliptic function	cn
cost, insurance, and freight	cif
cotangent	cot
coulomb	spell out
counter electromotive force	cemf
C_{tr}	transfer constant ($C_{tr} = k_{tr}/k_p$)
cubic	cu
cubic centimeter (liquid, meaning milliliter. ml)	cu, cm, cm^3
cubic centimeter	cm^3 cubic expansion coefficient $\propto$
cubic foot	cu ft
cubic feet per minute	cfm
cubic feet per second	cfs

Abbreviations	Symbols
cubic inch	cu in.
cubic meter	cu m or m^3
cubic micron	cu μ or cu mu or μ^3
cubic millimeter	cu mm or mm^3
cubic yard	cu yd
current density	spell out
cycles per second	spell out or c
cylinder	cyl
D	diffusion coefficient
D_{rot}	rotational diffusion coefficient
day	spell out
decibel	db
decigram	d.g.
decomposition, °C	T_{dc}
degree	deg or °
degree Celsius	°C
degree centigrade	C
degree Fahrenheit	F or °
degree Kelvin	K or none
degree of crystallinity	$\propto$
degree of polymerization	X
degree Réaumur	R
delta amplitude, an elliptic function	dn
depolymerization temperature	T_{dp}
density	ρ
diameter	diam
Dictionary of Architecture and Construction	DAC
diffusion coefficient	D
dipole moment	p
direct-current (as adjective)	d-c
dollar	$
dozen	doz
dram	dr
dynamic viscosity	η
E	energy (E_k = kinetic energy, E_p = potential energy, $E^{\ddagger}$ = energy of activation)
E	electronegativity
E	modulus of elasticity, Young's modulus ($E = \sigma_{ii}/\varepsilon_{ii}$)
E	general property

Abbreviations	Symbols
E	electrical field strength
e	elementary charge
e	parameter in the Q-e copolymerize-tion theory
e	cohesive energy density (always with an index)
edition	Ed.
Editor, edited	ed.
efficiency	eff
electric	elec
electric polarizability of a molecule	α
electrical current strength	I
electrical potential	V
electrical resistance	R or X
electromotive force	emf
electronegativity	E
elevation	el
energy	E
enthalpy	H
entropy	S
equation	eq
equivalent weight	equiv wt
et alii (and others)	et al.
et cetera	etc.
excluded volume	u
excluded volume cluster integral	β
exempli gratia (for example)	e.g.
expansion coefficient	α
external	ext
F	force
f	fraction (excluding molar fraction, mass fraction, volume fraction)
f	molecular coefficient of friction (e.g., f_s, f_D, f_{rot})
f	functionality
farad	spell out or f
Federal	Fed.
feet board measure (board feet)	fbm
feet per minute	fpm
feet per second	fps
flash point	flp

Abbreviations	Symbols
fluid	fl
foot	ft
foot-candle	ft-c
foot-Lambert	ft-L
foot-pound	ft-lb
foot-pound-second (system)	fps
foot-second (see cubic feet per second)	
fraction	$\int$
franc	fr
free aboard ship	spell out
free alongside ship	spell out
free on board	fob
freezing point	fp
frequency	spell out
fusion point	fnp
G	Gibbs energy (formerly free energy or free enthalpy) ($G = H - TS$)
G	shear modulus ($G = \sigma_{ij}$/angle of shear)
G	statistical weight fraction ($G_i = g_i/\Sigma_i\, g_i$)
g	gravitational acceleration
g	statistical weight
g	*gauche* conformation
g	parameter for the dimensions of branched macromolecules
G^m	molar Gibbs energy
gallon	gal
gallons per minute	gpm
gallons per second	gps
gauche conformation	g
Gibbs energy	G
grain	spell out
gram	g
gram-calorie	g-cal
greatest common divisor	gcd
H	enthalpy
H^m	molar enthalpy
h	height
h	Plank constant
haversine	hav

Abbreviations	Symbols
heat	Q
heat capacity	C
hectare	ha
henry	H
high pressure (adjective)	h-p
hogshead	hhd
horsepower	hp
horsepower-hour	hp-hr
hour	h or hr
hundred	C
hundredweight (112 lb)	cwt
hydrogen ion concentration, negative logarithm of	pH
hyperbolic cosine	cosh
hyperbolic sine	sinh
hyperbolic tangent	tanh
I	electrical current strength
I	radiation intensity of a system
i	radiation intensity of a molecule
ibidem (in the same place)	ibid.
id est (that is)	i.e.
inch	in.
inch-pound	in-lb
inches per second	ips
indicated horsepower	ihp
indicated horsepower-hour	ihp-hr
infrared	IR
inside diameter	ID
intermediate-pressure (adjective)	i-p
internal	int
International Union of Pure and Applied Chemistry	IUPAC
isotactic	it
J	flow (of mass, volume, energy, etc.), always with a corresponding index
joule	J
K	general constant
K	equilibrium constant
K	compression modulus ($p = -K\,\Delta V/V_o$)
k	Boltzmann constant

Abbreviations	Symbols
k	rate constant for chemical reactions (always with an index)
Kelvin	K (Not °K)
kilocalorie	kcal
kilocycles per second	kc
kilogram	kg
kilogram-calorie	kg-al
kilogram-meter	kg-m
kilograms per cubic meter	kg per cu m or kg/m^3
kilograms per second	kgps
kiloliter	Kl
kilometer or kilometer	km
kilometers per second	kmps
kilovolt	kv
kilovolt-ampere	kva
kilowatt	kw
kilowatthour	kwhr
Knoop hardness number	KHN
L	chain end-to-end distance
L	phenomenological coefficient
l	length
lambert	L
latitude	lat or ϕ
least common multiple	lcm
length	l
linear expansion coefficient	Y
linear foot	lin ft
liquid	liq
lira	spell out
liter	l
logarithm (common)	log
logarithm (natural)	log. or ln
kibgutyde	kibg. or λ
loss angle	δ
low-pressure (as adjuective)	l-p
lumen	1*
lumen-hour	1-hr*
luments per watt	lpw
M	"molecular weight" (IUPAC molar mass)
m	mass
mass	spell out or m
mass fraction	w

Abbreviations	Symbols
mathematics (ical)	math
maximum	max
mean effective pressure	mep
mean horizontal candlepower	mhcp
meacycle	mHz
megohm	MΩ
melting point, -temperature	mp, T_m
meter	m
meter-kilogram	m-kg
metre	m
mho	spell out
microsmpere	μa or mu a
microfarad	μf
microinch	μin.
micrometer (formerly micron)	μm
micromicrofarad	μμf
micromicron	μμ
micron	μ
microvolt	μv
microwatt	μw or mu w
mile	spell out
miles per hour	mph
miles per hour per second	mphps
milli	m
milliampere	ma
milliequivalent	meq
milligram	mg
millihenry	mh
millilambert	mL
milliliter or milliliter	ml
millimeter	mm
millimeter or mercury (pressure)	mm Hg
millimicron	mμ or m mu
million	spell out
million gallons per day	mgd
millivolt	mv
minimum	min
minute	min
minute (angular measure)	′

Abbreviations	Symbols
minute (time) (in astronomical tables)	m
mile	spell out
modal	m
modulus of elasticity	E
molar	M
molar enthalpy	H_m
molar Gibbs Energy	G_m
molar heat capacity	C_m
mole	mol
mole fraction	x
molecular weight	mol wt or M
month	spell out
N	number of elementary particles (e.g., molecules, groups, atoms, electrons)
N_L	Avogadro number (Loschmidt's number)
n	amount of a substance (mole)
n	refractive index
nanometer (formerly millimicron)	nm
National Association of Corrosion Engineers	NACE
National Electrical Code	NEC
newton	N
normal	N
number of elementary particles	N
Occupational Safety and Health Administration	OSHA
ohm	Ω
ohm-centimeter	ohm-cm
oil absorption	O.A.
ounce	oz
once-foot	oz-ft
ounce-inch	oz-in.
outside diameter	OD
osomotic pressure	Π
P	permeability of membranes
p	probability
p	dipole moment
$\mathbf{p}_i$	induced dipolar moment
p	pressure

Abbreviations	Symbols
p	extent of reaction
Paint Testing Manual	PTM
parameter	Q
partition function (system)	Q
parts per billion	ppb
parts per million	ppm
pascal	Pa
peck	pk
penny (pency – new British)	p.
pennyweight	dwt
per	diagonal line in expressions with unit symbols or (see Fundamental Rules)
percent	%
permeability of membranes	P
peso	spell out
pint	pt.
Planck's constant (in $E = h\nu$) (6.62517 +/− 0.00023 x 10^{-27} erg sec)	h
polymolecularity index	Q
potential	spell out
potential difference	spell out
pound	lb
pound-foot	lb-ft
pound-inch	lb-in.
pound sterling	£
pounds-force per square inch	psi
pounds per brake horsepower-hour	lb per bhp-hr
pounds per cubi foot	lb per cut ft
pounds per square foot	psf
pounds per square inch	psi
pounds per square inch absolute	psia
power factor	spell out or pf
pressure	p
probability	p
Q	quantity of electricity, charge
Q	heat
Q	partition function (system)
Q	parameter in the Q–e copolymerize-tion equation

Abbreviations	Symbols
Q, Q	polydispersity, polymolecularity in-dex ($Q = \overline{M_w}/\overline{M_n}$)
q	partition function (particles)
quantity of electricity, charge	Q
quart	qt
quod vide (which see)	q.v.
R	molar gas constant
R	electrical resistance
R_G	radius of gyration
R_n	run number
R_ϑ	Rayleigh ratio
r	radius
r_o	initial molar ratio of reactive groups in polycondensations
radian	spell out
radius	r
radius of gyration	R_G
rate constant	k
Rayleigh ratio	R_ϑ
reactive kilovolt-ampere	kvar
reactive volt-ampere	var
reference(s)	ref
refractive index	n
relaxation time	τ
resistivity	ρ
revolutions per minute	rpm
revolutions per second	rps
rod	spell out
root mean square	rms
S	entropy
S^m	molar entropy
S	solubility coefficient
s	sedimentation coefficient
s	selectivity coefficient in osmotic measurements)
Saybolt Universal seconds	SUS
secant	sec
second	s or sec
second (angular measure)	″
second-foot (see cubic feet per second)	

Abbreviations	Symbols
second (time) (in astronomical tables)	s
Second virial coefficient	A_2
shaft horsepower	shp
shilling	s
sine	sin
sine of the amplitude, an elliptic function	sn
society	Soc.
Soluble	sol
solubility coefficient	S
solubility parameter	δ
solution	soln
specific gravity	sp gr
specific heat	sp ht
specific heat capacity (formerly: specific heat)	c
specific optical rotation	$[\propto]$
specific volume	sp vol
spherical candle power	scp
square	sq
square centimeter	sq cm or cm^2
square foot	sq ft
square inch	sq in.
square kilometer	sq km or km^2
square meter	sq m or m^2
square micron	sq μ or μ^2
square root of mean square	rms
standard	std
Standard	Stnd.
Standard deviation	σ
Staudinger index	$[\eta]$
stere	s
syndiotactic	st
T	temperature
t	time
t	*trans* conformation
tangent	tan
temperature	T or temp
tensile strength	ts
threodiisotactic	tit
thousand	M
thousand foot-pounds	kip-ft
thousand pound	kip

Abbreviations	Symbols
ton	spell out
ton-mile	spell out
trans conformation	t
trans-tactic	tt
U	voltage
U	internal energy
U^m	molar internal energy
u	excluded volume
ultraviolet	UV
United States	U.S.
V	volume
V	electrical potential
v	rate, rate of reaction
v	specific volume always with an in-dex
vapor pressure	vp
versed sine	vers
versus	vs
volt	v or V
volt-ampere	va
volt-coulomb	spell out
voltage	U
volume	V or vol.
Volume (of a publication)	Vol
W	weight
W	work
w	mass function
watt	w or W
watthour	whr
watts per candle	wpc
week	spell out
weight	W or w
weight concentration*	c
work	y yield
X	degree of polymerization
X	electrical resistance
x	mole fractio y yield
yard	yd
year	yr
Young's	E
Z	collision number
Z	z fraction
z	ionic charge

Abbreviations	Symbols
z	coordination number
z	dissymmetry (light scattering)
z	parameter in excluded volume theory
α	angle, especially angle of rotation in optical activity
α	cubic expandion coefficient [$\alpha = V^{-1} (\partial V/\partial T)_p$]
α	expansion coefficient (as reduced length, e.g., α_L in the chain end-to-end distance or α_R for the radius of gyration)
α	degree of crystallinity (always with an index)
α	electric polarizability of a molecule
[α]	"specific" optical rotation
β	angle
β	coefficient of pressure
β	excluded volume cluster integral
Γ	preferential solvation
γ	angle
γ	surface tension
γ	linear expansion coefficient
δ	loss angle
δ	solubility parameter
δ	chemical shift
ε	linear expansion ($\varepsilon = \Delta l/l_o$)
ε	expectation
ε_r	relative permittivity (dielectric number)
η	dynamic viscosity
[η]	Staudinger index (called J_o in DIN 1342)
Θ	characteristic temperature, especial-ly theta temperature
θ	angle, especially angle of rotation
ϑ	angle, especially valence angle
κ	isothermal compressibility [$\kappa = V^{-1} (\partial V/\partial p)_T$]
κ	enthalpic interaction parameter in solution theory
λ	wavelength
λ	heat conductivity
λ	degree of coupling
μ	chemical potential
μ	moment
μ	permanent dipole moment
ν	mement, with respect to a reference value
ν	frequency
ν	kinetic chain length
ξ	shielding ratio in the theory of random coils
Ξ	partition function
Π	osmotic pressure
ρ	density
σ	mechanical stress (σ_{ii} = normal stress, σ_{ij} = shear stress)
σ	standard deviation
σ	hindrance parameter
τ	relaxation time
τ_i	internal transmittance (transmission factor) (represents the ratio of transmitted to absorbed light)
φ	volume fraction
φ(r)	potential between two segments separated by a distance r
Φ	constant in the viscosity-molecular-weight relationship
[Φ]	"molar" optical rotation
χ	interaction parameter in solution theory
ψ	entropic interaction parameter in solution theory
ω	angular frequency, angular velocity
Ω	angle
Ω	probability
Ω	skewness of a distribution

*(= weight of solute divided by volume of solvent); IUPAC suggests the symbol ρ for this quantity, which could lead to confusion with the same IUPAC symbol for density.

Notations

The abbreviations for chemicals and polymer were taken from the "Manual of Symbols and Terminology for Physicochemical Quantities and Units," *Pure and Applied Chemistry* **21***1) (1970), but some were added because of generally accepted use.

The ISO (International Standardization Organization) has suggested that all extensive quantities should be described by capital letters and all intensive quantities by lower-case letters. IUPAC doe not follow this recommendation, however, but uses lower-case letters for specific quantities.

The following symbols are used above or after a letter.

Symbols Above Letters

— signifies an average, e.g., $\overline{M}$ is the average molecular weight; more complicated averages are often indicated by $\langle \rangle$, e.g., $\langle R_G^2 \rangle$ is another way of writing $\overline{\left(R_G^2\right)}_z$

— stands for a partial quantity, e.g., $\tilde{v}_A$ is the partial specific volume of the compound A; V_A is the volume of A, wherea $\tilde{V}_A^m$xxx is the partial molar volume of A.

Superscripts

°	pure substance or standard state
∞	infinite dilution or infinitely high molecular weight
m	molar quantity (in cases where subscript letters are impractical)
(q)	the q order of a moment (always in parentheses)
‡	activated complex

Subscripts

Initial	State
1	solvent
2	solute
3	additional components (e.g., precipitant, salt, etc.)
am	amorphous
B	brittleness
bd	bond
cr	crystalline
crit	critical
cryst	crystallization
e	equilibrium

Initial	State
E	end group
G	glassy state
i	run number
i	initiation
i	isotactic diads
ii	isotactic triads
Is	heterotactic triads
j	run number
k	run number
m	molar
M	melting process
mon	monomer
n	number average
p	polymerization, especially propagation
pol	polymer
r	general for average
s	syndiotactic diads
ss	syndiotactic triads
st	start reaction
t	termination
tr	transfer
u	monomeric unit
w	weight average
z	z average
Prefixes	
at	atactic
ct	*cis*-tactic
eit	erythrodiisotactic
it	isotactic
st	syndiotactic
tit	threodiisotactic
tt	*trans*-tactic

Square brackets around a letter signify molar concentrations. (IUPAC prescribes the symbol c for molar councentrations, but to date this has consistently been used for the mass/volume unit.)

Angles are always given by °.

Apart from some exceptions, the meter is not used as a unit of length; the units cm and mm derived from it are used. Use of the meter in macromolecular science leads to very impractical units.

Mathematical Signs

Sign	Definition
Operations	
$+$	Addition
$-$	Subtraction
$\times$	Multiplication
$\cdot$	Multiplication
$\div$	Division
$/$	Division
$\circ$	Composition
$\cup$	Union
$\cap$	Intersection
$\pm$	Plus or minus
$\mp$	Minus or plus
Convolution	
$\oplus$	Direct sum, variation
$\ominus$	Various
$\otimes$	Various
$\odot$	Various
$:$	Ratio
$\amalg$	Amalgamation
Relations	
$=$	Equal to
$\neq$	Not equal to
$\approx$	Nearly equal to
$\cong$	Equals approximately, isomorphic
$<$	Less than
$<<$	Much less than
$>$	Greater than
$>>$	Much greater than
$\leq$	Less than or equal to
$\leq$	Les than or equal to
$\leqq$	Less than or equal to
$\geq$	Greater than or equal to
$\geq$	Grean than or equalt o
$\geqq$	Greater than or equal to
$\equiv$	Equivalent to, congruent to
$\not\equiv$	Not equivalent to, not congruent to
$\vert$	Divides, divisible by
$\sim$	Similar to, asymptotically equal to
$:=$	Assignment

Sign	Definition
$\in$	A member of
$\subset$	Subset of
$\subseteq$	Subset of or equal to
$\supset$	Superset of
$\supseteq$	Superset of or equal to
$\propto$	Varies as, proportional to
$\doteq$	Approaches a limit, definition
$\rightarrow$	Tends to, maps to
$\leftarrow$	Maps from
$\mapsto$	Maps to
$\hookrightarrow$ or $\hookleftarrow$	Maps into
$\Box$	d'Alembertian operator
Σ	Summation
Π	Product
$\int$	Integral
$\oint$	Contour integral
Logic	
$\wedge$	And, conjunction
$\vee$	Or, distunction
$\neg$	Negation
$\Rightarrow$	Implies
$\rightarrow$	Implies
$\Leftrightarrow$	If and only if
$\leftrightarrow$	If and only if
$\exists$	Existential quantifier
$\forall$	Universal quantifier
$\in$	A member o
$\notin$	Not a member of
$\vdash$	Assertion
$\therefore$	Hence, therefore
$\because$	Because
Radial units	
$'$	Minute
$''$	Second
$°$	Degree
Constants	
π	pi (≈ 3.14159265)
e	Base of natural logarithms (≈ 2.71828183)

Sign	Definition
Geometry	
⊥	Perpendicular
∥	Parallel
∦	Not parallel
∠	Angle
∢	Spherical angle
$\stackrel{v}{=}$	Equal angles
Miscellaneous	
i	Square root of -1
′	Prime
″	Double prime
‴	Triple prime
√	Square root, radical
$\sqrt[3]{\ }$	Cube root
$\sqrt[n]{\ }$	nth root
!	Factorial
!!	Double factorial
∅	Empty set, null set
∞	Infinity

Sign	Definition
∂	Partial differential
Δ	Delta
∇	Nabla, del
∇^2, Δ	Laplacian operator

English–Greek–Latin Numerical Prefixes

English	Greek	Latin
2	bis	di
3	tris	tri
4	tetrakis	tetra
5	pentakis	penta
6	hexakis	hexa
7	heptakis	hepta
8	octakis	octa
9	nonakis	nona
10	decakis	deca

Greek-Russian-English Alphabets

Greek letter		Greek name	English equivalent	Russian letter		English equivalent
A	α	Alpha	(ä)	А	а	(ä)
B	β	Beta	(b)	Б	б	(b)
				В	в	(v)
Γ	γ	Gamma	(g)	Г	г	(g)
Δ	δ	Delta	(d)	Д	д	(d)
E	ε	Epsilon	(e)	Е	е	(ye)
Z	ζ	Zeta	(z)	Ж	ж	(zh)
				З	з	(z)
H	η	Eta	(ā)	И	и	(i, ē)
Θ	θ	Theta	(th)	Й	й	(ē)
I	ι	Iota	(ē)	К	к	(k)
				Л	л	(l)
K	k	Kappa	(k)	М	м	(m)
Λ	λ	Lambda	(l)	Н	н	(n)
				О	о	(ô, o)
M	μ	Mu	(m)	О	о	(ô, o)
				П	п	(p)
N	ν	Nu	(n)	Р	р	(r)
Ξ	ξ	Xi	(ks)	С	с	(s)
				Т	т	(t)
O	o	Omicron	α	У	у	ōō
Π	π	Pi	(P)	Ф	ф	(f)
				Х	х	(kh)
P	ρ	Rho	(r)	Х	х	(kh)
				Ц	ц	(t$_s$)
Σ	σ	Sigma	(s)	Ч	ч	(ch)
T	τ	Tau	(t)	Ш	ш	(sh)
Υ	υ	Upsilon	(ü, ōō)	Щ	щ	(shch)
				Ъ	ъ	8
Φ	ø	Phi	(f)	Ы	ы	(ё)
X	χ	Chi	(H)	ь	ь	(ё)
Ψ	ψ	Psi	(ps)	Э	э	(e)
				Ю	ю	(ū)
Ω	ω	Omega	(ō)	Я	я	(yä)

English-Greek-Latin Numbers

English	Greek	Latin
1	mono	uni
2	bis	di
3	tris	tri
4	tetrakis	tetra
5	pentakis	penta
6	hexakis	hexa
7	heptakis	hepta
8	octakis	octa
9	nonakis	nona
10	decakis	deca

International Union of Pure and Applied Chemistry: Rules Concerning Numerical Terms Used in Organic Chemical Nomenclature (specifically as prefixes for hydrocarbons)

1	mono-or hen-	10 deca-	100 hecta-	1000 kilia-
2	di- or do-	20 icosa-	200 dicta-	2000 dilia-
3	tri-	30 triaconta-	300 tricta-	3000 trilia-
4	tetra-	40 tetraconta-	400 tetracta	4000 tetralia-
5	penta-	50 pentaconta-	500 pentactra	5000 pentalia-
6	hexa-	60 hexaconta-	600 hexacta	6000 hexalia-
7	hepta-	70 hepaconta-	700 heptacta-	7000 hepalia-
8	octa-	80 octaconta-	800 ocacta-	8000 ocatlia-
9	nona-	90 nonaconta-	900 nonactta-	9000 nonalia-

Source: IUPAC, Commission on Nomenclature of Organic Chemistry (N. Lorzac'h and published in *Pure and Appl. Chem* 58: 1693–1696 (1986))

Elemental Symbols and Atomic Weights

Source: International Union of Pure and Applied Chemistry (IUPAC) 2001 Values from the 2001 table *Pure Appl. Chem.*, **75**, 1107–1122 (2003). The values of zinc, krypton, molybdenum and dysprosium have been modified. The *approved name* for element 110 is included, see *Pure Appl. Chem.*, **75**, 1613–1615 (2003). The *proposed name* for element 111 is also included.

A number in parentheses indicates the uncertainty in the last digit of the atomic weight.

List of Elements in Atomic Number Order

At No	Symbol	Name	Atomic Wt	Notes
1	H	Hydrogen	1.00794(7)	1, 2, 3
2	He	Helium	4.002602(2)	1, 2
3	Li	Lithium	[6.941(2)]	1, 2, 3, 4
4	Be	Beryllium	9.012182(3)	
5	B	Boron	10.811(7)	1, 2, 3
6	C	Carbon	12.0107(8)	1, 2
7	N	Nitrogen	14.0067(2)	1, 2
8	O	Oxygen	15.9994(3)	1, 2
9	F	Fluorine	18.9984032(5)	
10	Ne	Neon	20.1797(6)	1, 3
11	Na	Sodium	22.989770(2)	
12	Mg	Magnesium	24.3050(6)	
13	Al	Aluminium	26.981538(2)	
14	Si	Silicon	28.0855(3)	2
15	P	Phosphorus	30.973761(2)	
16	S	Sulfur	32.065(5)	1, 2
17	Cl	Chlorine	35.453(2)	3
18	Ar	Argon	39.948(1)	1, 2
19	K	Potassium	39.0983(1)	1
20	Ca	Calcium	40.078(4)	1
21	Sc	Scandium	44.955910(8)	
22	Ti	Titanium	47.867(1)	
23	V	Vanadium	50.9415(1)	
24	Cr	Chromium	51.9961(6)	
25	Mn	Manganese	54.938049(9)	
26	Fe	Iron	55.845(2)	
27	Co	Cobalt	58.933200(9)	
28	Ni	Nickel	58.6934(2)	
29	Cu	Copper	63.546(3)	2
30	Zn	Zinc	65.409(4)	
31	Ga	Gallium	69.723(1)	
32	Ge	Germanium	72.64(1)	
33	As	Arsenic	74.92160(2)	
34	Se	Selenium	78.96(3)	
35	Br	Bromine	79.904(1)	
36	Kr	Krypton	83.798(2)	1, 3
37	Rb	Rubidium	85.4678(3)	1
38	Sr	Strontium	87.62(1)	1, 2
39	Y	Yttrium	88.90585(2)	
40	Zr	Zirconium	91.224(2)	1
41	Nb	Niobium	92.90638(2)	
42	Mo	Molybdenum	95.94(2)	1
43	Tc	Technetium	[98]	5
44	Ru	Ruthenium	101.07(2)	1
45	Rh	Rhodium	102.90550(2)	
46	Pd	Palladium	106.42(1)	1
47	Ag	Silver	107.8682(2)	1
48	Cd	Cadmium	112.411(8)	1
49	In	Indium	114.818(3)	
50	Sn	Tin	118.710(7)	1
51	Sb	Antimony	121.760(1)	1
52	Te	Tellurium	127.60(3)	1
53	I	Iodine	126.90447(3)	
54	Xe	Xenon	131.293(6)	1, 3
55	Cs	Caesium	132.90545(2)	
56	Ba	Barium	137.327(7)	
57	La	Lanthanum	138.9055(2)	1
58	Ce	Cerium	140.116(1)	1
59	Pr	Praseodymium	140.90765(2)	
60	Nd	Neodymium	144.24(3)	1
61	Pm	Promethium	[145]	5
62	Sm	Samarium	150.36(3)	1
63	Eu	Europium	151.964(1)	1
64	Gd	Gadolinium	157.25(3)	1
65	Tb	Terbium	158.92534(2)	
66	Dy	Dysprosium	162.500(1)	1
67	Ho	Holmium	164.93032(2)	
68	Er	Erbium	167.259(3)	1

At No	Symbol	Name	Atomic Wt	Notes
69	Tm	Thulium	168.93421(2)	
70	Yb	Ytterbium	173.04(3)	1
71	Lu	Lutetium	174.967(1)	1
72	Hf	Hafnium	178.49(2)	
73	Ta	Tantalum	180.9479(1)	
74	W	Tungsten	183.84(1)	
75	Re	Rhenium	186.207(1)	
76	Os	Osmium	190.23(3)	1
77	Ir	Iridium	192.217(3)	
78	Pt	Platinum	195.078(2)	
79	Au	Gold	196.96655(2)	
80	Hg	Mercury	200.59(2)	
81	Tl	Thallium	204.3833(2)	
82	Pb	Lead	207.2(1)	1, 2
83	Bi	Bismuth	208.98038(2)	
84	Po	Polonium	[209]	5
85	At	Astatine	[210]	5
86	Rn	Radon	[222]	5
87	Fr	Francium	[223]	5
88	Ra	Radium	[226]	5
89	Ac	Actinium	[227]	5
90	Th	Thorium	232.0381(1)	1, 5
91	Pa	Protactinium	231.03588(2)	5
92	U	Uranium	238.02891(3)	1, 3, 5
93	Np	Neptunium	[237]	5
94	Pu	Plutonium	[244]	5
95	Am	Americium	[243]	5
96	Cm	Curium	[247]	5
97	Bk	Berkelium	[247]	5
98	Cf	Californium	[251]	5
99	Es	Einsteinium	[252]	5
100	Fm	Fermium	[257]	5
101	Md	Mendelevium	[258]	5
102	No	Nobelium	[259]	5
103	Lr	Lawrencium	[262]	5
104	Rf	Rutherfordium	[261]	5, 6
105	Db	Dubnium	[262]	5, 6
106	Sg	Seaborgium	[266]	5, 6
107	Bh	Bohrium	[264]	5, 6
108	Hs	Hassium	[277]	5, 6
109	Mt	Meitnerium	[268]	5, 6
110	Ds	Darmstadtium	[281]	5, 6
111	Rg	Roentgenium	[272]	5, 6

At No	Symbol	Name	Atomic Wt	Notes
112	Uub	Ununbium	[285]	5, 6
114	Uuq	Ununquadium	[289]	5, 6
116	Uuh	Ununhexium		see Note above
118	Uuo	Ununoctium		see Note above

1. Geological specimens are known in which the element has an isotopic composition outside the limits for normal material. The difference between the atomic weight of the element in such specimens and that given in the Table may exceed the stated uncertainty.
2. Range in isotopic composition of normal terrestrial material prevents a more precise value being given; the tabulated value should be applicable to any normal material.
3. Modified isotopic compositions may be found in commercially available material because it has been subject to an undisclosed or inadvertent isotopic fractionation. Substantial deviations in atomic weight of the element from that given in the Table can occur.
4. Commercially available Li materials have atomic weights that range between 6.939 and 6.996; if a more accurate value is required, it must be determined for the specific material [range quoted for 1995 table 6.94 and 6.99].
5. Element has no stable nuclides. The value enclosed in brackets, e.g. [209], indicates the mass number of the longest-lived isotope of the element. However three such elements (Th, Pa, and U) do have a characteristic terrestrial isotopic composition, and for these an atomic weight is tabulated.
6. The names and symbols for elements 112-118 are under review. The temporary system recommended by J Chatt, *Pure Appl. Chem.*, **51**, 381–384 (1979) is used above. The names of elements 101-109 were agreed in 1997 (See *Pure Appl. Chem.*, 1997, **69**, 2471–2473) and for element 110 in 2003 (see *Pure Appl. Chem.*, 2003, **75**, 1613–1615). The proposed name for element 111 is also included.

List of Elements in Name Order

At No	Symbol	Name	Atomic Wt	Notes
89	Ac	Actinium	[227]	5
13	Al	Aluminium	26.981538(2)	
95	Am	Americium	[243]	5
51	Sb	Antimony	121.760(1)	1

At No	Symbol	Name	Atomic Wt	Notes
18	Ar	Argon	39.948(1)	1, 2
33	As	Arsenic	74.92160(2)	
85	At	Astatine	[210]	5
56	Ba	Barium	137.327(7)	
97	Bk	Berkelium	[247]	5
4	Be	Beryllium	9.012182(3)	
83	Bi	Bismuth	208.98038(2)	
107	Bh	Bohrium	[264]	5, 6
5	B	Boron	10.811(7)	1, 2, 3
35	Br	Bromine	79.904(1)	
48	Cd	Cadmium	112.411(8)	1
55	Cs	Caesium	132.90545(2)	
20	Ca	Calcium	40.078(4)	1
98	Cf	Californium	[251]	5
6	C	Carbon	12.0107(8)	1, 2
58	Ce	Cerium	140.116(1)	1
17	Cl	Chlorine	35.453(2)	3
24	Cr	Chromium	51.9961(6)	
27	Co	Cobalt	58.933200(9)	
29	Cu	Copper	63.546(3)	2
96	Cm	Curium	[247]	5
110	Ds	Darmstadtium	[281]	5, 6
105	Db	Dubnium	[262]	5, 6
66	Dy	Dysprosium	162.500(1)	1
99	Es	Einsteinium	[252]	5
68	Er	Erbium	167.259(3)	1
63	Eu	Europium	151.964(1)	1
100	Fm	Fermium	[257]	5
9	F	Fluorine	18.9984032(5)	
87	Fr	Francium	[223]	5
64	Gd	Gadolinium	157.25(3)	1
31	Ga	Gallium	69.723(1)	
32	Ge	Germanium	72.64(1)	
79	Au	Gold	196.96655(2)	
72	Hf	Hafnium	178.49(2)	
108	Hs	Hassium	[277]	5, 6
2	He	Helium	4.002602(2)	1, 2
67	Ho	Holmium	164.93032(2)	
1	H	Hydrogen	1.00794(7)	1, 2, 3
49	In	Indium	114.818(3)	
53	I	Iodine	126.90447(3)	
77	Ir	Iridium	192.217(3)	
26	Fe	Iron	55.845(2)	
36	Kr	Krypton	83.798(2)	1, 3
57	La	Lanthanum	138.9055(2)	1
103	Lr	Lawrencium	[262]	5
82	Pb	Lead	207.2(1)	1, 2
3	Li	Lithium	[6.941(2)]	1, 2, 3, 4
71	Lu	Lutetium	174.967(1)	1
12	Mg	Magnesium	24.3050(6)	
25	Mn	Manganese	54.938049(9)	
109	Mt	Meitnerium	[268]	5, 6
101	Md	Mendelevium	[258]	5
80	Hg	Mercury	200.59(2)	
42	Mo	Molybdenum	95.94(2)	1
60	Nd	Neodymium	144.24(3)	1
10	Ne	Neon	20.1797(6)	1, 3
93	Np	Neptunium	[237]	5
28	Ni	Nickel	58.6934(2)	
41	Nb	Niobium	92.90638(2)	
7	N	Nitrogen	14.0067(2)	1, 2
102	No	Nobelium	[259]	5
76	Os	Osmium	190.23(3)	1
8	O	Oxygen	15.9994(3)	1, 2
46	Pd	Palladium	106.42(1)	1
15	P	Phosphorus	30.973761(2)	
78	Pt	Platinum	195.078(2)	
94	Pu	Plutonium	[244]	5
84	Po	Polonium	[209]	5
19	K	Potassium	39.0983(1)	1
59	Pr	Praseodymium	140.90765(2)	
61	Pm	Promethium	[145]	5
91	Pa	Protactinium	231.03588(2)	5
88	Ra	Radium	[226]	5
86	Rn	Radon	[222]	5
75	Re	Rhenium	186.207(1)	
45	Rh	Rhodium	102.90550(2)	
111	Rg	Roentgenium	[272]	5, 6
37	Rb	Rubidium	85.4678(3)	1
44	Ru	Ruthenium	101.07(2)	1
104	Rf	Rutherfordium	[261]	5, 6
62	Sm	Samarium	150.36(3)	1
21	Sc	Scandium	44.955910(8)	
106	Sg	Seaborgium	[266]	5, 6
34	Se	Selenium	78.96(3)	
14	Si	Silicon	28.0855(3)	2

At No	Symbol	Name	Atomic Wt	Notes
47	Ag	Silver	107.8682(2)	1
11	Na	Sodium	22.989770(2)	
38	Sr	Strontium	87.62(1)	1, 2
16	S	Sulfur	32.065(5)	1, 2
73	Ta	Tantalum	180.9479(1)	
43	Tc	Technetium	[98]	5
52	Te	Tellurium	127.60(3)	1
65	Tb	Terbium	158.92534(2)	
81	Tl	Thallium	204.3833(2)	
90	Th	Thorium	232.0381(1)	1, 5
69	Tm	Thulium	168.93421(2)	
50	Sn	Tin	118.710(7)	1
22	Ti	Titanium	47.867(1)	
74	W	Tungsten	183.84(1)	

At No	Symbol	Name	Atomic Wt	Notes
112	Uub	Ununbium	[285]	5, 6
116	Uuh	Ununhexium		see Note above
118	Uuo	Ununoctium		see Note above
114	Uuq	Ununquadium	[289]	5, 6
92	U	Uranium	238.02891(3)	1, 3, 5
23	V	Vanadium	50.9415(1)	
54	Xe	Xenon	131.293(6)	1, 3
70	Yb	Ytterbium	173.04(3)	1
39	Y	Yttrium	88.90585(2)	
30	Zn	Zinc	65.409(4)	
40	Zr	Zirconium	91.224(2)	1

Pronounciation Symbols and Abbreviations

ə	Banana, collide, abut		ȯ	saw, all, gnaw, caught
ˈə, ˌə	Humdrum, abut		ü̇	fool
ᵊ	Immediately preceding \l\, \n\, \m\, \ŋ\, as in battle, mitten, eaten, and sometimes open \ˈō-pᵊm\, lock and key \-ᵊ ŋ-\; immediately following \l\, \m\, \r\, as often in French table, prisme, titre		u̇̇	took
			œ	French cœuf, German Hölle
			œ̄	French feu, German Höhle
			ȯi	coin, destroy
ər	further, merger, bird		p	pepper, lip
ˈə-, ˈə-r	As in two different pronunciations of hurry \ˈhər-ē, \ˈhə-rē\		r	red, car, rarity
			s	source, less
a	mat, map, mad, gag, snap, patch		sh	as in shy, mission, machine, special (actually, this is a single sound, not two); with a hyphen between, two sounds as in grasshopper \ˈgras-ˌhä-pər\
ā	day, fade, date, aorta, drape, cape			
ä	bother, cot, and, with most American speakers, father, cart		t	tie, attack, late, later, latter
á	father as pronounced by speakers who do not rhyme it with bother; French patte		th	as in thin, ether (actually, this is a single sound, not two); with a hyphen between, two sounds as in knighthood \ˈnīt-ˌh----d\
au̇	now, loud, out			
b	baby, rib		t̲h̲	then, either, this (actually, this is a single sound, not two)
ch	chin, nature \ˈnā-chər\			
d	did, adder		ü	rule, youth, union \ˈyün-yən\, few \ˈfyü\
e	bet, bed, peck		u̇	pull, wood, book, curable \ˈky u̇ r-ə-bəl\, fury \ˈfy----r-ē\
ˈē, ˌē	beat, nosebleed, evenly, easy			
ē	easy, mealy		ue	German füllen, hübsch
f	fifty, cuff		u̲e̲	French rue, German fühlen
g	go, big, gift		v	vivid, give
h	hat, ahead		w	we, away
hw	whale as pronounced by those who do not have the same pronunciation for both whale and wail		y	yard, young, cue \ˈkyü\, mute \ˈmyüt\, union \ˈyün-yən\
i	tip, banish, active		ʸ	indicates that during the articulation of the sound represented by the preceding character the front of the tongue has substantially the position it has for the articulation of the first sound of yard, as in French digne \dēnʸ\
ī	site, side, buy, tripe			
j	job, gem, edge, join, judge			
k	kin, cook, ache			
k̲	German ich, Buch; one pronunciation of loch		z	zone, raise
l	lily, pool		zh	as in vision, azure \ˈa-zhər\ (actually this is a single sound, not two).
m	murmur, dim, nymph			
n	no, own		\	reversed virgule used in pairs to mark the beginning and end of a transcription: \ˈpen\
ⁿ	Indicates that a preceeding vowel or diphthong is pronounced with the nasal passages open, as in French un bon vin blanc \œⁿ-bōⁿvaⁿ-bläⁿ\			
			ˈ	mark preceding a syllable with primary (strongest) stress: \ˈpen-mən-ˌship\
ŋ	sing \ˈsiŋ\, singer \ˈsiŋ-ər\, finger \ˈfiŋ-gər\, ink \ˈiŋk\		ˌ	mark preceding a syllable with secondary (medium) stress: \ˈpen-mən-ˌship\
ō	bone, know, beau		-	mark of syllable division

( )	indicate that what is symbolized between is present in some utterances but not in others: *factory* \ ▎fak-t(ə-)rē
÷	indicates that many regard as unacceptable the pronunciation variant immediately following: *cupola* \ ▎kyü-pə-lə, ÷- ▎lō\

Explanatory Notes and Abbreviations

(date)	date that word was first recorded as having been used
[. . .]	etomology and origin(s) of word
{. . .}	usage and/or languages, including French, German, Italian and Spanish
adj	adjective
adv	adverb
B.C.	before Christ
Brit.	Britain, British
C	centigrade, Celsius
c	century
E	English
Eng.	England
F	French, Fahrenheit
Fr.	France
fr.	from
G	German
Gr.	Germany
L	Latin
ME	middle English

n	noun
neut.	neuter
NL	new Latin
OE	old English
OL	old Latin
pl	plural
prp.	present participle
R	Russian
sing.	singular
S	Spanish
U.K.	United Kingdom
v	verb

Source: From *Merriam-Webster's Collegiate© Dictionary*, Eleventh Editioh, ©2004 by Merriam-Webster, incorporated, (www.Merriam-Webster.com). With permission.

Languages

French, German and Spanish translations are enclosed in {--} and preceded by F, G, I and S, respectively; and gender is designated by f-feminine, m-masculine, n-neuter. For example: **Polymer**--{F polymere m} represents the French translation "polymere" of the English word polymer and it is in the masculine case. These translations were obtain from multi-language dictionaries including: *A Glossary of Plastics Terminology in 5 Languages*, 5th Ed., Glenz, W., (ed) Hanser Gardner Publications, Inc., Cinicinnati, 2001. By permission).

J

J *n* (1) The SI abbreviation for ▶ Joule. (2) In most references to technical publications whose title contains the word "Journal", the abbreviation of that word. (Handbook of chemistry and physics, 52nd edn. Weast RC (ed). The Chemical Rubber, Boca Raton, FL).

J Acid *n* Fusion product of beta-naphthylamine-3,6-bisulfonic acid with caustic. Used in the manufacture of dyes. Also known as *2-Amino-5-Naphthol-7-Sulfonic Acid* and *6-Amino-1-Naphthol-3-Sulfonic Acid*.

Jack \jak\ *n* [ME *Jacke*, familiar term of address to a social inferiorm nickname for *Johan* John] (1548) (1) A blade having high and/or low butts used to actuate the movement of latch knitting needles. (2) Part of a dobby head designed to serve as a lever in the operation of the harness of a loom. (Elsevier's textile dictionary. Vincenti R (ed). Elsevier Science and Technology Books, New York, 1994).

Jacket \ja-kət\ *n* [ME *jaket*, fr. MF *jacquet*, dimin. of *jaque* short jacket, fr. *jacque* peasant, fr. the name *Jacques* James] (15c) (1) A woven or felted tubular sleeve for covering and shrinking on a machine roll. (2) A short coat. (3) In polymer manufacture, an external shell around a reaction vessel. For example, jacketed vessels are used when heat-transfer medium is circulated around the vessel. (Elsevier's textile dictionary. Vincenti R (ed). Elsevier Science and Technology Books, New York, 1994).

Jacquard \ja-ˌkärd\ *n* (1841) A system of weaving that utilizes a highly versatile pattern mechanism to permit the production of large, intricate designs. The weave pattern is achieved by a series of punched cards. Each card perforation controls the action of one warp thread for the passage of one pick. The machine may carry a large number of cards, depending upon the design, because there is a separate card for each pick in the pattern. Jacquard weaving is used for tapestry, brocade, damask, brocatelle, figured necktie and dress fabrics, and some floor coverings. A similar device is used for the production of figured patterns on some knit goods. (Elsevier's textile dictionary. Vincenti R (ed). Elsevier Science and Technology Books, New York, 1994).

Jalousie \ja-lə-sē\ *n* [F] (1766) A window with movable, horizontal glass slats angled to admit ventilation and keep out rain. The term is also used for outside shutters of wood constructed in this way. (Harris CM (2005) Dictionary of architecture and construction. McGraw-Hill, New York).

Jamb \jam\ *n* [ME *jambe*, fr. MF, LL *gamba*] (14c) Vertical face inside an opening to the full thickness of a wall; vertical slide members of a window frame, door frame or lining. (Harris CM (2005) Dictionary of architecture and construction. McGraw-Hill, New York).

Jamba Seed Oil *n* Nondrying oil, similar to rape oil, obtained from the seeds of *Eruca sativa*. Its main constituent acids are erucic, oleic, and linoleic. Approximate constants: sp gr, 0.916/15°C; iodine value, 100; saponification value, 173.

Japan \jə-ˈpan\ *n* (1688) Glossy black enamel, either air drying or baking, based on asphaltum and drying oil. Japan should not be confused with black enamels which are produced by pigmentation and do not have an asphaltum base. Also known as *Black Japan*. (Ash M, Ash I (1982–1983) Encyclopedia of plastics, polymers, and resins, vols I–III. Chemical Publishing, New York).

Japan Color *n* Paste containing pigment and a grinding japan vehicle, used for lettering and decorating.

Japan Drier Resinate-base liquid drier.

Japanese Lacquer *n* A glossy coating obtained by tapping the sap from the Japanese varnish tree (Langenheim JH (2003) Plant resins: Chemistry, evolution ecology and ethnobotany. Timber, Portland, OR; *Rhus vernicifera*) or sumac. (Weismantal GF (1981) Paint handbook. McGraw-Hill, New York).

Japan, Grinding *n* Rapid, hard-drying varnish suitable for use as a vehicle for japan colors; frequently contains shellac. (Weismantal GF (1981) Paint handbook. McGraw-Hill, New York).

Japanners' Brown *n* A brown pigment made by high temperature oxidation of ferrous hydroxide precipitated from a solution of an iron salt. (Kirk-Othmer encyclopedia of chemical technology: Pigments-powders. Wiley, New York, 1996).

Japanning Process of finishing with a baking black japan.

Japan Wax *n* (1859) Product obtained from the berries of trees indigenous to Japan and China. Although described as a wax, it differs from waxes in composition, in that it is a mixture of triglycerides. Mp, 52°C; sp gr, 0.987/15°C; saponification value, 215; iodine value, 11. (Langenheim JH (2003) Plant resins: Chemistry, evolution ecology and ethnobotany. Timber, Portland, OR; Ash M, Ash I (1982–1983) Encyclopedia of plastics polymers, and resins, vols I–III. Chemical Publishing, New York).

Jar Mill *n* A small BALL MILL utilizing a portable jar of porcelain or metal rather than a fixed cylinder for containing the material to be ground and the grinding media. After being charged and tightly closed, the jar is placed on a pair of rubber rollers – one driven, the other idling – and rotated for the desired time, typically overnight. (Perry's chemical engineer's handbook, 7th edn. Perry RH, Green DW (eds). McGraw-Hill, New York, 1997; Weismantal GF (1981) Paint handbook. McGraw-Hill, New York) ▶ Ball Mill.

Jasmine Oil \ˈjaz-mən-\ *n* [Arabic *yāsamīn*, from Persian] An essential oil in perfumery, fragranced, and flavoring, dextrorototary. Origin: any of numerous often climbing scrubs (genus *Jasminum*) of the olive olvie family that usually have extremely fragrant flowers. (Langenheim JH (2003) Plant resins: Chemistry, evolution ecology and ethnobotany. Timber, Portland, OR; Bailey's industrial oil and fat products. Shahidi F, Bailey AE (eds). Wiley, 2005; Merriam-Webster's collegiate dictionary, 11th edn. Merriam-Webster, Springfield, MA, 2004).

Jaspé *n* (1) A fabric used for suiting, draperies, or upholstery characterized by a series of faint stripes formed by dark, medium, and light yarns of the same color. (2) A term describing carpets having a faint striped effect. (Tortora PG, Merkel RS (2000) Fairchild's dictionary of textiles, 7th edn. Fairchild, New York).

Javalle Water \zha-ˈvel\ *n* Sodium hyprochlorite (NaHClO) dissolved in water, a disinfectant and bleaching solution. (Merriam-Webster's collegiate dictionary, 11th edn. Merriam-Webster, Springfield, MA, 2004).

J-Box *n* A J-shaped holding device used in continuous operations to provide varying amounts of intermediate material storage such as in wet processing of fabrics and in tow production. The material is fed to the top and pleated to fill the long arm before being withdrawn from the short arm. (Vigo TL (1994) Textile processing, dyeing, finishing and performance. Elsevier Science, New York).

JCT *n* Abbreviation for the ▶ Journal of Coatings Technology.

J-Cut *n* In tufting cut-pile carpet constructions, uneven cutting of the loops caused by poor adjustment of knives and hooks or excessive tension.

Jean \ˈjēn\ *n* [short for *jean fustian*, fr. ME *Gene* Genoa, Italy + *fustian*] (1577) Cotton twill fabric, similar to denim, but lighter and finer, in a 2/1 weave for sportswear and linings. (Fairchild's dictionary of textiles. Tortora PG (ed). Fairchild Books, New York, 1997).

Jeffamine® *n* Trade mark (Texaco) for polyoxypropyleneamines, grades are 3 diamine, 1 triamaine, liquid form, corrosive. Used as curing agent for epoxy resin systems in adhesives, elastomers, and foam formulations, and as an intermediate for textile and paper-treating chemicals. (Ash M, Ash I (1996) Handbook of paint and coating raw materials: Trade name products - chemical products dictionary with trade name cross-references. Ashgate, New York).

Jersey \ˈjər-zē\ *n* (1587) (1) A circular-knit or flat-knit fabric made with a plain stitch in which the loops intermesh in only one direction. As a result, the appearance of the face and the back of a jersey fabric is wholly different. (2) A tricot fabric made with a simple stitch, characterized by excellent drape and wrinkle recovery properties. (Elsevier's textile dictionary. Vincenti R (ed). Elsevier Science and Technology Books, New York, 1994).

Jet \ˈjet\ *n* [ME, fr. MF *jaiet*, fr. L *gagates*, fr. Gk *gagatēs*, fr. *Gagas*, town and river in Asia Minor] (14c) Term used to describe the blackness or intensity of the mass tone of black or near black surfaces. It is frequently used to describe dark blues, blue–black or black pigments. (Billmeyer FW, Saltzman M (1966) Principles of color technology. Wiley, New York).

Jet A device used to bulk yarns by introducing curls, coils, and loops that are formed by the action of a high velocity stream, usually of air or steam. (Vigo TL (1994) Textile processing, dyeing, finishing and performance. Elsevier Science, New York).

Jet Abrader *n* Device used for measuring the abrasion resistance of organic coating materials, in terms of the time required for a controlled jet of fine abrasive particles to abrade through the coating to the substrate.

Jet-Abrasion Test *n* A test for the abrasion resistance of coatings in which the time required for an air blast of fine abrasive particles to wear through the coating is measured (ASTM D 658, Section 06.01).

Jet Dyeing Machine *n* A high temperature piece dyeing machine that circulates the dye liquor through a Venturi jet, thus imparting a driving force to move the fabric. The fabric, in rope form, is sewn together to form a loop.

Jet Loom *n* A shuttleless loom that employs a jet of water or air to carry the filling yarn through the shed. ▶ Weft Insertion.

Jet Molding *n* (offset molding) A modification of injection molding designed for molding thermosets. An elongated nozzle or "jet" is attached to the front of the molding cylinder and is provided with a high-watt-density heating element and means for rapid cooling. It is also necessary to control cylinder temperatures carefully to prevent premature hardening of the resin. (Strong AB (2000) Plastics materials and processing. Prentice Hall, Columbus, OH).

Jet Printing *n* A process, wherein charged ink droplets are emitted from a nozzle and deflected vertically and horizontally by positively and negatively charged electrodes. The operation is analogous to electron beam tracing as in television tubes and, consequently, is extremely fast with speeds of over 1,200 words per minute easily attained. (Printing ink manual, 5th edn. Leach RH, Pierce RJ, Hickman EP, Mackenzie MJ, Smith HG (eds). Blueprint, New York, 1993).

Jet Spinning *n* For most purposes, similar to melt spinning of staple fiber. Hot-gas-jet spinning uses a directed jet of hot gas to "pull" molten polymer from a die lip and instantly draw it into fine fibers. (Kadolph SJJ, and Langford AL (2001) Textiles. Pearson Education, New York).

Jetting *n* In injection molding, a wriggly flow of resin from a small gate into the mold cavity, mistakenly referred to as turbulence and probably related to melt fracture. Jetting is the antithesis of the desired laminar flow forming a smooth flow front across the mold. It can cause strength problems in molded parts because of incomplete welding of the wormlike surfaces of the jet. (Strong AB (2000) Plastics materials and processing. Prentice Hall, Columbus, OH).

Jewelers' Rouge \ˈjü-ə-lərs ˈrüzh *esp. Southern* ˈrüj\ *n* Very fine and pure ferric oxide (Fe_2O_3) powder used for polishing metals and plastics. *Rouge paper* contains the same abrasive glued to paper. It also comes in cloth form, called *crocus cloth*. (Whittington's dictionary of plastics. Carley, James F (ed). Technomic, 1993).

Jig \ˈjig\ *n* [perhaps fr. MF *giguer* to frolic, fr. *gigue* fiddle, of Gr origin; akin to OHGr *gīga* fiddle; akin to ON *geiga* to turn aside] (ca. 1560) (1) A device for positioning component parts while they are being assembled or otherwise worked on, or for holding tools. (2) A clamping device used to secure a bonded assembly until the adhesive has set. (3) A restraining frame into which freshly molded parts are placed to prevent their warping during annealing or final cooling. (Merriam-Webster's collegiate dictionary, 11th edn. Merriam-Webster, Springfield, MA, 2004; Whittington's dictionary of plastics. Carley, James F (ed). Technomic, 1993).

Joggles \ˈjä-gəl\ *v* [frequentative of ^{1}*jog*] (1513) (keys) A term sometimes employed for matching inserts that exactly position the parts of a multi-piece mold. (Whittington's dictionary of plastics. Carley, James F (ed). Technomic, 1993).

Joining \ˈjói-niŋ\ *n* (14c) The process of assembling plastic parts by means of mechanical fastening devices such as rivets, screws, clamps, etc. ▶ Fabricate.

Joint \ˈjóint\ *n* [ME *jointe*, fr. OF, fr. *joindre*] (13c) The location where two separately made parts are joined with each other by adhesive bonding, welding, or fastening. ▶ Butt Joint, ▶ Lap Joint, and ▶ Scarf Joint. (Handbook of adhesives. Skeist I (ed). Van Nostrand Reinhold, New York, 1990).

Joint Filler *n* Sealant inserted between abutting ends of wallboard.

Joint, Scarf *n* A joint made by cutting away similar angular segments of two adherents and bonding the adherents with the cut areas fitted together. (Handbook of adhesives. Skeist I (ed). Van Nostrand Reinhold, New York, 1990) ▶ Joint, ▶ Lap.

Joint, Starved *n* A joint that has an insufficient amount of adhesive to produce a satisfactory bond. NOTE – This condition may result from too thin a spread to fill the gap between the adherents, excessive penetration of the adhesive into the adherend, too short an assembly time, or the use of excessive pressure. ▶ Starved Joint.

Joint Tape *n* Paper or paper-faced cotton tape, metal, fabric, glass mesh, or other material, sometimes embossed or perforated, which is fixed over the joints between wallboards, to conceal the joints and provide a smooth surface for painting.

Joist \ˈjóist\ *n* [ME *joiste*, fr. MF *giste*, fr. (assumed) VL *jacitum*, fr. L *jacēre*] (15c) A small rectangular sectional member arranged parallel from wall to wall in a building, or resting on beams or girders. They support a floor or the laths or furring strips of a ceiling.

Jolly Balance *n* Spring balance for determining the specific gravity of a solid by weighing it alternatively in air and water.

Joule \ˈjü(ə)l *also* ÷ˈjaü(ə)l\ *n* [James P. *Joule*] (1882) (J) The SI unit of work and energy, equal to 1 m N, that replaces a variety of not-quite-equal older joules as well as numerous calories, all of which equal 4.18–4.19 J. Six different ▶ British Thermal Units are all about equal to 1,055 J. (Serway RA, Faugh JS, Bennett CV (2005) College physics. Thomas, New York).

Joule-Thomson Effect *n* The cooling which occurs when a highly compressed gas is allowed to expand in such a way that no external work is done is known as the

Joule-Thomson effect. This cooling is inversely proportional to the square of the absolute temperature. (Serway RA, Faugh JS, Bennett CV (2005) College physics. Thomas, New York).

Journal of Coatings Technology *n* (JCT) The official journal of the Federation of Societies for Coatings Technology.

Journeyman Painter *n* One who has had at least 3 years' experience and schooling as an apprentice.

Juniper Gum \ˈjü-nə-pər-\ (Industrial gums: Polysaccharides and their derivatives. Whistler JN, BeMiller JN (eds). Elsevier Science and Technology Books, 1992) ▶ Gum Sandarac.

Juta Hycica Resin *n* Brazilian resin occasionally used in certain types of varnishes. It is a very hard and pale-colored material, insoluble in most of the common solvents. (Langenheim JH (2003) Plant resins: Chemistry, evolution ecology and ethnobotany. Timber, Portland, OR; Paint: Pigment, drying oils, polymers, resins, naval stores, cellulosics esters, and ink vehicles, vol 3. American Society for Testing and Material, 2001).

Jute (Burlap) \ˈjüt\ *n* [Bengali *jhuto*] (1746) A fiber obtained from the stems of several species of the plant *Corchorus* grown mainly in India and Pakistan. It is used in the form of fiber, yearn, and fabric for reinforcing phenolic and polyester resins. (Kadolph SJJ, Langford AL (2001) Textiles. Pearson Education, New York).

Jute Count *n* The weight in pounds of a spindle of 14,400 yd of yarn.

Jute Seed Oil *n* Obtained from the seed of *Corchorus capsularis*, grown chiefly in Indian, Egypt, and China. It contains linoleic and oleic acids. Iodine value, 103; sp gr, 0.921/15°C; saponification value, 185; acetyl value 27. (Langenheim JH (2003) Plant resins: Chemistry, evolution ecology and ethnobotany. Timber, Portland, OR; Paint: Pigment, drying oils, polymers, resins, naval stores, cellulosics esters, and ink vehicles, vol 3. American Society for Testing and Material, 2001).

K

k \ˈkā\ *n* (1) Abbreviation for SI prefix, ▶ Kilo-. (2) Symbol for ▶ Thermal Conductivity. (Ready RG (1996) Thermodynamics. Pleum, New York).

k′ Symbol for ▶ Huggins Constant. (Huggins ML (1958) Physical chemistry of high polymers. Wiley, New York).

K (1) Abbreviation for ▶ Kelvin. (2) Chemical symbol for potassium (Latin: kalium). (3) Symbol for ▶ Bulk Modulus.

Kakemono \ˌkä-ki-ˈmō-(ˌ)nō\ *n* [Japanese] (1890) A painting mounted on a margin of brocade; hung by its top when in use, and rolled up when not in use.

Kalrez® Dupont's trade name for fluoroelastomers made from tetrafluoroethylene, perfluorovinylmethyl ether, and a small percentage of crosslinkable monomer. These elastomers combine the rubbery properties of ▶ Viton with the thermal stability, chemical resistance, and electrical characteristics of tetrafluoroethylene resin. (Handbook of plastics, elastomers and composites. 4th edn. Harper CA (ed). McGraw-Hill, New York, 2002; Ash M, Ash I (1982–1983) Encyclopedia of plastics polymers, and resins, vols I–III. Chemical Publishing, New York).

Kalsomine ▶ Calcimine.

Kaolin \ˈkā-ə-lən\ *n* [F *kaolin*, fr. *Gaoling* hill in China] (ca. 1741) (china clay, bolus alba) A variety of ▶ Clay consisting essentially of the minerals *kaolinite*, *dickite*, and *nacrite* (all are $Al_2O_3 \cdot 2SiO_2 \cdot 2H_2O$). The name kaolin comes from the Chinese *kaoling*, meaning high hill, the name of the mountain in China which yielded the first kaolin sent to Europe. (Hibbard MJ (2001) Mineralogy. McGraw-Hill, New York).

Kaolinite \-lə-ˌnīt\ *n* (1867) $Al_2O_3 \cdot 2SiO_2 \cdot 2H_2O$. A finely divided crystalline form of hydrated aluminum silicate that occurs as monoclinic crystals with a basal cleavage, resulting chiefly from the alteration of feldspars under conditions of hydrothermal or pneumatolytic metamorphism. It is an important clay mineral. Density, 2.58 g/cm^3 (21.5 lb/gal); refractive index, 156; O.A., 32 to 55; Mohs hardness, 2.5. (Hibbard MJ (2001) Mineralogy. McGraw-Hill, New York; Solomon DH, Hawthorne DG (1991) Chemistry of pigments and fillers. Krieger, New York).

Kapok \ˈkā-ˌpäk\ *n* [Malay] (ca. 1750) Short, lightweight cotton-like fibers from the seed pod of trees of the family *Bombacabeae*. A very brittle fiber, it is generally not spun. It is used for stuffing cushions, mattresses, etc., and for life jackets because of its buoyancy and moisture resistance. (Kadolph SJJ, Langford AL (2001) Textiles. Pearson Education, New York; Elsevier's textile dictionary. Vincenti R (ed). Elsevier Science and Technology Books, New York, 1994).

Karaya Gum \kə-ˈrī-ə-\ *n* [Hindi *karāyal* resin] (1916) Dry exudation from *Sterculia urens*, which grows in India. It swells in water, and has some resemblance to gum tragacanth. (Industrial gums: Polysaccharides and their derivatives. Whistler JN, BeMiller JN (eds). Elsevier Science and Technology Books, 1992) Also known as *Gum Karaya*.

Karl Fischer Reagent A colored solution of iodine, sulfur dioxide, and pyridine in methanol. It reacts quantitatively with water, becoming colorless. It is used to determine small amounts of water in a wide range of materials, including many polymerics. (Goldberg DE (2003) Fundamentals of chemistry. McGraw-Hill Science/Engineering/Math, New York).

Kauri \ˈkau(-ə)r-ē\ *n* [Maori *kawri*] (1823) A fossil copal resin used in oleo-resinous varnishes found in New Zealand. (Langenheim JH (2003) Plant resins: Chemistry, evolution ecology and ethnobotany. Timber, Portland, OR; Paint: Pigment, drying oils, polymers, resins, naval stores, cellulosics esters, and ink vehicles, vol 3. American Society for Testing and Material, 2001).

Kauri-Butanol Value Volume in ml at 25°C (77°F) of a solvent, corrected to a defined standard, required to produce a defined degree of turbidity when added to 20 g of a standard solution of kauri resin in normal butyl alcohol. For kauri-butanol values of 69 and over, the standard is toluene and has an assigned value of 105. For kauri-butanol values under 60, the standard is a blend of 75% *n*-heptane and 25% toluene and has an assigned value of 40. Abbreviation is KB Value. (Paint: Pigment, drying oils, polymers, resins, naval stores, cellulosics esters, and ink vehicles, vol 3. American Society for Testing and Material, 2001; Flick EW (1991) Industrial synthetic resins handbook. Williams Andrews/Noyes, New York).

Kauri Glue Urea-formaldehyde resin, manufactured by BASF, Germany.

Kauri Reduction Test Test for measuring the flexibility of a varnish. (Paint: Pigment, drying oils, polymers, resins, naval stores, cellulosics esters, and ink vehicles, vol 3. American Society for Testing and Material, 2001).

Kautex Poly(vinyl chloride, manufactured by Kautex Werke, Germany.

Kautschin ▶ Dipentine.

K.B. Value A numerical measure of the solvent power of hydrocarbon solvents and oils using a kauri-butanol reagent. The values range from 20, which is a poor solvent, to a high of 105, which is an excellent solvent. (Paint: Pigment, drying oils, polymers, resins, naval stores, cellulosics esters, and ink vehicles, vol 3. American Society for Testing and Material, 2001) ▶ Kauri-Butanol Value.

Kel-F Elastomer Copolymer from vinylidene fluoride and trifluorochloroethylene. Manufactured by M. W. Kellog, US.

Kelvin \\ˈkel-vən\ *n* [William Thomson, Lord *Kelvin* (1824–1907) (1968) (K) The SI unit of both temperature and difference between temperatures, equal to 1/273.16 of the thermodynamic triple point of water, i.e., the temperature and pressure at which all three phases of water – ice, liquid, and vapor – are in equilibrium. A change or difference of 1 K is exactly equal to 1° difference on the Celsius (formerly centigrade) scale, and the temperature 0°C corresponds to 273.15 K. Symbol K; the name "degree of Kelvin" (Symbol °K) was discontinued by international agreement in 1967.

Kelvin Temperature Scale An absolute temperature scale, in which the unit is the *kelvin* (K), defined as $\frac{1}{273.16}$ of the temperature difference between absolute zero (0 K) and the triple point of water. K = °C + 273. This scale is used to describe the correlated color temperature of light sources and illuminants in color designations and color rendition. (Paint: Pigment, drying oils, polymers, resins, naval stores, cellulosics esters, and ink vehicles, vol 3. American Society for Testing and Material, 2001) Symbol K. ▶ Correlated Color Temperature and ▶ Kelvin. Kelvin is used extensively in gas laws and calculation for expansion of gases. Theoretically, all molecular motion ceases at 0 K or absolute zero temperature.

Keratin \\ˈker-ə-tᵊn\ *n* [ISV] (ca. 1849) (1) The protein derived from feathers, hair, hoofs, horns, etc of animals by calcinations. It is sometimes used as filler in plastics, particularly urea-formaldehyde molding compounds, in which it reduces brittleness and permits drilling and tapping. (2) A class of natural fibrous proteins occurring in vertebrate animals and man, characterized by their high content of several amino acids, especially crystine, arginine, and serine. They are generally harder than the fibrous collagen group of proteins. Keratins are insoluble in organic solvents but do absorb and hold water. The molecules contain both acidic and basic groups and are thus amphoteric. (Morrison RT, Boyd RN (1992) Organic chemistry, 6th edn. Prentice Hall, Englewood Cliffs, NJ).

Kernel \\ˈkər-nᵊl\ *n* [ME, fr. OE *cyrnel*, dim. of *corn*] (before 12c) All of an atom except for its valence shell of electrons; also called the *core*.

Kerosene \\ˈker-ə-ˌsēn\ *n* [Gk *kēros* + E *–ene* (as in camphene)] (1854) (*or* kerosene) A low viscosity oil distilled from petroleum or shale oil, used as a fuel, paint thinner, and alcohol denaturant. (Handbook of solvents. Wypych G (ed). Chemtec, New York, 2001).

Kerr Effect *n* When plane polarized light is incident on the pole of an electromagnet, polished so as to act like a mirror, the plane of polarization of the reflected light is not the same when the magnet is "on" as when it is "off". It was found that the direction of rotation was opposite to that of the currents exciting the pole from which the light was reflected. (Handbook of chemistry and physics, 52nd edn. Weast RC (ed). The Chemical Rubber, Boca Raton, FL).

Kerrolic Acid *n* $C_{15}H_{27}(OH)_4COOH$. Monobasic tetrahydroxy acid constituent of shellac.

Kersey \\ˈkər-zē\ *n* [ME, fr. *Kersey*, England] (14c) A heavily milled woolen fabric having a high lustrous nap and a "grainy" face, kersey is frequently used in overcoats.

Ketohexamethylene Syn: Cyclohexanone.

Ketone \\ˈkē-ˌtōn\ *n* [Gr *Keton*, alter. of *Aceton* acetone] (1851) An organic compound containing a carbonyl group (C=O) bound to two carbon atoms. The simplest one is acetone, $(CH_3)_2C=O$. It and the other lower ketones are widely used as solvents for vinyl and cellulosic resins, and as intermediates in the production of resins. (Odian GC (2004) Principles of polymerization. Wiley, New York).

Ketone-Based Resins *n* Consist of ether and ketone groups combined with phenyl rings in different sequences. The rigid ketone and phenyl groups produce high thermal stability. (Odian GC (2004) Principles of Polymerization. Wiley, New York).

Ketone Condensation Resins *n* Resins produced as the result of condensation of ketones and aldehydes, for example, the resin obtained from methyl ethyl ketone and formaldehyde. (Odian GC (2004) Principles of polymerization. Wiley, New York).

Ketones *n* A class of strong organic solvents used in gravure inks; for example, acetone, methyl ethyl ketone (MEK). (Handbook of solvents. Wypych G (ed). Chemtec, New York, 2001).

Kettle \\ˈke-tᵊl\ *n* [ME ketel, fr. ON ketill (akin to OE *cietel* kettle), both from a prehistoric Germanic word borrowed fr. L *catillus*, dimin. of *catinus* bowl] (13c) Reaction vessel for varnish or resin manufacture.

Kevlar® *n* DuPont's trade name for poly-(*p*-phenylene terephthalamide) fibers. ▶ Aramid.

K-Factor *n* A term sometimes used (incorrectly) for insulation value or used for ▶ Thermal Conductivity.

K Film *n* A chemical wood pulp made by the sulphate process, or paper or paperboard made from such pulp.

Kg SI abbreviation for ▶ Kilogram. (CRC handbook of chemistry and physics. Lide DR (ed). CRC, Boca Raton, FL, 2004 Version).

Khaki \ˈka-kē, ˈkä-, *Canad often* ˈkär-\ *n* [Hindi *khākī* dust-colored, fr. *khāk* dust, fr Persian] (1857) (1) A light yellowish brown. (2) A khaki-colored cloth of cotton, wool, or combinations of these fibers with manufactured fibers used primarily in military uniforms and workclothes. (Elsevier's textile dictionary. Vincenti R (ed). Elsevier Science and Technology Books, New York, 1994).

KHN ▶ Knoop Hardness Number.

Kibbled \ˈki-bəl\ *vt* (ca. 1790) Broken into small lumps, about ¼ in. diameter, e.g., kibbled glue.

Kick Board ▶ Baseboard.

Kidney Oil *n* Fraction boiling between 250°C and 270°C, obtained by the destructive distillation of rosin.

Kienle's Functionality Theory *n* Fundamental postulate which covers, among other things, the likelihood of reactions between compounds to form products of high molecular weight; and the relation between size and shape of reacting molecules and physical properties of the reaction products. (Fundamentals of polymer science: An introductory text. CRC, Boca Raton, FL, 1998; Concise dictionary of polymer science and engineering. Kroschwitz JI (ed). Wiley, New York, 1990).

Kier *n* A large metal tank, capable of being heated uniformly, used for wet processing.

Kier Boiling Process of boiling cellulosic materials in alkaline liquors in a kier at or above atmospheric pressure.

Kieselguhr \ˈkē-zəl-ˌgur\ *n* [Gr *Kieselgur*] (1875) Alternate name for ▶ Diatomite. ▶ Diatomaceous Silica.

Kiln \ˈkiln, ˈkil\ *n* [ME *kilne*, fr. OE *cyln*, fr. L *culina* kitchen, fr. *coquere* to cook] (before 12c) An oven, furnace, or heated enclosure used for processing a substance by burning, firing, or drying.

Kiln Dried *n* Lumber dried in chambers. Heat is controlled to prevent cracking and warping. Syn: Kiln Seasoned.

Kiln Seasoned ▶ Kiln Dried.

Kilo -\ˈkē-(ˌ)lō *also* ˈki-\ *combining form* [F, mod. of Gk *chilioi*] (k) The SI prefix meaning $\times 10^3$.

Kilogram \-ˌgram\ *n* [F *kilogramme*, fr. *kilo* + *gramme* gram] (1797) (kg) One of the basic units of SI, the mass of a particular platinum cylinder kept at the International Bureau of Weights and Measures in Paris. One avoirdupois pound = 0.4535924 kg. (CRC handbook of chemistry and physics. Lide DR (ed). CRC, Boca Raton, FL, 2004 Version).

Kinel Polyimide, manufactured by Rhone Poulenc, France.

Kinematics \ˌki-nə-ˈma-tiks\ *n plural but singular in construction* [F. *cinématique*, fr. Gk *kinēmat-*, *kinēma* motion, fr. *kinein* to move] (1840) A branch of dynamics that deals with aspects of motion apart from considerations of mass and force.

Kinematic Viscosity *n* (kinetic viscosity) The absolute (dynamic) viscosity of a fluid divided by the density of the fluid. The SI unit is m^2/s, but the cgs unit, the *stoke*, which equals 10^{-4} m^2/s, is still in wide use, as is its submultiple, the centistoke. (Goodwin JW, Goodwin J, Hughes RW (2000) Rheology for chemists. Royal Society of Chemistry, UK, August).

Kinetic Coefficient of Friction *n* The ratio of tangential force, which is required to sustain motion without acceleration of one surface with respect to another, to the normal force, which presses the two surfaces together. Also called coefficient of friction, coefficient of friction, kinetic.

Kinetic Energy *n* (1870) Energy associated with the motion of an object. An object of mass *m* moving at velocity v has kinetic energy $\frac{1}{2}$ mv². (Serway RA, Faugh JS, Bennett CV (2005) College physics. Thomas, New York).

Kinetic-Energy Correction *n* (1) In the Izod, Charpy and tensile-impact tests, a subtraction of the kinetic energy imparted to the broken-off part of the specimen. (2) In measurement of ▶ Dilute-Solution Viscosity a correction for the energy required to accelerate the liquid in the reservoir to its higher velocity in the capillary. A similar correction is theoretically needed in melt rheometry, but has so far been found to be much smaller than the errors of measurement. (Kamide K, Dobashi T (2000) Physical chemistry of polymer solutions. Elsevier, New York).

Kinetics \kə-ˈne-tiks *also* kī-\ *n* (ca. 1859) A branch of dynamics concerned with the relations between the movement of bodies and the forces acting upon them. (Connors KA (1990) Chemial kinetics. Wiley, New York).

Kinetic Theory *n* (1864) Either of two theories in physics based on the fact that the minute particles of a substance are in vigorous motion. The first theory is that the particles of a gas move in straight lines with high average velocity, continually encounter one another and thus change their individual velocities and directions, and cause pressure by their impact against the walls of a container. *Also known as the Kinetic Theory of Gases.* The second theory is that the temperature of a substance increases with an increase in either the average kinetic energy of the particles or the average potential energy of separation (as in fusion) of the particles or in both when heat is added. (Connors KA (1990) Chemial kinetics. Wiley, New York) Also known as the *Kinetic Theory of Heat.*

Kink \ˈkiŋk\ *n* [D; akin to MLGr *kinke* kink] (1678) (1) In fabrics, a place where a short length of yarn has spontaneously doubled back on itself. (2) In yarn, ▶ Snarl.

Kinking *n* The doubling back of yarn on itself to relieve torque imparted by twisting or texturing. ▶ Kink.

Kira-ye [Japanese] A print with a mica background.

Kirchhoff's Law (Emissivity) $1 = \varepsilon_R + \varepsilon_T + \varepsilon_A$, where ε_R is the energy reflected, E_T is the energy transmitted through the material and E_A is the energy absorbed by the material which is reemitted. The ability to reemit the energy is the emissivity. The common emissivity (ε) of a material is equal to $1 - \varepsilon_R$. This is an indicator of the property of a material to absorb radiation such as white light or infrared radiation and a perfect absorber of energy is a black body where $\varepsilon = 1$. (Driggers RC, Cox P, Edwards T (1998) Introduction to infrared and electro-optical systems. Artech House, Aylesford, UK, January; Handbook of infrared materials. Klocek P (ed). Marcel Dekker, New York, 1991).

Kirksite An alloy of aluminum and zinc, easily castable at relatively low temperatures, often used for molds for blow molding. Its high thermal conductivity hastens cooling.

Kling Test A method for determining the degree of fusion between flexible vinyl sheets, coated fabrics, and thin sections of cast or molded parts, by immersing the folded specimen in a solvent and observing the elapsed time at which disintegration commences. Useful solvent systems comprise methyl ethyl ketone, tetrahydrofuran, ethyl acetate, and carbon tetrachloride. The preferred solvent system is one that will initiate degradation within 5–10 min in a fully fused specimen. (Handbook of polyvinyl chloride formulating. Wickson EJ (ed). Wiley, New York, 1993).

Kneader \ˈnēd-\ *n* [ME *kneden*, fr. OE *cnedan*; akin to OHGr *knetan* to knead] (before 12c) A mixer with a pair of intermeshing blades, often S-shaped, used for working plastic masses of semi-dry or rubbery consistency. (Perry's chemical engineer's handbook, 7th edn. Perry RH, Green DW (eds). McGraw-Hill, New York, 1997).

Knee Break-Out Test *n* A method to evaluate the performance of fabrics, especially boys' wear, when subjected to abrasion, stretch, and impact forces under conditions which simulate ordinary wear at the knee. (Fairchild's dictionary of textiles. Tortora PG (ed). Fairchild Books, New York, 1997).

Kneeing *n* Abnormal behavior of a spinning threadline (especially in melt spinning) in which one or more filaments form an angle (knee).

Knife Coating *n* A method of coating a substrate in which the substrate, in the form of a continuous moving web, is coated with a plastic whose thickness is controlled by an adjustable knife or bar set at a suitable angle to the substrate. ▶ Spread Coating and ▶ Air-Knife Coating.

Knife Mark Doctor Mark.

Knife Test *n* Test for brittleness, toughness and tendency to ribbon, by cutting a narrow strip of the coating from the test panel, with a knife. It is not recommended as a test for adhesion, as other methods are more accurate. (Weldon DG (2001) Failure analysis of paints and coatings. Wiley, New York; Paint and coating testing manual (Gardner-Sward handbook) MNL 17, 14th edn. ASTM, Conshohocken, PA, 1995).

Knifing Filler *n* Filling composition suitable for application with a filling knife as distinct from one made for brush application.

Knit-de-Knit ▶ Texturing.

Knit Fabric *n* A structure produced by interlooping one or more ends of yarn or comparable material. ▶ Knitting.

Knit-Miss *n* A form of tricot knitting in which yarns on each bar of a two-bar machine are knit at alternate courses only. This type of knitting permits the use of heavy-denier yarns without creating undesirable bulkiness in the fabric. (Solomon DH, Hawthorne DG (1991) Chemistry of pigments and fillers. Krieger, New York).

Knitting *n* A type of knitting in which the yarns run lengthwise in the fabric. The yarns are prepared as warps on beams with one or more yarns for each needle. Examples of this type of knitting are tricot, milanese, and raschel knitting. (a) *Milanese Knitting*: A type of run-resistant warp knitting with a diagonal rib effect using several sets of yarns. (b) *Raschel Knitting*: A versatile type of warp knitting made in plain and Jacquard patterns; the latter can be made with intricate eyelet and lacy patterns and is often used for underwear fabrics. Raschel fabrics are coarser than other warp-knit fabrics, but a wide range of fabrics can be made. Raschel knitting machines have one or two sets of latch needles and up to thirty sets of guides.

(c) *Tricot Knitting*: A run-resistant type of warp knitting in which either single or double sets of yarn are used. ▶ Tricot. (d) *Weft Knitting*: A common type of knitting, in which one continuous thread runs crosswise in the fabric making all of the loops in one course. Weft knitting types are circular and flat knitting. (e) *Circular Knitting*: The fabric is produced on the knitting machine in the form of a tube, the threads running continuously around the fabric. (f) *Flat Knitting*: The fabric is produced on the knitting machine in flat form, the threads alternating back and forth a cross the fabric. The fabric can be given shape in the knitting process by increasing or decreasing loops. Full-fashioned garments are made on a flat knitting machine. (Kadolph SJJ, Langford AL (2001) Textiles. Pearson Education, New York; Fairchild's dictionary of textiles. Tortora PG (ed). Fairchild Books, New York, 1997; Vigo TL (1994) Textile processing, dyeing, finishing and performance. Elsevier Science, New York; Elsevier's textile dictionary. Vincenti R (ed). Elsevier Science and Technology Books, New York, 1994) ▶ Flat-Knit Fabric. {*knitted fabric* G Gewirke n, F tissu tricoté, tissu m, S género de punto, género m, I tessuto a maglia, tessuto m} (Fairchild's dictionary of textiles. Tortora PG (ed). Fairchild Books, New York, 1997).

Knockout \ˈnäk-ˌaút\ *n* (1887) Any part or mechanism of a mold whose function is to eject the molded article. (Strong AB (2000) Plastics materials and processing. Prentice Hall, Columbus, OH).

Knoop Hardness Number, KHN *n* The indentation hardness determined with a Knoop indenter, and calculated as follows:

$$KHN = \frac{L}{Ap} = \frac{L}{1^2 Cp}$$

Where L = load in kilograms applied to the indenter, 1 = measured length of long diagonal of the indention in millimeters, Cp = indenter constant relating 1^2 to Ap, and Ap = projected area of indention in square millimeters. (Handbook of physical polymer testing, vol 50. Brown R (ed). Marcel Dekker, New York, 1999; www.astm.org).

Knoop Hardness Test *n* An indentation hardness test using calibrated machines to force a rhombic-based pyramidal diamond indenter having specific edge angles, under specified conditions, into the surface of the material under test and to measure the long diagonal after removal of the load. (Handbook of physical polymer testing, vol 50. Brown R (ed). Marcel Dekker, New York, 1999; www.astm.org).

Knoop Indenter *n* Pyramidal diamond of prescribed dimensions used for testing the indentation hardness of organic coatings. In a more restricted sense, a type of diamond hardness indenter having edge angles of 172°, 30° and 130°. (Handbook of physical polymer testing, vol 50. Brown R (ed). Marcel Dekker, New York, 1999; www.astm.org).

Knoop Microhardness *n* A test employing a diamond indenter whose point is an obtuse pyramid that makes an indentation of length seven times its width. This tester makes smaller, shallower indentations than Brinell or Vickers, suiting it to testing hardnesses of surfaces, as in case-hardened steel, or coatings. The Knoop hardness number (KHN) is equal to 14.2 P/L^2, where P = the applied load in grams and L = the length in millimeters of the long axis of the indentation. (Handbook of physical polymer testing, vol 50. Brown R (ed). Marcel Dekker, New York, 1999; www.astm.org).

Knot Sealer *n* Solutions of various resins in alcohol, used to seal knots in new wood.

Knot Tenacity *n* (knot strength) The strength of a yarn specimen containing an overhand knot to measure, by comparison with the strength of the unknotted yearn, its sensitivity to compression or shearing. (Kadolph SJJ, Langford AL (2001) Textiles. Pearson Education, New York).

Knotting Fairly concentrated solution of shellac in alcohol, used for sealing knots in new wood. (Paint/coatings dictionary. Federation of Societies for Coatings Technology, Philadelphia, Blue Bell, PA, 1978).

Knuckle Area *n* In reinforced plastics, the area of transition between sections of different geometry in a filament-wound part. (Iaac MD, Ishal O (2005) Engineering mechanics of composite materials. Oxford University Press, UK).

Kodar Polyester, manufactured by Eastman, US.

Kodel-2 Polyester from terephthalic acid and 1,4-dimethylol cychlohexane, manufactured by Eastman, US.

Kodel-10 Poly(ethylene terephthalate), manufactured by Eastman, US.

Koettstorfer Number ▶ Saponification Value.

Kohinoor Test *n* A test for ▶ Scratch Hardness employing a series of pencils of different hardnesses.

Kohlrausch-Williams-Watts Equation *n* This empirical equation aids designers of load-bearing products made of plastics and reinforced plastics.

$$G(t) = G_o e^{-(t/t_o)^m}$$

Where $G(t)$ is the stress-relaxed modulus of the test piece at time t from application of the load, G_o is the modulus at the reference time t_o, usually in the range of a set of measurements on the logarithmic time scale, and m is a material-specific constant between 0.33 and 0.5 for many polymers and composites. (Handbook of physical polymer testing, vol 50. Brown R (ed). Marcel Dekker, New York, 1999; Shah V (1998) Handbook of plastics testing technology. Wiley, New York).

Kordofan Gum \ ˈkór-də-ˌfan-\ *n* Another name for gum acacia, derived from the Sudanese providence of Kordofan. (Industrial gums: Polysaccharides and their derivatives. Whistler JN, BeMiller JN (eds). Elsevier Science and Technology Books, 1992).

Kraemer Equation For dilute polymer solutions, an equation relating the inherent viscosity to intrinsic viscosity and concentration. It is

$$\eta_{inh} = (\ln \eta_r)/c = [\eta] + k''[\eta]^2 \cdot c$$

where η_r = the reduced viscosity, $[\eta]$ = the intrinsic viscosity and c = the concentration in g/dL. ▶ Huggins Equation and ▶ Dilute-Solution Viscosity. Huggins' constant, k', and Kraemer's, k'', are related by: $k' - k'' = 0.5$. Thus, since k' is often between 0.6 and 0.8, k'' will often lie between 0.1 and 0.3. (Kamide K, Dobashi T (2000) Physical chemistry of polymer solutions. Elsevier, New York; Huggins ML (1958) Physical chemistry of high polymers. Wiley, New York).

Kraemer-Sarnow Method *n* (K & S) Softening point. Method of determining the softening points of resins and aliphatic materials. (Paint: Pigment, drying oils, polymers, resins, naval stores, cellulosics esters, and ink vehicles, vol 3. American Society for Testing and Material, 2001; www.astm.org).

Kraftcord This yarn produced by tightly twisting plant fiber is sometimes used in carpet backings. (Kadolph SJJ, Langford AL (2001) Textiles. Pearson Education, New York; Elsevier's textile dictionary. Vincenti R (ed). Elsevier Science and Technology Books, New York, 1994).

Kraft Pulps *n* Pulps prepared in the alkaline liquor consisting of sodium hydroxide, sodium carbonate, and sodium sulfide. Also called *Sulfate Pulp*.

Kraft Yarn *n* A yarn made by twisting a strip of paper manufactured from kraft pulp.

Kralastic *n* ABS, manufactured by US Rubber, US.

Krebs-Stormer Viscometer *n* This is the most commonly used viscometer in the paint industry. Paint consistency is measured by its resistance to stirring by a two-vane paddle. The two vanes are offset to avoid channeling in viscous paints. Approximate weights are added to a platform on a string attached to a pulley which is connected to the paddle until a speed of 100 rpm is reached, generally measured by a stroboscopic timer. The grams required to reach this speed are then converted, by using a conversion table, to Krebs Units. (Paint and coating testing manual (Gardner-Sward handbook) MNL 17, 14th edn. ASTM, Conshohocken, PA, 1995). ▶ Viscometer.

Krebs Unit *n* An unit used in reporting viscosity measurements made with the weight-driven ▶ Stormer Viscometer used primarily for evaluating the viscosity of formulated paint and coating materials. A Krebs unit is the weight in grams that will turn a paddle-type rotor, submerged in the sample, 100 revolutions in 30 s.

Kroy® Shrinkproofing Process *n* Continuous process for shrinkproofing wool tops in which there is a direct chlorination step with no intervening chemical reaction followed by anti-chlorination and neutralization. Provides better hand and strength than conventional shrinkproofing. (Vigo TL (1994) Textile processing, dyeing, finishing and performance. Elsevier Science, New York).

Kryston *n* Polyester, manufactured by Goodrich, US.

Krytox *n* Perfluorinated polyether, manufactured by DuPont, US.

K.U ▶ Krebs Units.

Kubelka-Munk Equation *n* (Apply to a single wavelength of light).

$$R = \frac{1 - R_g(a - b \coth bSX)}{A - R_g + b \coth bSX} \quad (1)$$

$\coth$ = hyperbolic contangent;
abbreviation, $(e^u + 1/e^u)/(e^u - 1/e^u)$

$$SX = \frac{1}{b}\left(\operatorname{arc\,coth} \frac{a-R}{b} - \operatorname{arc\,coth} \frac{a-R_g}{b}\right) \quad (2)$$

arc coth = inverse hyperbolic contangent

where
R = reflectance over substrate of reflectance R_g
R_g = reflectance of substrate

$$a = \frac{S+K}{S} = \frac{1}{2}\left(\frac{1}{R_\infty} + R_\infty\right)$$

$$b = (a^2 - 1)^{1/2} = \frac{1}{2}\left(\frac{1}{R_\infty} + R_\infty\right)$$

S = unit scattering coefficient expressed in X−1
X = thickness (or product of thickness times concentration)
R_∞ = reflectance at infinite thickness (complete hiding)

K = unit absorption coefficient expressed in X−1.
When hiding in complete (substrate is completely obscured), the first equation reduces to

$$\frac{(1-R_\infty)^2}{2R_\infty} = \frac{K}{S}$$

and for a mixture of pigments

$$\left(\frac{K}{S}\right)_{MIX} = \frac{C_1 K_1 + C_2 K_2 + C_3 K_3 + \ldots}{C_1 S_1 + C_2 S_2 + C_3 S_3 + \ldots} \quad (3)$$

where the C's refer to relative concentrations of pigments in the mixture (Ec's = 100%) and subscripts identify the pigments. When all of the scattering comes from a single component (such as white pigment in a pasted color or the fiber in a textile), the denominator of the above equation is determined by the S of this component, and the equation becomes simply

$$\left(\frac{K}{S}\right)_{MIX} = C_1 \left(\frac{K}{S}\right)_1 + C_2 \left(\frac{K}{S}\right)_2 + C_3 \left(\frac{K}{S}\right)_3 + \ldots \quad (4)$$

The latter is referred to as the single constant K-M equation compared to the third one using separate K's and S's, which is called the two-constant equation. The above equations are those most widely used for computer color matching. (Colour physics for industry, 2nd edn. McDonald, Roderick, Society of Deyes and Colourists, West Yorkshire, England, 1997; Paint/coatings dictionary. Federation of Societies for Coatings Technology, Philadelphia, Blue Bell, PA, 1978).

Kubelka-Munk Theory *n* A theory describing the optical behavior of materials containing small particles which scatter and absorb radiant energy. It is widely used for color matching calculations. The mathematical equation describes the reflectance or transmittance in terms of an absorption coefficient, K, and a scattering coefficient, S. The K-M theory is based on an assumption of multiple scattering, that is, reflectances from one particle on to other particles before the reflected radiant energy is observed. This assumed behavior is in contrast to the Mie Theory which is based on an assumption of single isolated scattering of individual particles. (Colour physics for industry, 2nd edn. McDonald, Roderick, Society of Deyes and Colourists, West Yorkshire, England, 1997).

Kusters Dyeing Range *n* Continuous dye range for carpets. The unit wets the carpet, applies dyes and auxiliary chemicals by means of a doctor blade, fixes the dyes in a festoon steamer, and washes and dries the carpet in one pass through the range. An optional auxiliary unit may be installed to randomly drip selected dyes onto the background shade for special styling effects. This process is called TAK dyeing. (Vigo TL (1994) Textile processing, dyeing, finishing and performance. Elsevier Science, New York).

K-Value *n* An alternate name for ▶ Thermal Conductivity.

Kynol *n* Phenol-formaldehyde fiber, manufactured by Carborundium, US.

L

l \ˈel\ {*often capitalized, often attributive*} (before 12c) (also L) Symbol for length.

L (1) Abbreviation for SI-permitted (but discouraged) volume unit, the Liter, redefined in 1964 to be exactly 1 dm^3. (2) Symbol for length or magnetic inductance. (3) Abbreviation for ▶ Lightness, generally used in color order systems to describe the perception of the amount of light reflected or transmitted by materials. See ▶ Lightness and ▶ Luminance.

Labeling \ˈlā-b(ə-)liŋ\ *vt* (1601) To affix a precut, printed, flexible material to the surface of a product.

Labradorite \ˌla-brə-ˈdór-ˌīt\ *n* [*Labrador* Peninsula, Canada] (1814) Calcium sodium aluminum silicate, hardness 6 Mohs scale, used as a wear resistant filler in deck paints. (Herbst W, Hunger K (2004) Industrial organic pigments. Wiley)

Lac \ˈlak\ *n* [Persian *lak* & Hindi *lākh*] (1598) Resin secreted by the female of the insect, *Lacccifer iacca*, and deposited on the twigs of various species of trees in Indian and Indochina. After refining, the resin is known as shellac. (Langenheim JH (2003) Plant resins: Chemistry, evolution ecology and ethnobotany. Timber, Portland, OR)

Lace \ˈlās\ *n* [ME, fr. MF *laz*, fr. L *laqueus* snare] (14c) Ornamental openwork fabric, made in a variety of designs by intricate manipulation of the fiber by machine or by hand. (Fairchild's dictionary of textiles. Tortora PG (ed). Fairchild Books, New York, 1997).

Lace Stitch In this knitting stitch structure, loops are transferred from the needles on which they are made to adjacent needles to create a fabric with an open or a raised effect. (Fairchild's dictionary of textiles. Tortora PG (ed). Fairchild Books, New York, 1997)

L Acid See ▶ Cleve's Acid.

Lac Dye Red dye obtained by the maceration of lac.

Lacquer \ˈlac-kər\ *n* [Portuguese lacré sealing wax, fr. *laca* lac, fr. Arabic *lakk*, fr. Persian *lak*] (1592) (1) A solution of a film-forming natural or synthetic resin in a volatile solvent, with or without color pigment, which when applied to a surface forms an adherent film that hardens solely by evaporation of the solvent. The dried film has the properties of the resin used in making the lacquer. The word derives from the *lac* insect, which secreted the resinous substance from which *shellac* solutions were (and stilla are) made. Today most lacquers are made with cellulosic, alkyd, acrylic, and vinyl resins. (2) Finish on Chinese and Japanese lacquer ware. (Wicks ZN, Jones FN, Pappas SP (1999) Organic coatings science and technology, 2nd edn. Wiley-Interscience, New York) See ▶ Spirit Varnish.

Lactam *n* [ISV *lact*- + *am*ide] A cyclic amide obtained by removing one molecule of water from an amino acid. An example is ▶ Caprolactam.

Lactic Acid *n* (1790) (milk acid, α-hydroxypropionic acid) CH$_3$CHOHCOOH. A colorless or yellowish liquid with several applications in plastics. Reacted with glycerine, it forms an alkyd resin. It is a catalyst for vinyl polymerizations, and an additive for phenolic casting resins. It has a bp of 122°C/15 mmHg, mp of 18°C, and sp gr of 1.2 *Also known as Alpha-Hydroxypropionic Acid and Milk Acid.*

Lac Wax Wax obtained from lac consisting of myricyl and ceryl alcohols, free and combined with various fatty acids.

Ladder Polymers (double-stranded polymer) A polymer comprising chains made up of fused rings. Examples are cyclized (acid-treated) rubber and ▶ Polyimidazopyrrolone.

Laid-In Fabric A knit fabric in which an effect yarn is tucked in, not knitted into, the fabric structure. The laid-in yarns are held in position by the knitted yarns. See ▶ Axial Yarn.

Laitance \ˈlā-t°n(t)s\ *n* [F, fr. *lait* milk, fr. L *lact*-, *lac*] (ca. 1902) Milky white deposit on new concrete; efflorescence.

Lake A type of organic pigment prepared from water-soluble acid dyes, precipitated on an inert substrate by means of a metallic salt, tannin, or other reagent. Lakes were used in plastics at one time, but have been replaced by more permanent pigments.

Lake \ˈlāk\ *n* [F laque lac, fr. OP *laca*, fr. Arabic *lakk*] (1598) Special type of pigment consisting essentially of an organic soluble coloring mater combined more or less definitely with an inorganic base or carrier. It is characterized generally by a bright color and more or less pronounced translucency when made into an oil paint. Under this term are included two (or perhaps three) types of pigment: (a) the older original type composed of hydrate of alumina dyed with a solution of the natural organic color; (b) the more modern and far more extensive type made by precipitating from solution various coal-tar colors by means of a metallic salt, tannin, or other suitable reagent, upon a base or carrier either previously prepared or coincidently formed; and (c) a number combining both types in varying degree might be regarded as a third class.

Lake Bordeau B Red dyestuff produced from 2,3 hydroxy naphthoic acid and 2-naphthylamine-1-sulfonic acid. The calcium and manganese toners are known as Bordeaux R Toner and Maroon Toner B.B. respectively. The latter has good lightfastness.

Lake Dyes Dyes used for the making of lakes by combination with, or adsorption on, salts of calcium, barium, chromium, aluminum, phosphotungstic acid or phosphomolybydic acid.

Lake of Acid Yellow 1 Acid Yellow 1, Lake (10316). A nitro dye prepared by precipitation from an aqueous solution by a metal salt, i.e., aluminum or barium chloride.

Lake Red C Pigment Red 53 (15585). Red dyestuff derived from 6-chlor-3-toluidine-4-sulfonic acid, and β-naphthol. When laked with barium chloride, bronze scarlet is obtained.

Lake Red P Pigment produced by coupling diazotized *p*-nitroaniline-*o*-sulfonic acid with betanaphthol.

Laketine A transparent ink used for extending letterpress or lithographic inks.

Lallemantia Oil Drying oil obtained from the seeds of *Lallemantia iberica*, found in parts of Asia and Europe. It has sp gr of 0.934/20°C, iodine value of 190, and saponification value of 190. Its main constituents are linolenic and linolenic glycerides.

LALLS See ▶ Low-Angle Laser-Light Scattering.

Lambda Zero (λo) In dispersion staining, the wavelength at which both particle and liquid have the same refractive index.

Lambert \lam-bərt\ *n* [Johann H. *Lambert* † 1777 German physicist & philosopher] (1915) A deprecated unit of illumination equal to 1 lumen/cm^2. The SI unit is the *lux* (lx), equal to 1 lm/m^2, i.e., 10^{-4} lambert. See also ▶ Luminous Flux.

Lambert's Law of Absorption See Bouguer's Law.

Lambert's Law of Reflection The flux reflected per unit solid angle is proportional to the cosine of the angle measured from the normal (perpendicular) to the surface. If the reflected flux is isotropic, the surface is said to be a perfect Lambertian reflector or a perfect diffuser.

Lamé \lä-mā, la-\ *n* [F] (1922) A fabric woven with flat metal threads, usually silver or gold, that form either the background or the pattern.

Lamellae \lə-me-lə\ *n* [NL, fr. L, dim. of *lamina* thin plate] (1678) Thin, flat scales or plates.

Lamellae Structure Platelike single crystals that exist in most crystalline polymers. A thin, flat scale or part.

Lamina \la-mə-nə\ *n* [L] (ca. 1656) A single layer or ply within a laminate.

Laminar Flow Flow without turbulence, i.e., the movement of one layer of fluid past another layer with no eddying between them. See ▶ Reynolds Number. Most melt flow, even at high velocities, is laminar. With sudden changes in melt velocity, such as may occur at some extrusion-die entries and at injection-mold gates, laminar flow may be disrupted by ▶ Melt Fracture. Also see ▶ Jetting.

Laminar Scale Rust formation in heavy layers.

Laminate \la-mə-nāt\ (1665) (1, *n*) A product made by bonding together two or more layers of material or materials. The term most usually applies to preformed layers joined by adhesives or by heat and pressure. However, some authors apply the term to composites of plastic films, with other films, foils, and papers, even though they have been made by spread coating or by extrusion coating. In the reinforced-plastics industry, the term refers mainly to superimposed layers of resin-impregnated or resin-coated fabrics or fibrous reinforcements that have been bonded together, usually by heat and pressure, to form a single piece. (Natural fibers, plastics and composites. Wallenberger FT, Weston NE (eds), Springer, New York, 2003; Engineering plastics and composites. Pittance JC (ed), SAM International, Materials Park, OH, 1990; Handbook of adhesives. Skeist I (ed), Van Nostrand Reinhold, New York, 1990)

Laminated, Cross A laminate in which some of the layers of material are oriented at right angles to the remaining layers with respect to the grain or strongest direction in tension. NOTE — Balanced construction of the laminations about the center line of the thickness of the laminate is normally assumed. See also ▶ Laminated.

Laminated Fabric (Fabri composed of a high-strength reinforcing scrim or base fabric between two plies of flexible thermoplastic film. Usually open scrims are used to permit the polymer to flow through the interstices and bond during calendering.

Laminated Glass A structure consisting of two or more parallel sheets or shells of glass interleaved with, and bonded to, layers of tough, sticky plastic, typically polyvinyl butyral or polycarbonate. The former resin is used in the "safety-glass" windshields of US -made cars. The glass will splinter under a heavy blow, but it is very resistant to penetration and the shards stick to the interlayer.

Laminated, Parallel A laminate in which all the layers of material ore oriented approximately parallel with respect to the grain or strongest direction in tension.

Laminate, High Pressure Laminates molded and cured at pressures not lower than 1,000 psi, and commonly in the range of 1,200–200 psi.

Laminates Products made by bonding together two or more layers of material and materials.

Lamination *n* (ca. 1676) The process of preparing a laminate. Also, any layer in a laminate.

Lampblack \-ˈblak\ *n* (1598) Pigment Black 6 (77266). A bulky black soot obtained from the incomplete combustion of creosote or fuel oils, of duller and less intense blackness than ▶ Channel Black and other ▶ Carbon Blacks, and having a blue undertone and a small oil content.

Land (1) The horizontal bearing surface of a semipositive or flash mold by which excess material escapes. (2) The bearing surface along the top of the flights of an extruder screw. (3) The final shaping surface of an extrusion die, usually parallel to the direction of melt flow. (4, plural) The mating surfaces of any mold, adjacent to the cavity depressions that, when in contact, prevent the escape of material.

Land Area The area of those surfaces of a mold that contact each other when the mold is closed, measured in a plane perpendicular to the direction of application of the closing pressure.

Landed Force A force with a shoulder that seats on the land in a landed positive mold.

Land Width (flight width) Of an extruder screw, the distance across the tip of the flight, perpendicular to the flight faces.

Lane Length In an extrusion die, the distance across the land in the direction of melt flow between the lands.

Lanital Fiber from milk albumin. Manufactured by Snia Viscosa, Italy.

Lanolin \ˈla-nᵊl-ən\ *n* [L *lana* wood + ISV ^{3}-*ol* + ^{1}-*in*] (1885) Purified wool grease.

Lap *n* (1800) (1) Region where a coat extends over an adjacent fresh coat. The object of the painter is usually to effect a joint between the two coats without showing the lap. *v* (2) To place one coat of finishing material alongside another, partly extending over it, causing increased thickness where the two coats are present. (3) To overlap or partly cover one surface with another, as in shingling. (4) The length of the overlap, as the distance one tile extends over another.

Lap Joint *n* (1823) A joint made by placing one surface to be joined partly over another surface and bonding or fastening the overlapping portions. Compare ▶ Butt Joint and ▶ Scarf Joint. See ▶ Joint, ▶ Lap.

Lapis Lazuli \ˌlap-əs-ˈla-zə-lē, -ˈla-zhə-\ *n* [ME, fr. ML, fr. L *lapis* + ML *lazuli*, general of *lazuhum* lapis lazuli, fr. Arabic *lāzaward*] (15c) A rich blue semiprecious stone; either used decoratively or ground and powdered for use as an ultramarine pigment. (Kirk-Othmer encyclopedia of chemical technology: Pigments-powders. Wiley, New York, 1996)

Lapis Lazuli Blue Natural ultramarine blue. (Kirk-Othmer encyclopedia of chemical technology: Pigments-powders. Wiley, New York, 1996)

Lapping A term describing the movement of yarn guides between needles, at right angles to the needle bar, or laterally in relation to the needle bar, or laterally in relation to the needle bar during warp knitting.

Lapping Time See ▶ Wet-Edge Time.

Lap Siding See ▶ Clapboard.

Lap Winding A variant of ▶ Filament Winding consisting of convolutely winding a resin-impregnated tape onto a mandrel of the desired configuration. The process has been used for making large chemical- and heat-resistant, conical or hemispherical parts such as heat shields for atmospheric-reentry vehicles.

Lase \ˈlāz\ An acronym for load at specified elongation: the load required to produce a given elongation of a yarn or cord.

Laser \ˈlā-zər\ *n* [*l*ight *a*mplification by *s*timulated *e*mission of *r*adiation] (1960) Laser is an acronym coined from the bold-face letters in light application by stimulated emission of radiation. Early lasers were made of synthetic-ruby rod, silvered at one end face, semi-silvered at the other, and surrounded by a toroidal flashlamp. Today, the NdYAG (for neodymium-yttrium aluminum-garnet) laser, with higher productivity than the ruby laser, has taken over many of its jobs. Lasers have been useful in drilling, perforating, cutting, and welding operations with plastics and other materials. Medium-power CO_2 lasers are preferred for machining plastics because they produce light at a wavelength of 10.6 μm, which is completely absorbed by plastics. Lasers have been useful in the analysis of polymers by laser mass spectrometry, in initiating polymerizations, and in curing polymers. (Giambattista A, Richardson R, Richardson RC, Richardson B (2003) College physics. McGraw Hill Science/Engineering/Math, New York; Whittington's dictionary of plastics. Carley, James F (ed). Technomic, 1993; Saleh BEA, Teich MC (1991) Fundamentals of photonics. Wiley, New York)

Laser-Ionization Mass Spectrometry A technique for chemical analysis of transparent and opaque plastics, capable of detecting all the elements but hydrogen and helium.

Lashed-In Filling See ▶ Pulled-In Filling.

Lastrile Fiber from copolymers with 10–15% acrylonitrile and an aliphatic diene. Lastrile is a generic name.

Lastril Fiber A manufactured fiber in which the fiber-forming substance is a copolymer of acrylonitrile and a diene composed of at least 10% by weight, but not more than 50% by weight, of acrylonitrile [−CH_2−CH(CN)−] units (FTC definition).

Latch Needle One of the two types of knitting machine needles. The latch needle has a small terminal hook with a latch that pivots automatically in knitting to close the hook. The fabric loop is cast off. The latch then opens, allowing a new loop to be formed by the hook, and loop-forming and casting-off proceed simultaneously. Also see ▶ Spring Needle.

Latch Plate A plate used for retaining a removable mold core of relatively large diameter, or for holding insert-carrying pins on the upper part of a mold. Release of the pins or core is effected by moving the latch plate.

Latent Crimp Crimp in fibers that can be developed by a specific treatment. Fibers are prepared specially to crimp when subjected to specific conditions, e.g., tumbling in a heated chamber or wet processing.

Latent Heat *n* (ca. 1757) The quantity of heat necessary to change 1 g of liquid to vapor (latent heat of vaporization) without change of temperature; heat necessary for change of state being negative or positive heat (e.g., fusion, evaporation and melting). For example, the latent heat of fusion of ice water is 80 cal/g. (Ready RG (1996) Thermodynamics. Pleum, New York)

Latent Heat of Fusion See ▶ Heat of Fusion. (Ready RG (1996) Thermodynamics. Pleum, New York)

Latent Heat of Vaporization The quantity of heat required to change a unit mass (or sometimes a mole) of a liquid to its vapor, the two phases remaining in equilibrium at constant temperature, most commonly the *normal boiling point*, i.e., the temperature of boiling at a pressure of 101.325 kPa. The convenient SI units are J/g, kJ/kg or kJ/mol. Because high polymers decompose without boiling, their heats of vaporization cannot be directly measured.

Latent Solvent An organic liquid that has little or no solvent effect on a particular resin until it is activated by either heat or a mixture with a true solvent.

Latex \ˈlā-ˌteks\ *n* [NS *latic-*. *latex*, from L. fluid] (1835) (*pl* lattices or latexes) (1) An emulsion of a polymeric substance in an aqueous medium. (2) The sap of the *hevea* (rubber) tree and other plants, or emulsions prepared from the same. Latices of interest to the coatings and plastics industry are based mainly on styrene-butadiene copolymers, polystyrene, acrylics, and vinyl polymers and copolymers. (3) Fine dispersion of rubber or resin, natural or synthetic, in water; the synthetic is made by emulsion polymerization. Latex and emulsion are often used synonymously in the paint industry. Emulsified monomers once polymerized become solids or plasticized gel particles and not emulsions but aqueous suspensions. (Emulsion polymerization and emulsion polymers. Lovell PA, El-Aasser (eds), Wiley, New York, 1997); Martens CR (1964) Emulsion and water-soluble paints and coatings. Reinhold, New York; Vanderhoff JW, Gurnee EF (1956) Motion picture investigation of polymer latex phenomena. TAPPI 39(2):71–77; Vanderhoff JW, Tarkowski HL, Jenkins MC, Bradford EG (1966) Theoretical considerations of the interfacial forces involved in the coalescence of latex particles. J Macromol Chem 1(2):361–397) See also ▶ Hydrosol.

Latex Mechanical Stability The ability of latex to resist coagulation under influence of mechanical agitation.

Latex Paint A paint containing a stable aqueous dispersion of synthetic resin, produced by emulsion polymerization, as the principal constituent of the binder. Modifying resins may also be present. (Emulsion polymerization and emulsion polymers. Lovell PA, El-Aasser (eds), Wiley, New York, 1997; Martens CR (1964) Emulsion and water-soluble paints and coatings. Reinhold, New York)

Lath \ˈlath *also* ˈlath\ *n* [ME, fr. (ass.) OE *læthth-*; akin to OHGr *latta* lath, Welsh *llath* yard] (13c) One of a number of thin narrow strips of wood nailed to rafters, ceiling joists, wall studs, etc., to make a groundwork or key for slates, tiles or plastering. (Harris CM (2005) Dictionary of architecture and construction. McGraw-Hill, New York)

Lattice Energy The energy necessary to separate one mole of a crystalline solid into a gaseous collection of the units at the lattice points. (*Exception*: In the case of a metallic solid a gaseous collection of *atoms* is formed.) (Rhodes G (1999) Crystallography made crystal clear: A guide for users of macromolecular models. Elsevier Science and Technology Books, New York)

Lattice Pattern In filament winding, a pattern with a fixed arrangement of open voids producing a basket-weave effect. (Rhodes G (1999) Crystallography made crystal clear: A guide for users of macromolecular models. Elsevier Science and Technology Books, New York)

Laundry Blue Synthetic ultramarine blue. See ▶ Ultramarine Blue.

Lauric Acid \ˈlȯr-ik-, ˈlär-\ n [ISV, fr. L *laurus*] (1873) CH$_3$(CH$_2$)$_{10}$COOH. Fatty acid occurring in many vegetable fats as the glyceride, especially in cocoanut oil and laurel oil. It has a sp gr of 0.833, mp of 44°C, and refractive index of 1.4323. Also called *Dodecanoic Acid*.

Lauroyl Peroxide [CH$_3$(CH$_2$)$_{10}$C(O)–]$_2$. A peroxide used as an initiator in free-radical polymerizations of styrene, vinyl chloride, and acrylic monomers.

Lauryl Alcohol \ˈlȯr-əl-, ˈlär-\ n (1922) C$_{12}$H$_{26}$O A liquid mixture of this and other alcohols used in making detergents.

Lauryl Methacrylate H$_2$C=C(CH$_3$)COO(CH$_2$)$_{11}$CH$_3$. A monomer used in the production of acrylic resins.

Lawn \ˈlȯn\ n [ME, fr. *Laon*, France] (15c) A light, thin cloth made of carded or combed yarns, this fabric is given a crease resistant, crisp finish. Lawn is crisper than voile but not as crisp as organdy.

Law of Constancy of Interfacial Angles In all crystals of the same substance (in the absence of polymorphism), angles between corresponding faces are identical. (Rhodes G (1999) Crystallography made crystal clear: A guide for users of macromolecular models. Elsevier Science and Technology Books, New York)

Law of Mixtures (rule of mixtures) Properties of binary mixtures lie between the corresponding properties of the pure components and are proportional to the volume fractions v (sometimes weight or mole fractions) of the components. For example, the density ρ of a blend is given by:

$$\rho_m = \rho_1 v_1 + \rho_2 v_2 = \rho_2 + (\rho_1 - \rho_2)$$

The "law" works well for properties of unidirectional composites, such as modulus μ, but often fails for melt viscosities of blends, where maxima and minima above or below the viscosities of the neat resins are common. Another rule suggested by Arrhenius for mixture viscosities has the form:

$$\log \mu_m = x_i \log \mu_1 + x_2 \log \mu_2$$

where the x_i are mole fractions. While this rule has a different form than the preceding one and fits the mixture data of some systems, it, too, cannot provide maxima or minima. (Engineering plastics and composites. Pittance JC (ed), SAM International, Materials Park, OH, 1990)

Law of Rational Indices The lengths of intercepts of different crystal faces on any crystallographic axis are in ratios of small integers. (Rhodes G (1999) Crystallography made crystal clear: A guide for users of macromolecular models. Elsevier Science and Technology Books, New York)

Lay \ˈlā\ n (1590) (1) The length of twist produced by stranding singly or in groups, such as fibers or rovings; or the angle that such filaments make with the axis of the strand during a stranding operation. The length of twist of a filament is usually measured as the distance parallel to the axis of the strand between corresponding points on successive turns of the filament. (2) The term is also used in the packaging of glass fibers for the spacing of the roving bands in the package expressed as the number of bands per inch.

Lay-Flat Film Film that has been extruded as a wide, thin-walled, circular tube, usually blown, cooled, then gathered by converging sets of rollers and wound up in flattened form.

Lay-Flat Width In blown-film manufacture, half the circumference of the inflated film tube.

Laying Off Final light strokes of the brush during a painting operation. See ▶ Feathering.

Layup The tailoring and placing of reinforcing material – mat or cloth – in a mold prior to impregnating it with resin. The reinforcement is usually cut and fitted to the mold contours. See also ▶ Sheet-Molding Compound.

Layup Molding (hand-layup molding) A method of forming reinforced plastics articles comprising the steps of placing a web of the reinforcement, which may or may not be preimpregnated with a resin, in a mold or over a form and applying fluid resin to impregnate and/or coat the reinforcement, followed by curing of the resin and extraction of the cured article from the mold. When little or no pressure is used in the curing process, it is sometimes called *contact-pressure molding*. When pressure is applied during curing, the process is often named for the method of applying pressure, e.g., *vacuum-bag molding* or *autoclave molding*. A related process is ▶ Sprayup molding. (Strong AB (2000) Plastics materials and processing. Prentice Hall, Columbus, OH; Whittington's dictionary of plastics. Carley, James F (ed), Technomic, 1993)

LCP See Liquid Crystal Polymer.

LC Polymer Abbreviation for ▶ Liquid-Crystal Polymer.

LC$_{50}$ The concentration, in parts per million, of a substance in air that is lethal to 50% of the laboratory animals exposed to it.

LC$_{50}$ Test (Lethal Dose 50%). A toxicity test based on the results of animal experiments, the results of which mean that half the animals die from a given dosage of the substance tested.

LDPE Abbreviation for ▶ Low-Density Polyethylene. Also see ▶ Polyethylene.

L/D Ratio In an extruder, the ratio of the flighted length of the screw to its nominal diameter.

Lea (1) One-seventh of an 840-yard cotton hank, i.e., 120 yards. (2) A standard skein with 80 revolutions of 1.5 yards each (total length of 120 yards). It is used for strength tests. (3) A unit of measure, 300 yards, used to determine the yarn number of linen yarn. The number of leas in 1 lb is the yarn number. (Kadolph SJJ, Langford AL (2001) Textiles. Pearson Education, New York)

Leaching *v* (1796) The process of the extraction of a component from a solid material by treating the material with a solvent that dissolves the component of interest but not the remaining principle material. Natural leaching occurs when rain water dissolves soluble salts from soil. Some components of a paint coating leach from the dried film after application to a substrate when exposed to rain or when immersed in water.

Leaching Rate Milligrams of water-soluble material released per square centimeter per day from an immersed anti-fouling paint in seawater.

Leacril Poly(acrylonitrile), manufactured by ACSA, Italy.

Lead \\led\ *n* [ME *leed*, fr. OE *lēad*; akin to MHGr *lōt* lead] (before 12c) Element (Pb) found mostly in combination and used in pipes, cable sheaths, batteries, solder, and shields against radioactivity. Powders of this metal have also been employed as fillers in plastics. (Hibbard MJ (2001) Mineralogy. McGraw-Hill, New York)

Lead \\lēd\ Of an extruder screw, the distance parallel to the screw axis from any point on the screw thread to the corresponding point on the next turn of that threat, or pitch.

Lead Acetate *n* (1885) Pb(C$_2$H$_3$O$_2$)$_2$·3H$_2$O. Crystalline salt, soluble in water and used in the manufacture of lead pigments; also used in production of varnishes. (Kirk-Othmer encyclopedia of chemical technology: Pigments-powders. Wiley, New York, 1996; Gooch JW (1993) Lead based paint handbook. Plenum, New York)

Lead Azide *n* (1918) Pb(N$_3$)$_2$ A crystalline explosive compound used as a detonating agent.

Lead Carbonate *n* (1873) See ▶ Basic Lead Carbonate.

Lead Carbonate, Black See ▶ Carbonate White Lead.

Lead Chrome Green See ▶ Chrome Green.

Lead Chrome Pigment Any of a series of inorganic pigments including yellows, oranges, and greens, used in PVC, polyolefins, cellulosics, acrylics, and polyesters.

Lead Dioxide *n* (1885) PbO$_2$ A poisonous compound used as an oxidizing agent and as an electrode in batteries.

Lead Drier One of many organic lead salts which are soluble in paints and varnishes; used to speed the drying and hardening of the oil vehicle.

Leaded Zinc Oxide Mixed pigment comprising zinc oxide and basic lead sulfate. Several grades are possible and vary in the proportions of the individual ingredients. Leaded zinc oxides are either mechanical mixtures of the two ingredients, or are made by heating together roasted zinc ore, lead sulfide ore and coal. (Kirk-Othmer encyclopedia of chemical technology: Pigments-powders. Wiley, New York, 1996; Gooch JW (1993) Lead based paint handbook. Plenum, New York)

Leader-Pin Bushing See ▶ Guide-Pin Bushing.

Leading Flight Face (leader flight) The forward or front side of the flight of an extruder screw, the rear side being the *trailing flight face*.

Lead in Oil Basic carbonate white lead ground in linseed oil; formerly wide use, now replaced largely by titanium dioxide pigments (Gooch JW (1993) Lead based paint handbook. Plenum, New York).

Lead Monoxide *n* (ca. 1909) PbO A yellow to brownish red poisonous compound used in rubber manufacture and glassmaking.

$$Pb^{++} \quad O^{--}$$

Lead Naphthenate Lead soap of naphthenic acids. Used as a drier.

Lead Nitrate $Pb(NO_3)_2 \cdot 2H_2O$. Crystalline salt soluble in water and used in the manufacture of lead chromes.

Lead Octoate Lead soap of 2-ethylhexanoic acid. Used as a drier.

Lead Oxide *n* (ca. 1926) Either yellow lead oxide, PbO (litharge) or red lead oxide, Pb_3O_4. Both are used as pigments, though much less today in the US than formerly because of concerns about lead's toxicity and the need to keep it out of the environment. The oxides are sometimes used as fillers in radiation-shielding applications.

$$Pb^{++} \quad O^{--}$$

Lead Phosphite, Dibasic $2 \, PbO \cdot PbHPO_3 \cdot \frac{1}{2}H_2O$. Fine white acicular crystals. Pigment grade is the dibasic lead salt of phosphorous acid. Used as an anti-corrosive white pigment, and for its ability to eliminate bleed-through of cedar and redwood stains in latex paints. (Kirk-Othmer encyclopedia of chemical technology: Pigments-powders. Wiley, New York, 1996)

Lead-Restricted Paint Normally, a paint having a lead content below a given limit, as used for applications where the presence of lead may be harmful. (Gooch JW (1993) Lead based paint handbook. Plenum, New York)

Lead Salicylate (dibasic lead Salicylate) Pb $(C_6H_4OHCOO)_2 \cdot H_2O$. A white crystalline material formerly used as a heat stabilizer.

Lead Stabilizer Any of a large family of highly effective heat stabilizers that are limited to use in applications where toxicity, sulfur staining, and lack of clarity are not objectionable. Examples are: basic lead carbonate, basic lead sulfate complexes, basic silicate or white lead, coprecipitated lead silicate and silica gel, dibasic lead maleate.

Lead Stearate $Pb(C_{17}H_{35}COO)_2$. A white powder used as a vinyl-resin stabilizer and lubricant in extrusion compounds, and, earlier, in phonograph-record compounds. (Handbook of polyvinyl chloride formulating. Wickson EJ (ed), Wiley, New York, 1993)

Lead Sulfide *n* (ca. 1898) PbS. Mineral which occurs as galena and is one of the chief sources of lead. It can be prepared by passing hydrogen sulfide gas into an acid solution of lead nitrate. It is the compound responsible for darkening of paints when a lead-pigmented paint is exposed to sulfur-containing gases. (Hibbard MJ (2001) Mineralogy. McGraw-Hill, New York)

$$Pb^{++} \quad S^{--}$$

Lead Titanate $PbTiO_3$. A yellow buff-colored pigment which possesses good resistance to weather and chalking. Pale yellow solid; insoluble in water. Its sp gr is 7.52. Derivation: Interaction of oxides of lead and titanium at a high temperature. Contains lead sulfate and lead oxide as impurities.

$$^-O-Ti\begin{matrix}\diagup\!\!\!\!O\\ \diagdown O^-\end{matrix}$$
$$Pb^{++}$$

Lead White See ▶ Carbonate White Lead.
Leaf Green See ▶ Chromium Oxide Green.
Leafing A phenomenon produced by metallic pigments forming a layer parallel to the surface of the print and providing a high metallic luster.
Leafing The ability of an aluminum or gold bronze to provide a brilliant or sappearance. The flat pigment particles align themselves parallel with the coated surface providing a high reflection of light.
Leathercloth A term sometimes used, especially in Europe, for plastic-coated fabric with a leather-like texture.
Le Châtelier's Principle When a system at equilibrium experiences a stress, it will adjust, if possible, so as to minimize the effect of the stress.
Lecithin \ˈle-sə-thən\ n [ISV, fr. Gk *lekithos* yolk of an egg] (1861) A liquid, obtained in refinement of soya beans or cottonseed; used in paints to promote pigment wetting and to control pigment settling and flow properties. (Weismantal GF (1981) Paint handbook. McGraw-Hill, New York)
Lecyar Model A model for viscosities for polymer blends which do not obey simple mixture laws. The Lecyar model can accommodate a viscosity minimum, maximum, or both in the composition range from 0% to 100% of one polymer intimately blended with another. For a given shear rate and temperature it has the form:

$$\ln \eta_B = Am_1^3 + Bm_2^3 + Cm_1^2 m_2 + Dm_1 m_2^2$$

in which A and B are the natural logarithms of the neat resins 1 and 2, C and D are the natural logarithms of interaction viscosities that must be evaluated from measurements on blends, and m_1 and m_2 are the mass fractions of the two components ($m_1 + m_2 = 1$). (Shenoy AV (1996) Thermoplastics melt rheology and processing. Marcel Dekker, New York)
LEFM Abbreviation for ▶ Linear Elastic Fracture Mechanics.
Legging n The drawing of filaments or strings when adhesive-bonded substrates are separated.
Legs The stringy effect that is apparent when cemented surfaces are separated shortly after the bond is made. Long legs or strings are indicative often of a weak bond, whereas short legs indicate a strong bond. (Handbook of adhesives. Skeist I (ed). Van Nostrand Reinhold, New York, 1990)
Leguval Unsaturated polyester, manufactured by Bayer, Germany.
Leipsic Yellow See ▶ Chrome Yellow, Light and Primrose.
Leithner's Blue See ▶ Cobalt Blue.
Lemon Yellow n (1807) Any of two families of yellow pigments, one being mixtures of barium chromate with zinc carbonate, the other mixtures of lead chromate and lead carbonate.
Length The property of an ink whereby it can be stretched out into a long thread without breaking. Long inks have good flow in the fountain. (Printing ink manual, 5th edn. Leach RH, Pierce RJ, Hickman EP, Mackenzie MJ, Smith HG (eds), Blueprint, New York, 1993)
Leno Weave \ˈlē-(ˌ)nō-\ n [per. fr. F *linon* linen fabric, lawn, fr. MF *lin* flax, linen, fr. L *linum* flax] (1821) A weave in which the warp yarns are arranged in pairs with one twisted around the other between picks of filling yarn as in marquisette. This type of weave gives firmness and strength to an open weave fabric and prevents slippage and displacement of warp and filling yarns. (Kadolph SJJ, Langford AL (2001) Textiles. Pearson Education, New York)
Let Down (1) The process of paint manufacturing in which the pigment paste (mill base) is reduced (let down) by the addition of the remaining ingredients of the formula. (2) Reducing the intensity or depth of a colored pigment through the addition of a white, or sometimes colorless, pigment.
Let-Go An area in laminated glass over which an initial adhesion between interlayer and glass has been lost. (Wicks ZN, Jones FN, Pappas SP (1999) Organic coatings science and technology, 2nd edn. Wiley-Interscience, New York; Weismantal GF (1981) Paint handbook. McGraw-Hill, New York)
Let-Off A device used in coating by calendering or extrusion to suspend a coil or reel from which the material to be coated is fed to the coating machine.

Letterpress \ˈle-tər-ˌpres\ n (ca. 1765) The process of printing from an inked raised surface when the paper is impressed directly upon the surface. (Wijnekur FJM (1967) Elsevier's dictionary of the printing and allied industries in four languages. Research Institute for the Graphic and Allied Industries, TNO, Amsterdam, The Netherlands/Elsevier, New York)

Letterpress Printing The process used for paper is adapted to plastics by the use of special inks and transfer rolls, and possibly a modification of press speed. Flexible printing plates are usually employed, made of vinyl or rubber. It is the oldest method of printing flat plastic substrates and uses a press for relief printing from metal type or raised surfaces formed from wood, metal, or linoleum. (Printing ink manual, 5th edn. Leach RH, Pierce RJ, Hickman EP, Mackenzie MJ, Smith HG (eds), Blueprint, New York, 1993)

Letterset See ▶ Dry Offset.

Levapren Ethylene/vinyl acetate copolymer. Manufactured by Bayer, Germany.

Leveler Group of steel rollers that bends any edges that are not flat prior to coating. Also, a process in the aluminum industry that flattens strip so that the thickness is the same from one side to the other.

Leveling (1) The measure of the ability of a coating to flow out after application so as to obliterate any surface irregularities such as brush marks, orange peel, peaks, or craters which have been produced by the mechanical processes of applying or coating. (2) The property of a freshly spread aqueous polish to dry to a uniform and streak-free appearance. (Paint/coatings dictionary. Federation of Societies for Coatings Technology, Philadelphia, Blue Bell, PA, 1978; Martens CR (1968) Technology of paints, varnishes and lacquers. Reinhold, New York)

Levigation \ˌle-və-ˈgā-shən\ n [L *levigatus*, pp of *levigare* to make smooth, fr. *levis* smooth (akin to Gk *leios* smooth and perhaps to L *linere* to smear) + *-igare* (akin to *agere* to drive)] (1612) Process of mixing a pigment with water and washing. The percentage of salts and heavy particles is reduced or eliminated by controlled sedimentation. (Paint/coatings dictionary. Federation of Societies for Coatings Technology, Philadelphia, Blue Bell, PA, 1978)

Levorotatory \ˌle-və-ˈrō-tə-ˌtōr-ē\ *adj* (1873) Turning toward the left or counterclockwise: esp: rotating the plane of polarization toward the left – compare ▶ Dextrorotatory. (Morrison RT, Boyd RN (1992) Organic chemistry, 6th edn. Prentice Hall, Englewood Cliffs, NJ)

Lewis Acid \ˈlü-əs-\ n [Gilbert N. Lewis † 1946) American chemist] (1944) A substance that is capable of accepting an unshared pair of electrons from a base to form a covalent bond.

Lewis-Acid Catalyst See ▶ Friedel-Crafts Catalysts.

Lewis-Nielsen Equation An equation derived from the theory of mixtures that provides an estimate of the modulus of a short-fiber, thermoplastic composite. It is given below.

$$E_c = \frac{E_r(1 + 2AB\phi_f)}{1 - B\psi\phi_f} \text{ where } B = \frac{(E_f/E_r) - 1}{(E_f/E_r) + 2A}$$

$$\text{and } \psi = 1 + \frac{(1-V)\phi_f}{2V_r}$$

In these expressions, E_c, E_f, and E_r are the moduli of the composite, fiber, and resin, ϕ_f is the volume fraction of fiber, A is the average aspect ratio of the fiber, and Ψ is the maximum packing fraction. Application of the equation is limited to small strains. (Natural fibers, plastics and composites. Wallenberger FT, Weston NE (eds), Springer, New York, 2003; Murphy J (1998) Reinforced plastics handbook. Elsevier Science and Technology Books, New York; Engineering plastics and composites. Pittance JC (ed), SAM International, Materials Park, OH, 1990)

Lewis Structure A method of indicating the assignment of valence electrons in an atom, molecule, or ion by representing them as dots placed around symbols, which represent the cores, or kernels, of the atoms. (Whitten KW, Davis RE, Davis E, Peck LM, Stanley GG (2003) General chemistry. Brookes/Cole, New York)

Lexan® General Electric's trade name for their polycarbonate resins produced by reacting bisphenol A and phosgene, the first commercial PC resins. See also ▶ Polycarbonate Resin.

Leyden Blue See ▶ Cobalt Blue.

Li Chemical symbol for the element ▶ Lithium.

Licanic Acid $CH_3(CH_2)_3(CH=CH)_3-(CH_2)_4CO(CH_2)_2-COOH$, IUPAC name: 4-oxy-octadeca-9,11,13-trienoic acid. Constitutes 75–80% of the fatty acid of oiticica oil; mp, 75°C. (Bailey's industrial oil and fat products. Shahidi F, Bailey AE (eds), Wiley, New York; Paint: Pigment, drying oils, polymers, resins, naval stores, cellulosics esters, and ink vehicles, vol 3. American Society for Testing and Material, 2001)

Lieberman-Storch Method Procedure for the qualitative detection of rosin in vehicles. The rosin may be present as either free rosin (abietic acid), esterified rosin, or metal salts. In Europe, the Lieberman-Storch-Morawski test is more commonly used. (Paint: Pigment, drying oils, polymers, resins, naval stores, cellulosics esters, and ink vehicles, vol 3. American Society for Testing and Material, 2001)

Lift (1) The complete set of moldings produced in one cycle of a molding press.

Lifting Softening and raising or wrinkling of a previous coat by the application of an additional coating; often caused by the solvents.

Ligand \ˈli-gənd, ˈlī-\ n [L ligandus, gerundive of ligare] (1949) An atom, molecule, or ion bonded to the central atom in a complex.

Light \ˈlīt\ n [ME, fr. OE lēoht; akin to OHGr lioht light, L luc-, lux light, lucēre to shine, Gk leukos white] (before 12c) (1) Electromagnetic radiation of which a human observer is aware through the visual sensations that arise from the stimulation of the retina of the eye. This portion of the spectrum includes wavelengths from about 380 to 770 nm. Thus, it is incorrect to speak of ultraviolet "light" because the human observer cannot see radiant energy in the ultraviolet region. (2) Adjective meaning high reflectance, transmittance, or level of illumination as contrasted to dark, or low level of intensity. (Optics. Springer, New York, 2003; Johnson SF (2001) History of light and colour measurement: A science in the shadows. Taylor & Francis, UK; Colour physics for industry, 2nd edn. McDonald, Roderick (eds), Society of Dyers and Colourists, West Yorkshire, England, 1997; Moller KD (1997) Colour physics for industry, 2nd edn. McDonald, Roderick (eds), Society of Dyers and Colourists, West Yorkshire, England; Saleh BEA, Teich MC (1991) Fundamentals of photonics. Wiley, New York)

Light Absorbance (absorbance, absorptivity) The percentage of the total luminous flux incident upon a test specimen that is neither reflected from, nor transmitted through the specimen. Compare ▶ Light Reflectance and ▶ Light Transmittance.

Light End (1) The low boiling fraction in distillation. (2) See ▶ Fine End (1).

Light-Fastness ad (1950) The resistance of coated, printed or colored material to the action of sunlight or artificial light. Syn: ▶ Colorfastness.

Lightness n (before 12c) (1) Achromatic dimension necessary to describe the three-dimensionality of color, the others being hue and saturation. Sometimes the lightness dimension is called "brightness." In the Munsell Color Order System, the lightness dimension is called "value." (2) Perception by which white objects are distinguished from gray and light objects from dark ones.

Light Reflectance (reflectance, reflectivity) The fraction of the total luminous flux incident upon a surface that is reflected, generally a function of the color (wavelength) of the light. (Johnson SF (2001) History of light and colour measurement: A science in the shadows. Taylor & Francis, UK; Saleh BEA, Teich MC (1991) Fundamentals of photonics. Wiley, New York) See also ▶ Light Absorbance and ▶ Light Transmittance.

Light Resistance (light fastness, color fastness) The ability of a plastic material to resist fading, darkening, or degradation upon exposure to sunlight or ultraviolet light. Nearly all plastics tend to change color under outdoor conditions, due to characteristics of the polymeric material and/or pigments incorporated therein. Tests for light resistance are made by exposing specimens to natural sunlight or to artificial light sources such as the carbon arc, mercury lamp, germicidal lamp, or xenon arc lamp. ASTM D 4459 and D 4674 describe procedures for testing light fastness of plastics for indoor applications. (Degradation and stabilization of polymers. Zaiko GE (ed), Nova Science, New York, 1995) See also ▶ Artificial Weathering.

Light Scattering (Rayleigh, Mie, Frankhoffer) In a dilute polymer solution, light rays are scattered and diminished in intensity by a number of factors including fluctuations in molecular orientation of the polymer solute. Observations of the intensity of light scattered at various angles provide the basis for an important method of measuring weight-average molecular weights of high polymers. See also ▶ Low-Angle Laser-Light Scattering. (Kokhanovsky AA (2004) Light scattering media optics. Springer, New York; Kamide K, Dobashi T (2000) Physical chemistry of polymer solutions. Elsevier, New York; Berne BJ (2000) Dynamic light scattering: Applications to chemistry, biology and physics. Dover, New York; Brown W (1996) Light scattering: Principles and development. Oxford University Press, UK; Modern techniques for polymer characterization. Pethrick RA (ed), Wiley, New York, 1999; Elias (1977) Macromolecules, vol 1–2. Plenum, New York; Miller ML (1966) The structure of polymers. Reinhold, New York)

Light Source An object which emits light or radiant energy to which the human eye is sensitive. The emission of a light source can be described by the relative amount of energy emitted at each wavelength in the visible spectrum, thus defining the source as an illuminant, or the emission may be described in terms of its correlated color temperature. (Johnson SF (2001)

History of light and colour measurement: A science in the shadows. Taylor & Francis, UK; Saleh BEA, Teich MC (1991) Fundamentals of photonics. Wiley, New York) See ▶ Illuminant and ▶ Correlated Color Temperature.

Light Stabilizer An agent added to a plastic compound to improve its resistance to light-induced changes. See also ▶ Stabilizer and ▶ Ultraviolet Stabilizer.

Light Transmittance (luminous transmittance, light transmissivity) The ability of a material to pass incident light through it, whether specular or diffuse. ASTM (www.astm.org) prescribes several tests of this property in plastics. *Transmissivity* is the ratio of the intensity of the transmitted light to that of the unreflected incident light. See also ▶ Opacity. (Fox AM (2001) Optical properties of solids. Oxford University Press, UK)

Lignin \ˈlig-nən\ *n* (1822) The major non-carbohydrate constituent of wood and woody plants, functioning in nature as a binder to hold the matrix of cellulose fibers together. Lignins are obtained commercially from by-products of coniferous woods, for example by treating wood flour with a derivative of lignosulfonic acid. They are used as extenders in phenolic resins, and sometimes as reactants in the production of phenol-formaldehyde resins. (Hoadley RB (2000) Understanding wood. The Taunton, Newtown, CT)

Lignin Plastic A plastic based on lignin resins.

Lignin Resin A resin made by heating lignin or by reaction of lignin with chemicals or resins, the lignin being in greatest amount of mass.

Lignostone Compressed wood. Manufactured by Röchling, Germany.

Ligroin (ligroine, benzine) Any of several saturated petroleum-naphtha fractions boiling in the range 60–135°C (68–275°F), used as solvents. The term *benzine* is depreciated due to confusion with *benzene*, and should not be used.

Lime \ˈlīm\ *n* [ME, fr. OE *līm*; akin to OHGr *līm* birdlime, L *limus* mud, slime, and perhaps to L *linere* to smear] (before 12c) A caustic highly infusible solid that consists of calcium oxide often together with magnesium oxide, that is obtained by calcining forms of calcium carbonate (as shells or limestone). Syn: ▶ Calcium Oxide.

Lime Blue Mixture of ultramarine and terra alba. Another type of lime blue is made from methylene blue by adsorption on natural earth.

Limed Rosin Commercial calcium resinate made by the direct interaction of lime and rosin.

Lime Green See ▶ Green, Lime.

Lime Putty See ▶ Putty.

Lime Red Lake produced by adsorbing magenta on a natural earth.

Limestone See ▶ Calcium Carbonate.

Limewashing Coating with limewash made from hydrated lime or by slaking quick lime, to which tallow is sometimes added. Syn: ▶ Whitewash, ▶ Whitening.

Lime Yellow Lake produced by adsorbing auramine or other yellow dyestuff on a natural earth.

Limiting Oxygen Index A relative measure of flammability that is determined as follows. A sample is ignited in an oxygen/nitrogen atmosphere. The oxygen content is adjusted until the minimum required to sustain steady burning is found. The higher the value, the lower the flammability. (Troitzsch J (2004) Plastics flammability handbook: Principle, regulations, testing and approval. Hanser-Gardner, New York; Tests for comparative flammability of liquids, UI 340. Laboratories Incorporated Underwriters, New York, 1997).

Limiting Viscosity Number The IUPAC term for ▶ Intrinsic Viscosity.

Limonite \ˈlī-mə-ˌnīt\ *n* [Gr *Limonit*, fr. Gk *leimōn* wet meadow; akin to Gk *limnē* pool] (1823) A native hydrous ferric oxide of variable composition that is a major ore of iron. See ▶ Iron Oxides.

Lineal Density The meaning of Denier. Mass per unit length expressed as grams per centimeter, pounds per foot, or equivalent units. It is the quotient obtained by dividing the mass of a fiber or yarn by its length.

Linear Elastic Fracture Mechanics (LEFM) A theory of fracture, applicable to brittle plastics (and other brittle materials), based on the assumption that the material is Hookean up to the point of fracture, with yielding restricted to a small volume near crack tips in the stressed material.

Linear Expansion See ▶ Coefficient of Thermal Expansion.

Linear Expansion Coefficient *n* The change in specimen length resulting from a specified change in temperature per specimen length at a reference temperature per said change in temperature.

Linear Low-Density Polyethylene (LLDPE) The original low-density polyethylene (LDPE), produced at high pressure, has a highly branched structure. Using Ziegler-Natta catalysts and low pressure, with a small percentage of 1-butene or other comonomer, one can produce a more linear PE with density between 0.919 and 0.925 g/cm^3. LLDPE films have the gloss and clarity of LDPE films, but are stronger, so can be blown thinner to carry design loads. Because of higher melt viscosity, screw modifications are usually necessary for processing

LLDPE in extruders designed for LDPE. (Odian GC (2004) Principles of polymerization. Wiley, New York; Physical properties of polymers handbook. Mark JE (ed), Springer, New York, 1996)

Linear Polyethylenes *n* Linear polyethylenes are polyolefins with linear carbon chains. They are prepared by copolymerization of ethylene with small amounts of higher alfa-olefins such as l-butene. Linear polyethylenes are stiff, tough and have good resistance to environmental cracking and low temperatures. Processed by extrusion and molding. Used to manufacture film, bags, containers, liners, profiles and pipe.

Linear Polymer A polymer in which the molecules form long chains without branches or crosslinking. The molecular chains of a linear polymer may be intertwined, but the forces tending to hold the molecules together are physical rather than chemical and therefore can be weakened by heating. Linear polymers are thermoplastic. (Odian GC (2004) Principles of polymerization. Wiley, New York; Physical properties of polymers handbook. Mark JE (ed), Springer, New York, 1996)

Linear Unsaturated Polyesters See ▶ Polyester.

Linear Viscoelasticity ▶ Viscoelasticity is characterized by a linear relationship between stress, strain, and strain rate.

Line Etching A print made up of lines or pigmented areas and lighter spaces free from shading.

Linen \ˈli-nən\ *adj* [ME, fr. OE *līnen*, fr. *līn* flax, fr. L *linum* flax; akin to Gk *linon* flax, thread] (before 12c) Cellulosic fibers derived from the stem of the flax plant or a fabric made from these fibers. Linen fibers are much stronger and more lustrous that cotton; they yield cool, absorbent fabrics that wrinkle easily. Fabrics with linen-like texture and coolness but with good wrinkle resistance can be produced from manufactured fibers and blends.

Linen Lea The number of 300-yard hanks contained in 1 lb.

Liner (1) A continuous, usually flexible coating on the inside surface of a filament-wound pressure vessel, used to protect the laminate from chemical attack or to prevent leakage under stress. (2) In extruders and injection molders, the hard-alloy interior surface of the cylinder. Decades ago, some of these were separately fabricated and pressed into the steel cylinders. Today they are centrifugally cast into cylinders.

Linet A French-made lining fabric of unbleached linen.

Lining Fabrics (1) Muslin or canvas used underneath fine wallpapers to avoid small cracks possibly opening up in a plaster wall and showing through. (2) Fabric that is used to cover inner surfaces, especially when the inner surface is of a different material than the outer. May refer to garment lining, lining for boxes, coffins, etc. Generally of smooth, lustrous appearing fabrics, but also of felt and velvet. Both manufactured fibers and natural fibers are used.

Lining Paper Plain paper applied before the wallpaper. Assures a smoother surface and better adhesion.

Lining Tool (British) A small flat fitch with a slanting edge, used for painting lines with the help of a run. Syn: ▶ Liner.

Linoleates \lə-ˈnō-lē-ˌāt\ *n* (ca. 1865) Generally the salts or soaps of linseed fatty acid. Cobalt, lead, and manganese linoleates are widely used as driers in printing ink.

Linoleic Acid \ˌli-nə-ˈlē-ik-\ *n* [Gk *lin*on flax + ISV *oleic* (*acid*)] (1857) $C_{18}H_{32}O_2$. *cis*, *cis*-9,12 Octadecadienoic acid. An 18-carbon, straight-chain fatty acid with two double bonds that may be in the 9 and 12 or 9 and 11 positions. It is found in nature as its glyceryl trimester in many vegetable oils and is a starting material for some plasticizers for plastics. It has a mol wt of 280.44, bp of 230°C, iodine value of 181.1.

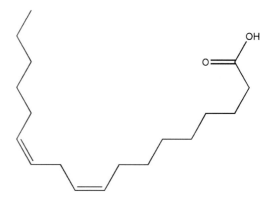

Linolein Glyceride of linoleic acid. It is one of the constituents of linseed oil which induces the drying properties.

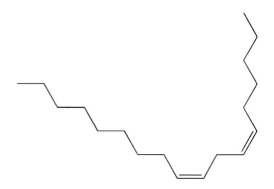

Linolenate Driers Certain metal salts of, and soaps of, linseed fatty acids.

Linolenic Acid \li-nə-ˈlē-nik-\ *n* [ISV, irreg. fr. *linoleic*] (1887) $CH_3CH_2CH=CHCH_2CH=CHCH_2CH=CH(CH_2)_7CO–OH$. *cis,cis,cis*-9,12,15-Octadecatrienoic acid. Tripoly-unsaturated fatty acid component of linseed and other drying oils, It has a bp of 230°C/17 mmHg, an acid value of 201.6, and iodine value of 273.7. (Paint: Pigment, drying oils, polymers, resins, naval stores, cellulosics esters, and ink vehicles, vol 3. American Society for Testing and Material, 2001)

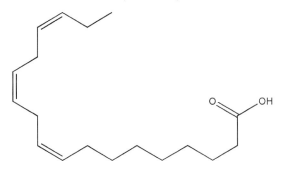

Linoleum and Oilcloth Varnishes Special highly flexible and elastic varnishes.

Linoleum, Floor and Wall Covering Made from oxidized linseed oil or combinations of drying oils, wood flour and/or ground cork, resins and pigment, rolled out and compressed onto an asphalt saturated felt, burlap, or other backing. Heat, which fuses and sets the oils and resins to form strong binding agents, is applied to the mixture during compression.

Linon A Jour A gauze-like linen fabric used as dress goods.

Linoxyn Semisolid, highly oxidized linseed oil; used in the manufacture of linoleum.

Linseed Oil *n* (15c) Drying oil from seeds of the flax plant *Linum usitatissimum*, a mixture of glyceryl esters of linolenic (25%), oleic (5%), linoleic (62%), stearic (3%), and palmitic (5%) acids. The oil is refined by treatments which remove water and mucilaginous material and is then described as refined oil, according to the method of treatment. Further processing produces boiled oil, blown oil or bodied oil. This best known and most widely used oil in the paint industry if characterized by its relatively short drying time. Its high degree of unsaturation, to which its good drying characteristics can be partially ascribed, is due to the presence of large percentages of linolenic and linoleic triglycerides. Many years ago the oil was obtained from seed by mechanical pressure, including both hydraulic presses and later expellers. In recent years the more modern solvent extraction is used. Oils thus obtained show lower percentages of impurities and better overall quality. Linseed oil responds very readily to a variety of refining techniques and is used in the paint industry both as a drying oil and as an ingredient in a wide array of modified resins of many varieties. (Paint: Pigment, drying oils, polymers, resins, naval stores, cellulosics esters, and ink vehicles, vol 3. American Society for Testing and Material, 2001)

Lint \ˈlint\ *n* [ME] (14c) Particles and short fibers that fall from a textile product during the stresses of use.

Lint Ball See ▶ Balling Up.

Lintel \ˈlin-təl\ *n* (ME, fr. MF, fr. LL *limitaris* threshold, fr. L, constituting a boundary mfr. *limit*-, limes boundary] (14c) Horizontal beam supported at each end and spanning an opening. Usually supports the structure above it.

Linters *n* (ca. 1889) Short fibers that adhere to cotton seeds after ginning. Used in rayon manufacture, as fillers for plastics, and as a base for the manufacture of cellulosic plastics.

Lipophilic Livering \ˌlī-pə-ˈfi-lik-\ An increase in the consistency of a paint resulting in a rubbery or coagulated mass.

Liquid Chromatography (LC or HPLC) Any chromatographic process in which the moving phase is a liquid, in contrast to the moving gas phase of gas chromatography. Materials which can be run through a liquid chromatograph are liquids or solutions. An example of a liquid chromatograph is the Series 200 LC Plus Diode Array System by PerkinELmer (Polymers: Polymer characterization and analysis. Kroschwitz JI (ed), Wiley, New York, 1990)

TCLC plus system

Liquid Crystal *n* (1891) A liquid in which the molecules are oriented parallel to each other resulting in birefringence and interference patterns visible in polarizing light. (Collins PJ (1997) Introduction to liquid crystals: Chemistry and physics, vol 1. Taylor & Francis, New York)

Liquid-Crystal Polymer (LC polymer, liquid-crystalline polymer, mesomorphic polymer) A polymer capable of forming regions of highly ordered structure (*mesophase*) while in the liquid (melt or solution) phase. The degree of order is somewhat less than that of a regular solid crystal. Four types have been identified: rodlike, including aromatic polyamides, esters, azomethines, and benzobisoxazoles; helical, mostly natural materials such as polypeptides; side-chain (*comb polymers*); and block copolymers with alternating rigid and flexible units. These polymers are described as *nematic*, in which the mesogens (ordered regions) show no positional order, only long-range order; *Cholesteric* or *chiral*, a modified nematic phase in which the orientation direction changes from layer to layer in a helical pattern; and *smectic*, in which the mesogens have both long-range order and 1- or 2-dimensional positional order. Liquid-crystal polymers are difficult to get into the molten condition because the solid crystals generally decompose before melting. The most commercially successful ones to date are those processed in solution, e.g., poly(*p*-phenylene terephthalamide) (Kevlar). LC polymers are also classified as *lyotropic* and *thermotropic*. Lyotropic ones show their liquid-crystalline character only in solution, while thermotropic ones can show it in the melt without the presence of a solvent. (Collins PJ (1997) Introduction to liquid crystals: Chemistry and physics, vol 1. Taylor & Francis, New York) (2) Aromatic polyester polymers that are extemely unreactive, chemically inert and fire resistant (Plat 2007). Liquid crystallinity in polymers may occur either by dissolving a polymer in a solvent (lyotropic liquid-crystal polymers) or by heating a polymer above its glass or melting transition point (thermotropic liquid-crystal polymers). Liquid-crystal polymers are present in melted/liquid or solid form (Callister 2007). In solid form the main example of lyotropic LCPs is the commercial aramid known as Kevlar. Chemical structure of this aramid consists of linearly substituted aromatic rings linked by amide groups. In a similar way, several series of thermotropic LCPs have been commercially produced by several companies (e.g., Vectra).

A high number of LCPs, produced in the 1980s, displayed order in the melt phase analogous to that exhibited by nonpolymeric liquid crystals. Processing of LCPs from liquid-crystal phases (or mesophases) gives rise to fibers and injected materials having high mechanical properties as a consequence of the self-reinforcing properties derived from the macromolecular orientation in the mesophase. Today, LCPs can be melt-processed on conventional equipment at high speeds with excellent replication of mold details. In fact, the high ease of forming of LCPs is an important competitive advantage against other plastics, as it offsets high raw material cost (Harper 2000).

A unique class of partially crystalline aromatic polyesters based on *p*-hydroxybenzoic acid and related monomers, liquid-crystal polymers are capable of forming regions of highly ordered structure while in the liquid phase. However, the degree of order is somewhat less than that of a regular solid crystal. Typically LCPs have a high mechanical strength at high temperatures, extreme chemical resistance, inherent flame retardancy, and good weatherability. Liquid-crystal polymers come in a variety of forms from high temperature to injection moldable compounds. LCP can be welded, though the lines created by welding are a weak point in the resulting product. LCP has a high Z-axis coefficient of thermal expansion. LCPs are exceptionally inert. They resist stress cracking in the presence of most chemicals at elevated temperatures, including aromatic or halogenated hydrocarbons, strong acids, bases, ketones, and other aggressive industrial substances. Hydrolytic stability in boiling water is excellent. Environments that deteriorate the polymers are high-temperature steam, concentrated sulfuric acid, and boiling caustic materials. Because of their various properties, LCPs are useful for electrical and mechanical parts, food containers, and any other applications requiring chemical inertness and high strength. LCP is particularly attractive for microwave frequency electronics due to low relative dielectric constants, low dissipation factors, and commercial availability of laminates. References: Plat NA (2007) Liquid crystal polymers. Springer, New York; Callister (2007) Materials science and engineering - an introduction, pp 557–558; Harper CA (ed) Modern plastics handbook, ISBN 0-07-026714-6, 2000.

Properties of liquid crystal polymers

Specific gravity	1.38–1.95
Modulus (E)	8,530–17,200 MPa
Tensile strength	52.8–185 MPA
Elongation	0.26–6.2%
Izod impact strength, Notched	21.0–82.5 kJ/m^2

Liquid Crystal Polymer Resins Self-reinforcing plastics because of their densely packed fibrous polymer chains.

(Odian GC (2004) Principles of polymerization. Wiley, New York)

Liquid Driers Solutions of soluble driers in volatile organic solvents, usually hydrocarbons.

Liquid Emulsion Polymer Coating Synthetic resin emulsion (latex) which produce hard, tough coatings having a high luster. Usually nonbuffable or semibuffable, and referred to and sold as self-polishing floor wax.

Liquid Injection Molding (LIM) A process of injection-molding thermosetting resins in which the uncured resin components are metered, mixed, and injected at relatively low pressures through nozzles into mold cavities, the curing or polymerization taking place in the mold cavities. The process is most widely used with resins that cure by addition polymerization such as polyesters, epoxies, silicones, alkyds, diallyl phthalate, and (occasionally) urethanes.

Liquid-Junction Potential A voltage produced across the junction between two dissimilar liquids.

Liquid Reaction Molding (LRM) Older Syn: ▶ Reaction Injection Molding.

Liquid Resin (Deprecated) See ▶ Tall Oil.

Liquid Rosin Syn: ▶ Tall Oil.

Liquid Solvent Wax Mixture of waxes and other ingredients in a solvent base. Must be polished for luster and are occasionally colored. *Sometimes called, simply, Solvent Wax.*

Liquid Water Emulsion Wax Dispersion of wax and other modifying materials in water.

Liquifying Stress The ease with which concentrated inks can be liquefied by remilling after storage.

Liquor Ratio In wet processing the ratio of the weight of liquid used to the weight of goods treated.

Lisle Yarn *n* [*Lisle* Lille, F] (1851) A high-quality cotton yarn made by plying yarns spun from long combed staple. Lisle is singed to hive it a smooth finish.

Lithium \li-thē-əm\ *n* [NL, fr. *lithia* oxide of lithium, fr. Gk *lithos*] (1818) (Li) Element number 3, the least dense of all the metals (density = 0.534 g/cm^3), with valence of +1, and highly reactive. Lithium aluminum hydride (LiAlH$_4$) is an important catalyst in organic reductions and lithium is a component of many greases, e.g., the high-temperature lubricant, ▶ Lithium Stearate.

Lithium Carbonate *n* (1873) Li$_2$CO$_3$. A crystalline salt used in the glass and ceramic industries.

Lithium Fluoride *n* (1944) LiF. A crystalline salt used in making prisms and ceramics and as a flux.

Lithium Niobate *n* [*niob*ium + ^{1}-*ate*] (1966) LiNbO$_3$. A crystalline material whose physical properties change in response to pressure or the presence of an electric field and which is used in fiber optics and as a synthetic gemstone.

Lithium Stearate LiOOCC$_{17}$H$_{35}$. A white crystalline material used as a lubricant in plastics.

Lithographed Paper Wallpaper made by the printing process used for billboards and posters. Lithography exploits the affinity of color for the greasy material in which the design is put on the roller, a transfer process.

Lithography \li-*thä-grə-fē\ *n* [Gr *Lithographie*, fr. *lith-* + *-graphie* –graphy] (1813) A process of planographic printing involving two different areas on the plate, one receptive to ink, the other receptive to fountain solution. (Printing ink manual, 5th edn. Leach RH, Pierce RJ, Hickman EP, Mackenzie MJ, Smith HG (eds), Blueprint, New York, 1993; Printing ink handbook. National Association of Printing Ink Manufacturers, 1976) Also see ▶ Stereolithography for three-dimensional formation of objects.

Lithographic Chalk A greasy crayon, composed of soap, wax, oil and lampblack, for drawing designs on a lithographic plate prior to etching.

Lithographic Inks Inks used in the lithographic process. The principal characteristic of a lithographic ink is its ability to resist excessive emulsification by the fountain solution.

Lithographic Stones Slabs of limestone, t3 or 4 in. thick, the surfaces of which are smoothed or grained, for use in lithographic printing.

Lithographic Varnishes Heat-bodied refined linseed oils used as vehicles for printing and lithographic inks.

Lithol Red C$_{20}$H$_{13}$N$_2$O$_4$SM*. (M* = Na, Ba or Ca). Pigment Red 49 (15630). Pigment made by combining the intermediates, tobias acid and beta-naphthol. This type of red is available as sodium, barium, and calcium toners, also lakes; the sodium is the lightest shade, the barium is what may be termed a medium shade, and the calcium lithols are deep reds and maroons.

Lithol Rubine Pigment Red 57 (15850). Azo pigment made by diazotizing paratoluidine-meta-sulfonic acid, and coupling with 3-hydroxy-2-naphthoic acid. Pigment is used as the calcium salt. Most frequently encountered in the resinated form, it is characterized by its distinctly brilliant deep masstone and its blue intense tint.

Lithopone \❙li-thə-❙pōn\ *n* [ISV *lith-* + Gk *ponos* work] (ca. 1884) BaSO$_4$·ZnS. Pigment White 5 (77115). (Charlton white, Orr's white, zinc baryta) A mixed pigment obtained by the interaction (*metathesis*) of equimolar solutions of barium sulfide and zinc sulfate, from which precipitate barium sulfate and zinc sulfide, both white.

Lithrage PbO. Oxide of lead made by controlled heating of metallic lead. Pure litharge has sp gr of 9.53, mp of 888°C, and mol wt of 223.21. Used as a raw material in the manufacture of pigments and driers and infrequently as a catalyst in pains.

$$Pb^{++} \quad O^{--}$$

Litre \❙lē-tər\ [*variant* of liter] The volume of a kilogram of water at 4°C, equal to 1.06 quarts or 61.02 in.^{3}

Little Joe A one- or two-station dry offset proof press.

Live Centers Point at the very center at each end of the roll. This is usually a center point of the journal at which the roll turns. The roll is driven from this point and is considered concentric from the same point.

Live Edge See ▶ Wet Edge.

Live-Feed Molding See Multi-Live –Feed Molding.

Livering The progressive, irreversible increase in consistency of a pigment-vehicle combination. Livering in the majority of cases arises from a chemical reaction of the vehicle and the solid dispersed material, but it may also result from polymerization of the vehicle.

Living Polymers A polymerization reaction in which there is no termination, and the polymer chains continue to grow as long as there are monomer molecules to add to the growing chain.

Living Ring See ▶ Revolving Spinning Ring.

LLDPE Abbreviation for ▶ Linear Low-Density Polyethylene.

LLDPE Resin See ▶ Linear Low Density Polyethylene.

Ln (1) Abbreviation for ▶ Lumen. (2) Abbreviation for Natural Logarithm, i.e., logarithm to the base e (= 2.71828...).

Load Cell An instrument, most often part of a machine for testing mechanical properties and some rheometers, that senses the force applied to the specimen (or piston).

Locking Pressure The pressure applied to an injection or transfer mold to keep it closed during molding.

Locking Ring A slotted plate in an injection or transfer mold that locks the parts of the mold together and prevents the mold from opening while the plastic is being injected.

Loft The properties of firmness, resilience, and bulk of a fiber batting, yarn, fabric, or other textile material.

Logarithm \❙lȯ- gə-❙ri-thəm\ *n* The exponent that indicates the power to which a number is raised to produce a given number. A common logarithm (log) has a base of 10 and a natural (ln) or Napierian logarithm has a base of e (irrational number 2.71828...). Conversion of log to ln: ln N = 2.303 log N.

Logarithmic Decrement \❙lȯ-gə-❙rith-mik ❙de-krə-mənt\ (Δ) In a damped, vibrating system, the natural logarithm of the ratio of the amplitude of any oscillation to the amplitude of the succeeding oscillation. Where damping is mild, the ratio of the amplitude of any oscillation to the amplitude of the succeeding oscillation. Where damping is mild, the ratio of amplitudes several vibrations apart will usually give a more accurate estimate. The equations are:

$$\Delta = \ln(A_i/A_{i+1}) \text{ and } \Delta = (1/n)\cdot\ln(A_i/A_{i+n})$$

Logarithmic Viscosity Number The IUPAC term for ▶ Inherent Viscosity.

Log normal Distribution (logarithmic normal distribution) A statistical probability-density function, characterized by two parameters, that can sometimes provide a faithful representation of a polymer's molecular-weight distribution or the distribution of particle sizes in ground, brittle materials. It is a variant of the familiar normal or Gaussian distribution in which the logarithm of the measured quantity replaces the quantity itself. It's mathematical for is

$$f(x) = \frac{1}{\sqrt{2\pi}\beta} x^{-1} e^{-(\ln x - \alpha)/2\beta^2} dx$$

or

$$f(\ln x) = \frac{1}{\sqrt{2\pi}\beta} e^{-(\ln x - \alpha)/2\beta^2} d\ln x$$

α and β are the mean and standard deviation of ln *x*. Anti-ln α is called the *log mean* of *x*. To test the suitability of this distribution, one plots the cumulative percent of members having weights or sizes below *x*, versus *x*, on lognormal probability paper and looks for linearity in the plot.

Logotype (or Logo) \❙lȯ-gə-❙tīp, ❙lä-\ *n* (ca. 1816) Name of a company or product in a unique design used as a trademark in advertising.

Logu Sluggish, low snap or recovery. A condition formed in poorly cured or overloaded vulcanizers.

London Dispersion Forces (London Forces) *n* [fr. Fritz London. (1930) Identifies weak intermolecular forces based on transient dipole interactions. One of *van der Waals forces*, also called dispersion forces, but distinct from dipole-dipole forces. These forces arise from momentary fluctuations in the electron charge cloud density in a atom or molecule. Changes in symmetry of the electron cloud that cause a momentary dipole moment and attractive/repulsive charges. The larger a molecule is and the more electrons it has, the more polarizable it will be, and thus the larger the London forces can be. Molecular shape and other factors are also important. (Whitten KW, Davis RE, Davis E, Peck LM, Stanley GG (2003) General chemistry. Brookes/Cole, New York)

Lone Pair A pair of electrons which belongs to only one atom and hence is not shared (VB model).

Long-Chain Branching In a polymer's structure, the presence of arms (branches) off the main chain that are about as long as the main chain. In making low-density polyethylene, a typical molecule may contain 50 short branches and only one or zero long branch, yet the presence of long branches greatly broadens the molecular-weight distribution. Polymers containing long branches tend to be less crystalline than the corresponding polymers without long branches.

Long-Fiber-Reinforced Thermoplastic A palletized thermoplastic resin for injection molding, usually nylon 6/6 or polypropylene, produced by pultrusion from continuous-filament glass yarn, and cut to lengths of 9–13 mm (about three times the length of short-fiber pellets). Nylon containing 50 weight percent longer-fiber glass is about 15% stronger and stiffer than its short-fiber mate, with double the notched-Izod impact strength.

Longo A colloquialism used in the filament-winding industry, designating an article that is wound longitudinally or with a low-angle helix.

Long Oil High ratio of oil to resin in a medium. (1) *Long oil alkyd* – an alkyd resin containing more than 60% of oil as a modifying agent. (2) *Long oil varnish* – an oleoresinous varnish, other than alkyd, containing more than 25 gal of oil per 100 lb of resin. A long oil varnish is usually slower drying, tougher and more elastic than a short oil varnish. See also ▶ Short Oil Varnish and ▶ Medium Oil Varnish.

Long Staple A long fiber. In reference to cotton, long staple indicates a fiber length of not less that 1–1/8 in. In reference to wool, the term indicates fiber 3–4 in. long suitable for combing.

Loom \ lüm\ *n* [ME *lome* tool, loom, fr. OE *gelōma* tool; akin to MD al*lame* tool] (15c) A machine for weaving fabric by interlacing a series of vertical, parallel threads (the warp) with a series of horizontal, parallel threads (the filling). The warp yarns from a beam pass through the heddles and reed, and the filling is shot through the "shed" of warp threads by means of a shuttle or other device and is settled in place by the reed and lay. The woven fabric is then wound on a cloth beam. The primary distinction between different types of looms is the manner of filling insertion. The principal elements of any type of loom are the shedding, picking, and beating-up devices. In shedding, a path is formed for the filling by raising some warp threads while others are left down. Picking consists essentially of projecting the filling yarn from one side of the loom to the other. Beating-up forces the pick, that has just been left in the shed, up to the fell of the fabric. This is accomplished by the reed, which is brought forward with some force by the lay. (Kadolph SJJ, Langford AL (2001) Textiles. Pearson Education, New York)

Loom Barré A repeated unevenness in the fabric, usually running from selvage to selvage, and caused by uneven let-off or take-up or by a loose crank arm.

Loom-Finished A term describing fabric that is sold in the condition in which it comes from the loom.

Loom Fly Waste fibers that are inadvertently woven into a fabric.

Looped Filling A woven-in loop caused by the filling sloughing off the quill or by the shuttle rebounding in the box.

Looped Pile A pile surface made of uncut looped yarns.

Looped Year See ▶ Kink.

Loop Elongation The maximum extension of a looped yarn at maximum load, expressed as a percentage of the original gauge length. (Kadolph SJJ, Langford AL (2001) Textiles. Pearson Education, New York)

Looping *v* (1832) Generally, a method of uniting knit fabrics by joining two courses of loops on a machine called a looper.

Looping Bar A bar inserted in the bottom of an extrusion métier around which the dried filaments pass as they leave the spinning cabinet.

Loop Pile Carpet construction in which the tufts are formed into loops from the supply yarn.

Loop Selvage A weaving defect at the selvage of excessive thickness or irregular filling loops that extend beyond the outside selvages.

Loop Tenacity The strength of a compound strand formed when one strand of yarn is looped through another strand, then broken. It is the breaking load in grams divided by twice the measured yarn denier or decitex. Loop tenacity, when compared with standard tenacity measurements, is an indication of the brittleness of a fiber.

Loop Test A simple test (ASTM 3291) for evaluating the compatibility of vinyl resin plasticizers based on the fact that a material under compressive stress will exude plasticizer more rapidly. A specimen in sheet form is folded double, forming a loop with internal radius equal to the sheet thickness. At intervals, the bend of the loop is reversed 360° and the former inside surface of the loop is examined for evidence of plasticizer spewing.

Loopy Yarn See ▶ Textured Yarns.

Loose Edge See ▶ Slack Selvage.

Loose End See ▶ Loose End.

Loose Filling A fabric defect that is usually seen as short, loose places in the filling caused by too little tension on the yarn in the shuttle or by the shuttle rebounding in the box. Loose filling can often be felt by an examiner when passing a hand over the surface of the fabric.

Loose Pick See ▶ Slack Pick.

Loose Punch A male portion of a mold constructed so that it remains attached to the molding when the press is opened, to be removed from the part after demolding.

Lopac Copolymer from methacrylonitrile and styrene or α-methyl styrene (9:1). Manufactured by Monsanto, US.

Loss Angle The inverse tangent of the electrical dissipation factor.

Loss Compliance The "imaginary" part of the complex compliance. See ▶ Compliance and ▶ Complex Modulus.

Loss Dielectric A loss of energy evidenced by the rise in heat of a dielectric placed in an alternating electric field. It is usually observed as a frequency-dependent conductivity.

Loss Factor The product of a power factor and dielectric constant of a dielectric material.

Loss Modulus "G" – the component of applied shear stress which is 90 degrees out of phase with the shear strain, divided by the strain.

Loss of Drier See ▶ Drier Dissipation.

Loss of Gloss A paint defect in which a dried film of paint loses gloss, usually over a period of several weeks.

Lost End An end on a section or tricot beam that has been broken at some stage in warping and has not been repaired by a knot.

Lost-Wax Process See ▶ Investment Casting.

Lot \lät\ *n* [ME, fr. OE *hlot*; akin to OHGr *hlōz*] (before 12c) A unit of production or a group of other units or packages that is taken for sampling or statistical examination, having one or more common properties and being readily separable from other similar units.

Lot Number The number used by the manufacturer to identify an entity of production.

Louvers \lü-vər-\ *n* [ME *lover*, fr. MF *lovier*] (14c) Slats placed at an angle, as in shutters.

Lovibond Color System A system of color specification by means of numbers proportional to the optical densities of three glass filters (yellow, red, and blue) required to modify a standard light source (such as daylight or incandescent lamp light) to produce a color match. The carefully calibrated glasses should be used as described by the manufacturer, Tintometer, Ltd., in order that results in different laboratories will agree. Lovibond specifications have long been used as the basis for describing the colors of edible oils and vegetable oils used for paint vehicles.

Lovibond Tintometer Optical comparison instrument manufactured by Tintometer, Ltd. (UK) for use of Lovibond glasses. See ▶ Lovibond Color System.

Low-Angle Laser-Light Scattering A technique for determining weight-average molecular weights of polymers in solution. The low angle – 2–10° – reduces the number of measurements needed and simplifies their interpretation, as compared with conventional, wide-angle light scattering. (Kamide K, Dobashi T (2000) Physical chemistry of polymer solutions. Elsevier, New York; Berne BJ (2000) Dynamic light scattering: Applications to chemistry, biology and physics. Dover, New York)

Low-Density Polyethylene (LDPE) This term is generally considered to include polyethylenes having densities between 0.915 and 0.925 g/cm^3. In LDPE, the ethylene molecules are linked in random fashion, with many side branches, mostly short ones. This branching prevents the formation of a closely knit pattern, resulting in material that is relatively soft, flexible, and tough, and which will withstand moderate heat. See also ▶ High-Density Polyethylene and ▶ Polyethylene.

Lower Explosive Limit Lower limit of flammability or explosibility of a gas or vapor at ordinary ambient temperatures expressed in percent of the gas vapor in air by volume. (Tests for comparative flammability of liquids, UI 340. Laboratories Incorporated Underwriters, New York, 1997) See ▶ Explosive Limits.

Low-Pressure Injection Molding A term sometimes used for the process of injecting a fluid material such as a vinyl plastisol into a closed mold, using a grease gun or similar low-pressure equipment.

Low-Pressure Laminate Various definitions place the upper limit of pressure for this term at from 6.9 MPa down to pressures obtained by mere contact of the piles. According to ASTM D 883, the upper limit is 1.4 MPa (200 psi). The Decorative Board Section of the National Electrical Manufacturers' Association (NEMA) has recommended abandonment of the term "low-pressure laminate" in favor of *decorative board* in the case of ". . .a product resulting from the impregnation or coating of a decorative web of cloth, paper, or other carrying media with a thermosetting resin and consolidation of one or more of these webs with a cellulosic substrate under heat and pressure of less than 500 lb/in.^{2}" This includes all boards that were formerly called low-pressure melamine and polyester laminates, but not vinyls. See also ▶ Contact-Pressure Molding and ▶ Laminate.

Low-Pressure Molding Molding or laminating in which the pressure is 1.4 MPa (200 psi) or less (ASTM D 883).

Low-Pressure Resin See ▶ Contact-Pressure Resin.

Low Rows A carpet defect characterized by rows of unusually low pile height across the width of the goods.

Low-Temperature Flexibility All plastics that are flexible at room temperature become less so as they are chilled, finally becoming brittle at some low temperature. This property is often measured by torsional tests over wide ranges of temperature, from which apparent moduli of elasticity are calculated. See also ▶ Brittleness Temperature and ▶ Clash-Berg Point. Some relevant ASTM tests are D 1043, D 3295, D 3296, D 3374 (Section 07.02), and D 1055 (Section 09.01).

L-Sealer A heat-sealing device, used in packaging, that seals a length of flat, folded film on the edge opposite the fold and simultaneously seals a strip across the width at 90° from the edge seals. The article to be packaged may be inserted between the two layers of folded film prior to sealing. When it is desired to sever the continuous length of sealed compartments into individual packages, a heated wire or knife is incorporated between two sealing bars that form the bottom of the **L**. These bars then make the top seal of the filled bag and the bottom seal of the next bag to be filled.

Luana A fabric characterized by a crosswise rib effect, usually made with a filament yarn warp and a spun yarn filling.

Lubricant \ˈlü-bri-kənt\ *n* (ca. 1828) A substance that tends to make surfaces slippery, reduce friction, and prevent adhesion.

Lubricant Bloom See ▶ Bloom. The term *lubricant bloom* should only be used when the exudation is known to be caused by a lubricant contained in the plastic compound or applied to it during processing.

Lucite® \ˈlü-ˌsīt\ Poly(methyl methacrylate) DuPont's trade name for methacrylate-ester monomers and polymers (*acrylic*), including PMMA and several other resins, and for certain products made from such resins.

Lumband Oil This oil is obtained from the nuts of the tree, *Aleurites moluccana*. Although a product of an Aleurites tree, it contains no elaeostearin. It dries somewhat better than soybean oil. *Also called Candlenut Oil.*

Lumen \ˈlü-mən\ *n* [NL *lumin-*, *lumen*, fr. L, light, air shaft, opening] (1873) (lm) The SI unit of ▶ Luminous Flux.

Luminance \ˈlü-mə-nən(t)s\ *n* (1880) The luminous intensity of light reflected or transmitted by a material in a given direction per unit of projected area of the material, as viewed from that direction.

Luminescent *n* (1889) Emitting light not due to high temperature, usually caused by excitation by rays of a shorter wavelength. (Saleh BEA, Teich MC (1991) Fundamentals of photonics. Wiley, New York)

Luminescent Pigment A pigment that produces striking effects in darkness or light. (Dainth J (2004) Dictionary of chemistry. Oxford University Press, UK; Solomon DH, Hawthorne DG (1991) Chemistry of pigments and fillers. Krieger, New York) See ▶ Fluorescent Pigment and ▶ Phosphorescent Pigment.

Luminosity Curve, CIE Graph representing the luminous intensity of wavelengths in the visible spectrum relative to the maximum intensity at the same wavelength. The relative luminosity curve is described by the Y color matching function in the CIE System. The shape of the curve depends on the area of the retina being stimulated and on the intensity of the incident light. Thus, the CIE luminosity curves for the 2° and 10° observers are slightly different, and those for daylight vision, i.e., photopic vision, and for dark adapted vision, i.e., scotopic vision, are extremely different.

Luminous \ˈlü-mə-nəs\ *adj* [ME, fr. L *luninosus*, fr. *lumin-*, *lumen*] (15c) (1) Adjective used to imply dependence on the spectral response characteristic of the Standard Observer defined in the CIE System. Thus, the luminous reflectance or the luminous transmittance is described by the Y tristimulus value in the CIE System. The adjective is applied to many measures of light, such as intensity, density, etc., and always indicates the measures are weighted for the relative luminous sensitivity of the human observer. (2) Material emitting or spearing to be emitting, visible radiant energy.

Luminous Directional Reflectance Reflectance of a surface for specified directions of illumination and view is the ratio of the brightness of the surface to the brightness that an ideally diffusing, perfectly white, surface would have if illuminated and viewed in the same manner. (Saleh BEA, Teich MC (1991) Fundamentals of photonics. Wiley, New York)

Luminous Energy *n* (ca. 1931) Energy transferred in the form of visible radiation. (Saleh BEA, Teich MC (1991) Fundamentals of photonics. Wiley, New York)

Luminous Flux *n* (1925) The total visible energy emitted by a source per unit time. The SI unit is the *lumen* (lm), defined as the luminous flux emitted in a solid angle of one steradian (sr, the solid central angle that cuts out of a spherical surface a square whose side is equal to the radius) by a point source having a uniform intensity of one candela. Therefore, 1 lm = 1 cd sr. See ▶ Flux, Luminous.

Luminous Paint *n* (ca. 1889) Paint which exhibits fluorescence. (1) *Fluorescent paint* – contains pigments which are capable of absorbing energy from the blue or ultraviolet end of the spectrum and reemitting it in the form of light in the visible wavelengths. A fluorescent paint ceases to "glow" if the activating source is removed. (2) *Phosphorescent paint* – contains pigments (phosphors) which absorb energy at one wavelength and emit it over a period of time, in the form of light at a longer wavelength in the visible spectrum. It differs from a fluorescent paint in that it continues to glow after the stimulating source has been removed. (3) *Radioactive* or *self-luminous paint* – normally, this is a phosphorescent paint containing a portion of radioactive compounds, and in such a paint the phosphor is permanently activated by absorbing energy from the bombardment by the radioactive rays and emits light in the visible spectrum. (Paint/coatings dictionary. Federation of Societies for Coatings Technology, Philadelphia, Blue Bell, PA, 1978)

Luminous Transmittance Syn: ▶ Light Transmittance.

Luparen Poly(propylene), manufactured by BASF, Germany.

Luphen Phenoplast, manufactured by BASF, Germany.

Lupolen Poly(ethylene) (high pressure), manufactured by BASF, Germany.

Luran Copolymer from styrene/acrylonitrile, manufactured by BASF, Germany.

Luster \ˈləs-tər\ *n* [MF *lustre*, from OI *lustro*, fr. *lustrare* to brighten, fr. L. to purify ceremonially, fr. *lustrum*] (ca. 1522) (1) Type of surface reflectance, or gloss, where the ratio of specular reflectance to diffuse reflectance is relatively high, but not so high as that from a perfect specular reflector (mirror). (2) Another term for gloss.

Lustering *v* (1582) The finishing of yarn or fabric by means of heat, pressure, steam, friction, calendering, etc., to produce luster.

Lusterless *adj* (ca. 1522) An adjective describing a non-glossy or non-reflecting surface with respect to illumination of the surface, dull in appearance.

Lute n. (lo͞ot) Also called **luting**, a mixture of cement and clay used to seal the joints between pipes, etc. 2. (Medicine / Dentistry) Dentistry a thin layer of cement used to fix a crown or inlay in place on a tooth; vb (Miscellaneous Technologies / Building) (tr) to seal (a joint or surface) with lute [via Old French ultimately from Latin *lutum* clay]. Generally, a material placed around a conduit for physically securing and/or thermally insulating for a gas or liquid. Some modern polymeric composite lutens have been developed.

Lux \ˈləks\ *n* [L, light] (1889) (lx) The SI unit of illuminance, defined as the illuminance produced by a luminous flux of one lumen uniformly distributed over a surface of 1 m². That is, 1 lx = 1 lm/m².

Lycra \ˈlī-krə\ Elastomer from segments of polyether and polyurethane, manufactured by DuPont.

Lyocell Fiber A manufacturing cellulose fiber made by direct dissolution of wood pulp in an amine oxide solvent, *N*-methylmorpholine-*N*-oxide. The clear solution is extruded into a dilute aqueous solution of amine oxide, which precipitates the cellulose in the form of filaments. The fiber is then washed before it is dried and finished. The solvent spinning process for making lyocell fiber is considered to be environmentally friendly because the non-toxic spinning solvent is recovered, purified, and recycled as an integral part of the manufacturing process. No chemical intermediates are formed, the minimal waste in not hazardous, and energy consumption is low. Wood pulp is a renewable resource, and the fiber is biodegradable. CHARACTERISTICS: Lyocell fiber is stronger than other cellulosic fibers. It is inherently absorbent, having a water Imbibition of 65–75%. Lyocell retains 85% of its dry tenacity when wet, making it stronger when wet than cotton. The fiber has a density of 1.15 g/cm³. END USES: Lyocell fiber is suitable for blending with cotton or other manufactured fibers. Because of its molecular structure, lyocell has the tendency to develop surface fibrils that can be beneficial in the manufacture of hydroentabled and other nonwovens, and in specialty papers. For apparel uses, the fiber's unique fibrillation characteristic has enabled the development of fabrics with a soft luxurious hand. The degree of fibrillation is controlled by cellulose enzyme treatment.

Lyophilic \ˌlī-ə-ˈfi-lik\ *adj* (1911) Describing a substance that easily forms colloidal suspensions. Such ability when the suspending medium is water is called *hydrophilic.* A PVC plastisol is an example of a lyophilic suspension.

Lyophobic \ˌlī-ə-ˈfō-bik\ *adj* (1911) Characterizing a material which exists in the colloidal state without any significant affinity for the medium.

Lyotropic See ▶ Liquid-Crystal Polymer.

Lyotropin Polymer Polymers that decompose before melting but that form liquid crystals in solution under appropriate condition. They can be extruded from high concentration dopes to give fibers of high modulus and orientation for use in advanced composites, tire cord, ballistic protective devices, etc.

M

m \\em\ *n* (1) Abbreviation for ▶ Meter (1). (2) Abbreviation for the SI Prefix ▶ Milli-. (3) (usually italicized) Abbreviation for chemical positional prefix ▶ Meta-.

M *n* (1) Abbreviation for prefix ▶ Mega-. (2) Symbol for molecular weight. (3) Symbol for being moment.

mA *n* Abbreviation for Milliampere.

M.A.C. *n* Maximum allowable concentrations of solvent vapors, also known as *Threshold Limit Values*. These values refer to air-borne concentrations of substances and represent conditions to which it is believed that nearly all workers may be repeatedly exposed, day after day, without adverse effect.

MacAdam Color Difference Equation *n* A color difference equation developed by David MacAdam which is now used as modified by Hugh Davidson and Fred Simon to incorporate the effect of lightness on the chromaticity differences:

$$\Delta E = [1/K(g_{11}\overline{\Delta x^2} + 2g_{12}\Delta x \Delta y + g_{22}\overline{\Delta y^2} + G\overline{\Delta Y^2})]^{1/7}$$

where g_{11}, $2G_{11}$, and g_{22} are constants depending on the chromaticity coordinates, x and y, and K and G are constants depending on the luminous reflectance or transmittance, Y. This color difference is frequently calculated from charts prepared by Simon and Goodwin, which have the required constants built-in.

MacAdam Limits *n* The theoretical limit or gamut of colors which can be obtained at various limits of luminance (Y). Thus, the gamut of colors which can be obtained theoretically decreases steadily as the luminance (Y) increases.

Machinability \mə-₁shē-nə-⌐bi-lə-tē\ *vt* (ca. 1864) (1) In fabricating materials by such operations as drilling, lathe-turning, and milling, the ease with which the material is removed.

Machine-Printing *n* The method by which the bulk of modern wallpapers are produced. Machine-printing employs a rotary press and a series of cylinders or rollers to turn out wallpaper at high speeds. Raw paper stock is first given a coating of the ground color by a special machine, after which the paper proceeds in a continuous web to the rotary press where the top colors are applied, and it is then festooned on specially heated drying racks.

Machine Shot Capacity See ▶ Shot Capacity.

Machine Twist *n* A hard-twist sewing thread, usually of 3-ply construction spun with S twists and plied with Z twist, especially made for use in sewing machines.

Machining of Plastics *n* Many of the machining operations used for metals are applicable to rigid plastics, with appropriate variations in tooling and speeds [see ▶ Machinability (1), above]. Among such operations are blanking, boring, drilling, grinding, milling, planning, ▶ Punching, routing, ▶ Sanding, sawing, shaping, tapping, threading, and turning.

Mach Number *n* (N_{Ma}) The ratio of a fluid velocity or the relative velocity of an object moving through a fluid to the velocity of sound in the fluid. All fluids (liquids and gases) have Mach mumbers.

Macrolattice *n* A repeating structure in very small microfibrils of alternating crystalline and amorphous regions. Yarn properties are thought to be governed by morphology at the macrolattice scale.

Macromolecule \₁ma-krō-⌐mä-li-₁kyü(ə)l\ *n* [ISV] (ca. 1929) The large ("giant") molecules that make up high polymers, both natural and synthetic. Each macromolecule may contain hundreds of thousands of atoms. See ▶ Polymer.

Macromonomers *n* High molecular weight monomers. Also called *Macromens*.

Macroscopic \₁ma-krə-⌐skä-pik\ *adj* [ISV macr- + scopic (as in *microscopic*] (1872) Visible to the naked eye, as opposed to *microscopic*.

Macroscopic Properties *n* See ▶ Thermodynamic Properties.

Madder Lake *n* Lightfast, nonbleeding, red-colored pigment prepared from the coloring matter of madder root.

Madder Lakes *n* A class of solvent-resistant and lightfast pigments; generally dirty in appearance.

Madras \⌐ma-drəs; mə-⌐dras, -⌐dräs\ *n* [*Madras*, India] (ca. 1830) A lightweight, plain weave fabric with a striped, checked, or plaid pattern. True madras is "guaranteed to bleed."

Magdala Red *n* $C_{30}H_{21}N_4Cl$. Red dyestuff. *Known also as Naphthalene Red*.

Magnesia \mag-⌐nē-shə, -zhə\ *n* [NL, fr. *magnes carneus*, a white earth, literally, flesh magnet] (1755) MgO. (1) Magnesium oxide. (2) Sometimes used incorrectly in the printing ink industry to mean magnesium carbonate. Syn: ▶ Magnesium Oxide.

$$Mg^{++} \quad O^{--}$$

Magnesite \⌐mag-nə-₁sīt\ *n* (1815) $MgCO_3$. Mineral, magnesium carbonate, principally used as a filler or extender.

Jan W. Gooch, *Encyclopedic Dictionary of Polymers*, DOI 10.1007/978-1-4419-6247-8,
© Springer Science+Business Media LLC 2011

Magnesite Floor *n* Hard composition floors in which magnesium oxychloride is the binder. This binder is formed in laying the floor, when magnesium oxide is combined with a strong solution of magnesium chloride. Fillers which may be added to this binder are: asbestos, cork, sand, wood flour, marble dust, talc, leather, etc. This great variety of fillers produces magnesite floors having variable porosity, resiliency, appearance and durability.

Magnesium Carbonate *n* (1903) (magnesia alba, precipitated magnesium carbonate) $MgCO_3$. A white powder of low density, prepared by metathesis, used as a filler or modifier in phenolic resins. This carbonate also occurs naturally as *magnesite*.

Magnesium Carbonate, Precipitated *n* Chemically, this is the same as magnesite, but physically it has a much better color, in bulk being a very intense white. It is usually a very fine light powder of rather high oil absorption.

Magnesium Chloride *n* (ca. 1910) A bitter deliquescent salt $MgCl_2$ used especially as a source of magnesium metal.

Magnesium Glycerophosphate *n* $MgPO_4C_3H_5(OH)_2$. A colorless powder, derived by the action of glycerophosphoric acid on magnesium hydroxide, used as a stabilizer for plastics.

Magnesium Hydrogen Phosphate Trihydrate *n* (dibasic magnesium phosphate, magnesium monohydrogen *ortho*phosphoric acid with magnesium oxide, used as a nontoxic stabilizer for plastics.

Magnesium Hydroxide *n* (ca. 1909) $Mg(OH)_2$. Used as a thickening agent for polyester resins. Its action is slower than that of magnesium oxide.

Magnesium Hydroxychloride Cement *n* (Sorel cement, magnesium oxychloride cement) A mixture of magnesium chloride and magnesium oxide that reacts with water for form a solid mass, presumed to be magnesium hydroxychloride, $Mg(OH)Cl$. It has been useful as an intumescent coating for urethane foams and other materials such as polystyrenes, nylons, acetals, polyesters, and silicones.

Magnesium Oxide *n* (ca. 1909) (magnesia, periclase) A white powder used as filler and as a thickening agent in polyester resins. It occurs naturally as the mineral *periclase*, but it is usually made in purer form by calcining magnesium hydroxide or carbonate.

Magnesium Phosphate, Dibase *n* See ▶ Magnesium Hydrogen Phosphate Trihydrate.

Magnesium Phosphate, Monobasic *n* (magnesium dihydrogen phosphate) $Mg(H_2PO_4)_2 \cdot 2H_2O$. A white, hygroscopic, crystalline powder derived by reacting phosphoric acid with magnesium hydroxide. It is used as a flame retardant and stabilizer for plastics.

Magnesium Phosphate, Tribasic *n* $Mg_3(PO_4)_2 \cdot 8H_2O$ or $\cdot 4H_2O$. A fine, soft white powder derived by reacting magnesium oxide and phosphoric acid at a high temperature, used as a nontoxic stabilizer.

Magnesium Silicate, Fibrous *n* $3MgO \cdot 2SiO_2 \cdot 2H_2O$. A fibrous chrysotile mineral white to gray powder, chemically inert used as extender and/or filler in paints and caulks. Pigment grades are used for their high temperature resistance, high oil absorption and water demand. Density, 2.48–2.56 g/cm³ (20.7–21.3 lb/gal); O.A., 50–180. Syn: ▶ Asbestos, ▶ Chrysotile and ▶ Fibrous Asbestos.

Magnesium Silicate, Nonfibrous *n* $3MgO \cdot 4SiO_2 \cdot H_2O$. Pigment White 26 (77718). A hydrated magnesium silicate extender of filler of wide range of composition. Soft white, gray or yellow shade. Natural product (talc) used in paint, rubber, ceramics, paper and roofing compounds. Density, 2.7–2.8 g/cm³ (22.5–23.3 lb/gal); O.A., 30–50; particle size, 0.5–2.5 μm.

Magnesium Soap *n* A magnesium salt of a fatty acid, e.g., ▶ Magnesium Stearate, precipitated by an inorganic magnesium salt from a solution of sodium or potassium soaps. See also ▶ Soap, Metallic.

Magnesium Soaps Saponification products of magnesium and various fatty acids.

Magnesium Stearate *n* $Mg(OOCC_{17}H_{35})_2$. A white, soft powder used as a lubricant and stabilizer.

Magnetic Field Due to a Current *n* The intensity of the magnetic field in oersted at the center of a circular conductor of radius *r* in which a current *I* in absolute electromagnetic units is flowing,

$$H = \frac{2\pi I}{r}$$

If the circular coil has *n* turns the magnetic intensity at the center is,

$$H = \frac{2\pi nI}{r}$$

The magnetic field in a long solenoid of *n* turns per centimeter carrying a current *I* in absolute electromagnetic units

$$H = 4\pi nI$$

If *I* is given in amperes the above formulae becomes,

$$H = \frac{2\pi I}{10r}, H = \frac{2\pi nI}{10r}, H = \frac{4\pi I}{10}$$

Magnetic Field Due to a Magnet *n* At a point on the magnetic axis prolonged, at a distance *r* cm from the center of the magnet of length 2*l* whose poles are +*m* and −*m* and magnetic moment *M*, the field strength is oersted is,

$$H = \frac{4mlr}{(r^2 - l^2)^2}$$

If *r* is large compared with *l*,

$$H = \frac{2M}{r^3}$$

At a point on a line bisecting the magnet at right angles, with corresponding symbols,

$$H = \frac{2ml}{(r^2 + l^2)^{\frac{3}{2}}}$$

For large value of r,

$$H = \frac{M}{r^3}$$

Magnetic Field Intensity or Magnetizing Force *n* Is measured by the force acting on unit pole. Unit field intensity, the oersted is that field which exerts a force of one dyne on unit magnetic pole. The field intensity is also specified by the number of lines of force intersecting unit area normal to the field, equal numerically to the field strength in oersted. Magnetizing force is measured by the space rate of variation of magnetic potential and as such its unit may be the **gilbert per**

centimeter. The gamma (γ) is equivalent to 0.00001 oersted. Dimensions,

$$\left[\varepsilon^{\frac{1}{2}}\ m^{\frac{1}{2}}\ l^{\frac{1}{2}}\ t^{-2}\right];\ \left[\mu^{-\frac{1}{2}}\ m^{\frac{1}{2}}\ l^{-\frac{1}{2}}\ t^{-1}\right]$$

Magnetic Filler *n* Any permanently magnetizable material in powder form that may be incorporated into plastics to produce molded or extruded-strip magnets. Major ones in use are Alnico, rare earths, and, most used in plastics, hard ▶ Ferrite. The figures shown below are a scanning electron micrograph of an acicular magnetic particle employed in the coated surface of the following figure of magnetic media (recording tape).

Acicular ferritic magnetic particles

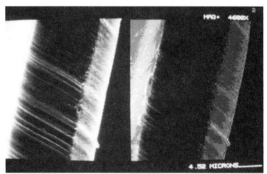
Magnetic coating on PET film cross-section

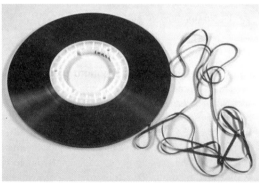
Magnetic storage tape reel

Audio-magnetic tape cassette

Magnetic Flux *n* (1896) Through any area perpendicular to a magnetic field is measured as the product of the area by the field strength. The units of magnetic flux, the **Maxwell** is the flux through a square centimeter normal to a field of one gauss. The line is also a unit of flux. It is equivalent to the Maxwell. Dimensions,

$$\left[e^{-\frac{1}{2}}\ m^{\frac{1}{2}}\ l^{\frac{3}{2}}\right];\ \left[\mu^{\frac{1}{2}}\ m^{\frac{1}{2}}\ l^{\frac{1}{2}}\ t^{-1}\right]$$

Magnetic Induction *n* Results when any substance is subjected to a magnetic field is measured as the magnetic flux per unit area taken perpendicular to the didrection of the flux. The unit is the Maxwell per square centimeter or its equivalent, the gauss. Dimensions,

$$\left[e^{-\frac{1}{2}}\ m^{\frac{1}{2}}\ l^{\frac{3}{2}}\right];\ \left[\mu^{\frac{1}{2}}\ m^{\frac{1}{2}}\ l^{-\frac{1}{2}}\ t^{-1}\right]$$

If a substance of permeability of μ is placed in a magnetic field *H* the magnetic induction in the substance,

$$\mathbf{M} = \mu H$$

If *I* is the magnetic moment for unit volume, or intensity of magnetization,

$$\mathbf{M} = H + 4\pi I.$$

The susceptibility,

$$k = \frac{I}{H},\quad \mu = 1 + 4\pi\kappa$$

Magnetic Inks *n* Inks made with pigments which can be magnetized after printing. The printed characters can later recognized by electronic reading equipment.

Magnetic Moment *n* (1865) The magnetic moment of a magnet is measured by the torque experienced when it is at right angles to a uniform field of unit intensity. The value of the magnetic moment is given by the product of the magnetic pole strength by the distance between the poles. Unit magnetic moment is that possessed by a magnet formed by two poles of opposite sign and of unit strength, one centimeter apart. Dimensions,

$$\left[\mu^{\frac{1}{2}}\ m^{\frac{1}{2}}\ l^{-\frac{1}{2}}\ t^{-1}\right];\ \left[e^{-\frac{1}{2}}\ m^{\frac{1}{2}}\ l^{\frac{3}{2}}\right]$$

If the poles are separated by a distance which is great compared with the dimensions of the magnet, the magnetic moment of a magnet of length 1 whose poles have values of $+m$ and $-m$ is, m = ml.

Magnetic Permeability *n* A property of materials modifying the action of magnetic poles placed therein and modifying the magnetic induction resulting when the material is subjected to a magnetic field or magnetizing force. The permeability of a substance may be defined as the ratio of the magnetic induction in the substance to the magnetizing field to which it is subjected. The permeability of a vacuum is unity. Dimensions $[e^{-1} l^{-2} t^2]$; $[\mu]$.

Magnetic Pole or Quantity of Magnetism *n* Two unit quantities of magnetism concentrated at points unit distance apart in a vacuum repeal each other with unit force. If the distance involved in 1 cm and the force 1 dyne, the quantity of magnetism at each point is one cgs unit of magnetism. Dimensions,

$$\left[e^{-\frac{1}{2}}\ m^{\frac{1}{2}}\ l^{\frac{1}{2}}\right],\ \left[\mu^{\frac{1}{2}}\ m^{\frac{1}{2}}\ l^{\frac{3}{2}}\ t^{-1}\right]$$

Magnetic Potential or Magnetomotive Force *n* At a point is measured by the work required to bring until positive pole from an infinite distance (zero potential) to the point. The unit is the *gilbert*, that magnetic potential against which an erg of work is done when unit magnetic pole is transferred. Dimensions,

$$\left[e^{\frac{1}{2}}\ m^{\frac{1}{2}}\ l^{\frac{3}{2}} t^{-2}\right],\ \left[\mu^{-\frac{1}{2}} m^{\frac{1}{2}}\ l^{\frac{1}{2}}\ t^{-1}\right]$$

Magnetic Quantum Number, m_l *n* (1923) A quantum number which indicates the orbital occupied by an electron.

Magnetic Separator *n* A device that removes tramp iron and steel from a stream of mainly nonmagnetic material, such as reground plastic or mixed wastes, by passing the stream close to strong magnets. Some design parameters for magnetic separators are given in Section 21 of Chemical Engineers' Handbook, Sixth edition, R. H. Perry and D. W. Green, editors, McGraw-Hill, and the two preceding editions.

Magnetite Black \ mag-nə- tīt blak\ *n* (1851) Fe_3O_4. Magnetic iron oxide. See ▶ Black Iron Oxide.

Magnification, Empty *n* A higher magnification than necessary to resolve detail.

Magnification, Maximum Useful (MUM) *n* The maximum magnification necessary to resolve detail. Magnification in excess of MUM gives no additional resolving power. It can usually be estimated as being 1,000 times the NA of the objective.

Magnifying Power *n* The magnifying power of an optical instrument is the ratio of the angle subtended by the image of the object seen through the instrument to the angle subtended by the object when seen by the unaided eye. In the case of the microscope or simple magnifier the object as viewed by the unaided eye is supposed to be a distance of 25 cm (10 in.).

Mahlstick \ mól-\ *variant of maulstick* Long stick, padded at one end, on which a painter can rest his hand to steady if when working.

Mahogany Sulfonates *n* Soaps, the sodium salts of sulfonic acids from petroleum refining sludge; used in synthetic resin production, as are sorbitan oleates and laurates, polyoxyethylene esters.

Maintenance Paints *n* Coatings used to maintain manufacturing plants, offices, stores and other commercial structures, hospitals and nursing homes, schools and universities, government and public buildings, and both building and nonbuilding requirements in such areas as public utilities, railroads, roads, and highways, and including industrial paint, other than the original coatings, the primary function of which is protection. Residential maintenance is excluded. See also ▶ Industrial Maintenance Paints.

Makeready *n* The preparation and correction of the printing plates, before starting the printing run, to insure uniformly clean impressions of optimum quality. All preparatory operations preceding a production run.

Makrolon Polycarbonate from bisphenol A and phosgene base units. Manufactured by Bayer, Germany.

Malachite \ ma-lə- kīt\ *n* [alt of ME *melochites*, fr. L *molochites*, fr. Gk *molchitēs*, fr. *molochē*, *malchē* mallow] (1656) $CuCO_3 \cdot Cu(OH)_2$. Basic carbonate of copper which occurs naturally. The color varies from a bright emerald to a dark green.

Malachite Green *n* (1) Bluish-green dyestuff made from dimethyl aniline and benzaldehyde. It is often sold in the form of its oxalate. (2) A green lake pigment produced for a basic dye, used in the manufacture of printing inks.

Maleic Acid \mə- lē-ik-, - lā-\ *n* [F, *acide maléique*, alter. of *acide malique* malic acid, fr. its formation by dehydration of malic acid] (1857) $COOH(CH)_2COOH$. Dibasic acid used in the manufacture of synthetic resins.

Maleic Anhydride *n* (1857) (2,5-furandione) A compound crystallizing as colorless needles, obtained by passing a mixture of benzene and air over a heated vanadium pentoxide catalyst, and having the structure shown below. It has many applications in plastics, including the production of alkyd, polyester, and vinyl-copolymer resins, and as a curing agent for thermosetting resins such as phenolics and urea's. About half the maleic anhydride produced in the US is used in the manufacture of unsaturated polyester resins, to which it imparts fast curing and high strength. Used in manufacturing synthetic resins and maleinized oils, MP, 56°C; bp, 202°C; acid value, 1.143.

Maleic Anhydride Value See ▶ Diene Value or Number.
Maleic Ester Resin *n* A synthetic resin made from maleic acid or maleic anhydride and a polyhydric alcohol.
Maleic Resin *n* A resin made from a natural resin and maleic anhydride or maleic acid.
Maleic Resins *n* A class of resins obtained from the condensation of maleic anhydride with rosin, terpenes, etc.
Maleic Value Another name for diene value.
Maleinized Oil *n* Oil which has been reacted, through its double bonds, with maleic anhydride.
Mallory Fatigue Test *n* A test to measure the endurance properties of tire cord.
Maltese Cross *n* (1877) A dark shadow, having the shape of a maltese cross, seen in polymer (e.g., polyethylene) spherulities when viewed under a polarizing microscope.
MAN Abbreviation for ▶ Methacrylonitrile.
Mandrel \\ˈman-drəl\\ *n* [prob. mod. of F *mandrin*] (1665) (1) The core around which paper, fabric, or resin-impregnated fibrous glass is wound to form pipes or tubes. (2) In extrusion, an extension of the core of a pipe or tubing die, internally cooled by circulating water or other fluid, that guides and cools the internal surface of the tube as it emerges from the die proper. The mandrel is an important determiner of the final internal diameter of the tube.
Mandrel Test *n* Test for determining the flexibility and adhesion of surface coatings, so named because it involves the bending of coated metal panels around mandrels.
Manganese Black \\ˈmaŋ-gə-ˌnēz\\ MnO_2. Manganese dioxide. A black pigment. Principal uses are as a drier and as a colorant for ceramics.

Manganese Brown *n* There are two types of manganese brown: (1) Burnt turkey umber; (2) A brown oxide pigment, made artificially as a by-product from chlorine manufacture.
Manganese Dioxide *n* (1882) MnO_2 A dark insoluble compound used especially as an oxidizing agent, as a depolarizer of dry cells, and in making glass and ceramics.

$$O^{--}\ Mn^{+4}\ O^{--}$$

Manganese Driers *n* (1) Material containing chemically combined manganese used to accelerate the oxidation and polymerization of an ink film. (2) These include manganese dioxide, the hydrated oxide, manganese acetate, sulfate and borate. The organic driers are salts of various organic acids such as naphthenic or 2-ethyl hexoic. Manganese driers are characterized by their reddish-brown colors and their surface drying activity.
Manganese Green *n* Strong green pigment, with good alkali resistance, prepared by roasting manganese dioxide and barium hydroxide together under oxidizing conditions.
Manganese Violet See ▶ Mineral Violet.
Manifold \\ˈma-nə-ˌfōld\\ *n* (1855) A pipe or channel with several inlets or outlets. With reference to blow molding, extrusion, and injection molding, a manifold is a piping or distribution system that receives the outflow of the extruder or molder and divides or distributes it to feed several blow-molding heads or injection nozzles.
Manila \\mə-ˈni-lə\\ *adj* (1834) Fiber obtained from the leaf stalks of the abaca plant. It is generally used for cordage.
Manila Copal *n* Natural resins, two types of which are used in varnish manufacture, in which they are described as hard and soft manilas. The hard type requires running and is used to some extent in oil varnishes. The soft type is readily soluble in industrial alcohol and forms the basis of spirit paper and other air-drying varnishes; the native name for the soft type, obtained by tapping, is Melengket.
Manjak *n* Intense black, naturally-occurring asphaltum, obtained from Barbados. It differs from other asphaltums used in the trade by reason of its unusual staining power and difficult solubility. Prolonged high temperature treatment is necessary in order to effect a reasonable solution in drying oils, and a substantial amount of mineral matter always remains undissolved. It is used alone, or with gilsonite, in black bituminous finishes of many types. Syn: ▶ Glance Pitch.
Man-Made Fiber Syn: ▶ *Synthetic* Fiber.

Mannich Reaction *n* The condensation of ammonia or a primary or secondary amine with formaldehyde and a compound containing at least one hydrogen atom of pronounced activity. The active hydrogen replaced by an aminomethyl or substituted aminomethyl group. This reaction has been employed in producing "mannich polyols" for use in making urethane foams.

Mannite \ˈma-ˌnīt\ *n* [F, fr. *manna*, fr. LL] (1830) See ▶ Mannitol.

Mannitol \ˈma-nə-ˌtól\ *n* [ISV] (1879) $C_6H_8(OH)_6$. A hexahydric alcohol which has been used in the production of synthetic oils and alkyd resins. Bp, 278°C/1 mm Hg; mp, 166°C. *Known also as Mannite.*

Manufactured Fiber *n* A class name for various genera of fibers (including filaments) produced from fiber-forming substances which may be: (1) polymers synthesized from chemical compounds, e.g., acrylic, nylon, polyester, polyethylene, polyurethane, and polyvinyl fibers; (2) modified or transformed natural polymers, e.g., alginic and cellulose-based fibers such as acetates and rayons; and (3) minerals, e.g., glasses. The term manufactured usually refers to all chemically produced fibers to distinguish them from the truly natural fibers such as cotton, wool, silk, flax, etc.

Manufactured Unit *n* A quantity of finished adhesive or finished adhesive component, processed at one time. NOTE – The manufactured unit may be a batch or a part thereof.

Marble \ˈmär-bəl\ *n* [ME, fr. OF *marbre*, fr. L *marmor*, fr. Gk *marmaros*] (12c) (1) Limestone that has crystallized to varying extent, often with veined inclusions, and occurring in many colors. Its preponderant constituent is ▶ Calcium Carbonate. (2) A smooth round sphere of any hard nonmetal in the size range from about 0.7–2.5 cm.

Marble Flour See ▶ Calcium Carbonate, Natural.

Marbling, Marbleizing *n* Imitating with finishing materials, as in antiquing, the figure and texture of polished marble or other decorative stones, usually by stippling or mottling in conjunction with graining, scratching and spattering.

March, Nonconditional See Nonconditional Match.

Margaric Acid See Daturic Acid.

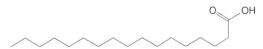

Marine Borers *n* Mollusks and crustaceans that attack submerged wood in salt and brackish water.

Marine Coatings *n* Paints and varnishes specifically formulated to withstand water immersion and exposure to marine atmosphere. See also ▶ Spar Varnish.

Marine Varnishes See ▶ Marine Coatings and ▶ Spar Varnish.

Marker *n* In the floor coverings industry, a distinctive threadline in the back of a carpet that enables the installer to assemble breadths of carpet so that the pile lays in one direction or so that patterns match.

Mark-Houwink Equation *n* Also referred to as Kuhn-Mark-Houwink-Sakurada equation; allows prediction of the viscosity average molecular weight M_v for a specific polymer in a dilute solution of solvent by $[\eta] = KM_v^a$ where K is a constant for the respective material and *a* is a branching coefficient; K and *a* (sometimes a^η) can be determined by a plot of log $[\eta]$ vs. log M_v^a and the slope is *a* and intercept on the Y axis is K. (Kamide K, Dobashi T (2000) Physical chemistry of polymer solutions. Elsevier, New York; Physical properties of polymers handbook. Mark JE (ed). Springer, New York, 1996; Elias HG (1977) Macromolecules, vol. 1–2, Plenum, New York)

Marking Nut Oil *n* Oil that resembles cashew nut shell liquid in that it is phenolic and quite unlike the glyceride vegetable oils. *Known also as DHOBI Marking Nut Oil or Bhilawan Oil.*

Marl A yarn made from two rovings of contrasting colors drafted together, then spun. Provides a mottled effect.

Marlex Poly(ethylene), manufactured by Phillips, US.

Marouflage *v* To glue a canvas to a wall which is to be covered by a mural painting.

Marquardt Index *n* In an infrared-absorption study of the cure advancement of a phenolic resin, the Marquardt index is the numerical difference in percent transmission between the absorption peaks at 12.2 and 13.3 μm. As the resin cure progresses, the intensity of the 13.3-μm absorption increases more rapidly than that of the initially stronger 12.2-μm peak.

Marquetry \ˈmär-kə-trē\ *n* [MF *marqueterie*, fr. *marqueter* to checker, inlay, fr. *marque* mark] (1563) Decorative inlay.

Marquisette \ˌmär-kwə-ˈzet\ *n* [*marquise* + -*ette*] (1908) A lightweight, open-mesh fabric made of cotton, silk, or manufactured fibers in a leno, doup, or gauze weave. Marquisettes are used for curtains, dresses, mosquito nets, and similar end uses.

Mar Resistance *n* The resistance of a glossy plastic surface to abrasive action. It is measured (ASTM D 673) by abrading a specimen to a series of degrees, then measuring the gloss of the abraded spots with a glossmeter and comparing the results to that of the unbraided area of the specimen. See also ▶ Gloss.

Married Fiber Clump *n* A defect that occurs in converter top. It consists of a group of unopened, almost coterminous fibers with the crimp in register.

Martens Heat-Deflection Temperature *n* The temperature at which, under four-point loading, a bar of polymer deflects by a specified amount. For amorphour polymers, the Martens temperature is about 20°C below the glass-transition temperature. Compare ▶ Deflection Temperature.

Martius Yellow *n* Calcium derivative of naphthalene yellow.

Mask \ˈmask\ *n* [MF *masque*, fr. OIt *maschera*] (1534) A stencil used for spray-painting plastics, consisting of a relatively thin sheet shaped to fit the part to be painted with openings for areas to be painted.

Masking *n* Temporarily covering that part of a surface to which it is not desired to apply a coating.

Masking Tape *n* Adhesive backed paper tape used to mask or protect parts of a surface not to be finished.

Masonry \ˈmā-sən-rē\ *n* (13c) The art of the mason in shaping, arranging and uniting stone, brick, building blocks, tile and similar materials, to form walls and other parts of a building.

Masonry Conditioner *n* A solvent-based, pigmented primer coating formulated to have great penetrating power so as to prepare masonry (especially chalky stucco) to receive finish coats. Particularly important under latex paints.

Masonry Paint *n* An alkali-resistant coating, usually a latex paint, used for masonry substrates.

Mass \ˈmas\ *n* [ME *masse*, fr. MF, fr. L *massa*, fr. Gk *maza*; akin to Gk *massein* to knead] (15c) (1) Quantity of matter, whose unit, the kilogram, is one of seven base units of the SI system. The term is often confused with *weight* in everyday use, probably because, when weighed on an equal-arm balance, the mass being determined is compared with standard masses, ordinarily referred to as "weights". Although the kilogram-force (*kilopond*) has long been used and is still being used, it has no place in the SI system. (2) Units of mass – the gram is $\frac{1}{1000}$ the quantity of matter in the International Prototype Kilogram; one of the three fundamental units of the cgs system. The British standard of mass is the pound, of which a standard is preserved by the government. The United States standard mass is the avoirdupois pound defined as ½.20462 kg. (Giambattista A, Richardson R, Richardson RC, Richardson B (2003) College physics. McGraw Hill Science, New York; Kricheldorf HR, Swift G, Nuyken O, Huang SJ (2004) Handbook of polymer synthesis. CRC, Boca Raton, FL)

Mass-Action Expression, Q The product of the concentrations or partial pressures (or, better, activities) of the products in a reaction, divided by those of the reactants. Each term is raised to an exponential power corresponding to the coefficient written before the corresponding substance or species in the balanced equation. Pure solids and liquids are omitted, as are substances present in large excess, and therefore almost constant concentration.

Mass-Action Law *n* For a homogeneous reacting system, the rate of chemical reaction is proportional to the active masses of the reacting substances, the molecular concentration of a substance in a gas or liquid being taken a its active mass.

Mass Action, Law of *n* At a constant temperature the product of the active masses on one side of a chemical equation when divided by the product of the active masses on the other side of the chemical equation is a constant, regardless of the amounts of each substance present, at the beginning of the action. At constant temperature the rate of the reaction is proportional to the concentration of each kind of substance taking part in the reaction.

Mass by Weighing on a Balance with Unequal Arms *n* If W_1 is the value for one side, W_2 the value for the other, the true mass,

$$W = \sqrt{W_1 W_2}$$

Mass Color *n* The color, when viewed by reflected light, of a pigment-vehicle mixture of such thickness as to obscure completely the background. *Sometimes called Over-Tone or Mass-Tone.*

Mass Defect *n* (ca. 1923) Difference between atomic mass and mass number of a nuclide.

Mass Dyeing See Spin Drying.

Mass-Energy Equivalence *n* The equivalence of a quantity of mass and a quantity of energy when the two quantities are related by the equation $E = mc^2$. The conversion factor c^2 is the square of the velocity of light. The relationship was developed from **relativity theory**, but has been experimentally confirmed.

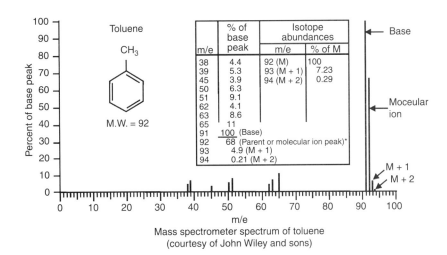

Mass spectrometer spectrum of toluene
(courtesy of John Wiley and sons)

Mass (Fiber) Strength *n* The force per unit of lineal density required to break a fiber. The SI measure is newtons per (kilogram per meter), or N m/kg. Long used in the staple-fiber industry has been the unit gram-force per denier. 1 g_f/denier = 88,259 N m/kg.

Massicot (Massocot) *n* Another name for lead monoxide.

Mass Number *n* (1923) The total number of nucleons (protons and neutrons) in an atom.

Mass Polymerization See ▶ Bulk Polymerization.

Mass Spectrometry (Spectroscopy) *n* (1943) (MS) Mass spectrometry is an analytical technique in which a material (e.g., a polymer) is pyrolyzed, the fragment molecules are injected into a vacuum chamber where they are ionized with an electron gun, accelerated in an electric field, and forced through a magnetic field, the paths of the more massive molecules deflecting (curving) less than the lighter ones. A detector registers the mass number and ion count at each mass number and from this information develops a spectrum. An analyst can determine the composition of the original polymer from his interpretation of the spectrum fragments. The MS method may be supplemented by ▶ Gas Chromatography which can identify the types of chemical structures in the fragments (Polymers: Polymer characterization and analysis. Kroschwitz JI (ed) Wiley, New York, 1990; Willard HH, Merritt LL, Dean JA (1974) Instrumental methods of analysis. D. Van Nostrand, New York) An example of a MS spectrum of toluene is shown.

Mass Tone *n* The color produced by a single color dispersed full strength in a suitable vehicle.

Masstone *n* (1) A pigment-vehicle mixture which contains a single pigment only. (2) Occasionally, this term is used more loosely to describe a pigment-vehicle mixture which contains no white pigment. See ▶ Mass Color.

Masstone Color *n* The color of a masstone paint applied at complete hiding.

Masterbatch A term used in the rubber industry for rubber compounds containing high percentages of pigments and/or other additives, to be added in small amounts to batches during compounding. The term is often used in the plastics industry for ▶ Color Concentrate.

Master Curve *n* The curve one gets by applying the principle of ▶ Time-Temperature Equivalence to viscoelastic data on, say, relaxation modulus or creep.

Mastic \ˈmas-tik\ *n* [ME *mastik*, fr. L *mastiche*, fr. Gk *mastichē*, prob. back-formation fr. *mastichan*] (14c) (1) A solid resinous material obtained from the mastic tree (*Pistacia lentiscus*) and used in adhesives and lacquers. (2) *Asphalt mastic*, a composition of mineral matter with resin and solvent. (3) Any pasty material used as a waterproof coating or as cement for setting tile.

Masticate \ˈmas-tə-ˌkāt\ *v* [LL *masticatus*, pp of *masticare*, fr. Gk *mastichan* to gnash the teeth; akin to Gk *masasthai* to chew] (1649) To work rubber on a mixing mill or in an internal mixer until it becomes soft and plastic. Synonymous with breakdown.

Mastication *n* Intense shearing of unvulcanized rubber by working in a roll mill or internal mixer to reduce its molecular weight preparatory to compounding and molding.

Mastication of Resins *n* Process of hot working of resins which is believed to reduce molecular complexity and to confer solubility. It has been used on natural copals, and advantages of this treatment are that thermal cracking is avoided and pale colors maintained. Patented methods include treatment between rotating rollers, sometimes in the presence of solvents.

Mat *n* A fabric or felt of glass or other reinforcing fibrous material cut to the contour of a mold, for use in reinforced-plastics processes such as matched-die molding, hand layup, or contact-pressure molding. The mat is usually impregnated with resin just before or during the molding process.

Matched-Die Molding *n* A reinforced plastic manufacturing process in which close-fitting metal matching male and female molds are used to form the part using pressure, temperature, and time cycle.

Matched-Mold Thermoforming *n* A sheet-thermoforming process in which the heated plastic sheet is shaped between male and female halves of a matched mold. The molds may be of metal or inexpensive materials such as plaster, wood, epoxy resin, etc., and must be vented to permit the escape of air as the mold closes. See ▶ Sheet Thermoforming.

Matching, Color *n* Act of making one material appear to match another color. If the achieved match is dependent on the conditions of illumination and viewing, the match is termed conditional or metameric. If the achieved is independent of the quality of the illuminant viewer, or viewing conditions, the match is termed nonconditional or nonmetameric.

Matelassé *n* A soft, double or compound fancy-woven fabric with a quilted appearance. Heavier types are used as draperies and upholsteries. Crepe matelassé is used for dresses, wraps, and other apparel. Matelassé is usually woven on a Jacquard loom.

Maximum Permissible Stress See ▶ Allowable Stress and ▶ Factor of Safety.

Maxwell \maks-wel\ *n* [James Clerk *Maxwell*] (1900) The cgs emu magnetic flux is the flux through a square centimeter normal to a field at 1 cm from a unit magnetic pole.

Maxwell Model *n* (Maxwell element) A concept useful in modeling the deformation behavior of viscoelastic materials. It consists of an elastic spring in series with a viscous dashpot. When the ends are pulled apart with a definite force, the spring deflects instantaneously to its stretched position then motion is steady as the dashpot opens. See also ▶ Voigt Model. A simple combination of these two types provides a fair analogic representation of real viscoelastic behavior under stress.

Maxwell's Rule *n* A law stating that every part of an electric circuit is acted upon by a force tending to move it in such a direction as to enclose the maximum amount of magnetic flux.

MBK *n* Abbreviation for ▶ Methyl Butyl Ketone

MBS *n* Abbreviation for Methacrylate-Butadiene-Styrene Resin. These are mixtures of PMMA and butadiene-styrene copolymers, formulated in a variety of types with markedly different characteristics according to their composition and molecular weight. MBS resins can be processed by all the usual thermoplastics processes.

Mc *n* Abbreviation for Megacycle, one million cycles, loosely used to mean 1 MHz, one million cycles *per second*.

MC *n* Methyl cellulose.

MD *n* (1) Abbreviation for Machine Direction. (2) Abbreviation for Methylene Dianiline, little used because of its carcinogenicity.

MDI *n* Abbreviation for Diphenylmethane-4,4′-Diisocyanate. See ▶ Diisocyanate.

MDPE *n* Abbreviation for Medium-Density Polyethylene. See ▶ Polyethylene.

Measling *n* The appearance of spots or stars under the surface of the resin portion of an epoxy/glass-fiber laminate (from *measles*).

Mechanical Adhesion *n* See ▶ Adhesion and ▶ Specific.

Mechanical Equivalent of Heat *n* A conversion factor that transforms work or kinetic energy into heat. Probably the best known one is 788 ft-lb per British thermal unit; others are 2,545 Btu per horsepower-hour, 4.186×10^7 ergs/cal, and 3,413 Btu/(kW h). In SI there is no need for such factors because work, heat and electrical energy are all measured in joules. 1 J = 1 m-N = 1 W-s.

Mechanical Finishing *n* Changing the appearance or physical properties of a fabric by a mechanical process such as calendering, embossing, bulking, compacting, or creping.

Mechanical Grease Forming *n* A method of ▶ Sheet Thermoforming used with acrylic sheet when excellent optics are imperative and the shape desired cannot be produced by ▶ Free Forming. The mold surface is covered with a 1–2-mm-thick layer of felt soaked with melted grease that must be cleaned off the sheet after forming.

Mechanically Foamed Plastic *n* A cellular plastic in which the cells have been produced by gases introduced by physical means. See also ▶ Cellular Plastic.

Mechanical Loss *n* Loss in energy, dissipated as heat, that result when a material is subjected to an oscillatory load or displacement.

Mechanical Properties *n* Those properties of a martial that are associated with elastic and inelastic reaction when force is applied, or that involve the relationship between stress and strain.

Mechanical Property *n* Any property of a material that defines its response to a particular mode of stress or strain. Such properties include elastic moduli, strength and ultimate strain in several modes, impact strength, abrasion resistance, creep, ductility, coefficient of friction, hardness, cyclic fatigue strength, tear strength, and machinability. Many ASTM tests in Section 08 are devoted to the mechanical properties of plastics.

Mechanical Resins *n* Thermosetting resins prepared by condensation of formaldehyde with melamine. Have good hardness, scratch and fire resistance, clarity, colorability, rigidity, dielectric properties, and tensile strength, but poor impact strength. Molding grades are filled. Processed by compression, transfer, and injection molding, impregnation, and coating. Used in cosmetic containers, appliances, tableware, electrical insulators, furniture laminates, adhesives, and coatings.

Mechanical Spectrometer *n* An instrument (Rheometrics, Inc) capable of applying an alternating tensile/compressive (or flexural or torsional) deformation of constant amplitude to a plastic specimen in the frequency range from 0.002 to 80 Hz and measuring the variation of force so caused and the phase angle between the deformation and the force. For this information one can calculate the "real" and "imaginary" parts of the various moduli. (See ASTM, www.astm.org)

Mechanism *n* The set of steps (elementary processes) which together comprise an overall reaction.

Mechanisms *n* Step-by-step pathway from reactants to products showing which bonds break and which bonds form in what order.

Media *n* Aggregate used to effect dispersion in certain types of production equipment, such as ball, pebble, and sand mills. The media vary in size and composition. Some examples are: steel balls, natural stones or pebbles, synthetic ceramic balls, glass beads, and sand.

Media Mill *n* Any mill using any one of the various types of grinding media, e.g., sand, steel ball, pebble, etc.

Median \ˈmē-dē-ən\ *n* The value in an arrayed set of repeated measurements that divides the set into two equal-numbered groups. If the sample size is odd, the medium is the middle value. The median is a useful measure of the center when the distribution is strongly skewed toward low or high values. Compare ▶ Arithmetic Mean.

Medium \ˈmē-dē-əm\ *n* [L, fr. neuter of *medius* middle] (1593) In paints or enamels, the continuous phase in which the pigment is dispersed; thus, in the liquid paint in the can, it is synonymous with vehicle, and in the dry film it is synonymous with binder.

Medium *n* (**Art**) In a general sense, the particular material with which a work of art is executed: oils, water color, chalks, lithographic stone, pen and ink, etc. It may also refer to the liquid with which powdering pigments are ground to make artist's paint, and in a more restricted sense, to the liquid used to render such paint more fluid and workable.

Medium Oil Varnish *n* Varnish of medium oil content usually containing from 18 to 25 gal of oil per 100 lb of resin. See ▶ Long Oil.

Medium Yellow *n* A pigment based on pure, monoclinic lead chromate.

Mega- *adj combining form* [Gk, fr. *megas* large] (M) The SI prefix meaning $\times 10^6$.

Megahertz \ˈme-gə-ˌhərts, -ˌherts\ *n* [ISV] (1941) A unit of vibrational frequency equal to 10^6 cycles per second, i.e., 10^6 Hz.

Megapoise *n* One million poises. This unit is used for materials of very high viscosity, e.g., asphalts. See ▶ Viscosity and ▶ Poise.

MEK *n* Abbreviation for ▶ Methyl Ethyl Ketone

MEKP *n* Abbreviation for ▶ Methyl Ethyl Ketone Peroxide.

Melamine \ˈme-lə-ˌmēn\ *n* [Gr *Melamin*] (ca. 1835) (2,4,6-Triamino-1,3,5-Triazine) $C_3N_3(NH_2)_3$. A cyclic unsaturated compound, derived from cyanuric acid, with the structure shown below. It reacts with formaldehyde to give a series of heat reactive resins. Melamine's main use is for ▶ Melamine-Formaldehyde Resins.

Melamine-Formaldehyde Resins *n* (melamine resin) Any of a group of thermosetting resin of the amino-resin family, made by reacting melamine with formaldehyde. The lower-molecular-weight, uncured melamine resins are water-soluble syrups, used for impregnating paper, laminating, etc. High-molecular-weight resins, usually cellulose-filled, are powders widely used, from 1950–1970, for plastic tableware.

Melamine/Phenolic Resin *n* A mixture of melamine- and phenol-formaldehyde resins that combines the dimensional stability and ease of molding of phenolics with the wider range of colorability of the melamine resins.

Melamine Resins *n* (1939) Any of the class of thermosetting resins formed by the interaction of melamine and formaldehyde.

Melan, Melamin *n* Melamine/formaldehyde prepolymers. Manufactured by Henkel, Germany.

Melatrope *n* The center of rotation of the isogyres in biaxial interference figures representing the point of emergence of rays that, in the crystal, travel along the optic axes.

Melbrite *n* Melamine/formaldehyde resin. Manufactured by Montedison, Italy.

Melded Fabric *n* A nonwoven fabric of a base fiber and a thermoplastic fiber. The web is hot-calendered or embossed at the softening point of the thermoplastic fiber to form the bond.

Meldola Blue *n* Methylene blue type of dye. *Known also as New Blue or Naphthol Blue.*

Melengket *n* Native name for soft Manila copal obtained by tapping.

Melissic Acid *n* $CH_3(CH_2)_{28}COOH$. Monobasic fatty acid constituent of beeswax. MP, 90°C.

Melt \ˈmelt\ *n* (1854) A material, solid at room temperature that has been heated to a molten condition.

Melt-Bead Sealing See Extruder-Bead Sealing.

Melt Blend See ▶ Biconstituent Fiber.

Melt Blowing *n* The formation of a Nonwoven by extruding molten polymer through a die then attenuating and breaking the resulting filaments with hot, high-velocity air or steam. This results in short fiber lengths. Short fibers are then collected on a moving screen where they bond during cooling.

Melt Coating See ▶ Extrusion Coating.

Melt-Draining Screw see ▶ Solids-Draining Screw.

Melt-Dyed See ▶ Dyeing, Mass Colored.

Melt Extruder *n* A short extruder, typically of constant channel depth and lead throughout, designed to receive a molten feed and raise its pressure for extrusion through a die, such as a pelletizing die.

Melt Flow *n* The rate of extrusion of molten resin through a die of specified length and diameter. The conditions of the test (e.g., temperature and load) should be given. Frequently, however, the manufacturer's data lists only the value, not the condition as well.

Melt-Flow Index *n* (MFI, melt index) The rate of flow, in grams per 10 min, of a molten resin through an orifice 2.096 mm in diameter and 8.000 mm long at a specified temperature and weight of piston pressing on the melt. Numerous combinations of temperatures and weights are listed in ASTM (www.astm.org) various thermoplastics. This single-point flow measurement is useful in controlling production quality and resin purchasing, but most of the MFI conditions are at much lower shear than those prevailing in commercial processing, so MFI is not a reliable guide to processing behavior. MFI is inversely related to viscosity and decreases rapidly as the molecular weight in a resin family increases.

Melt Fracture *n* In extrusion, the distortion of the extrudate as it emerges from a die. The effect ranges from minor, regular ridges and valleys at 45° or 90° to the axis of the extrudate to violent wriggling and curling and, at its most extreme, breaking up of the extrudate into fragments

Melt Index *n* The amount, in grams, of a thermoplastic polymer which can be forced through an orifice of 0.0825 in. diameter when subjected to a force of 2,160 gf in 10 min at 190°C.

Melting Point *n* (melting range) In pure compounds, the temperature at which the transition from solid to liquid occurs, requiring heat input. Polymers, being broad mixtures of homologs, melt over a substantial range of temperature, the shorter chains melting first with rising temperature, the longer ones later. Crystalline polymers have narrower, more distinct melting ranges than amorphous polymers. See also ▶ Heat of Fusion.

Melting Zone *n* In a well-designed extruder screw, the section, to be coincident with the ▶ Transition Section, in which most, if not all, of the melting of the feedstock occurs. The pumping section, in which the plastic is presumed to be fully melted, is sometimes called the *melt zone*.

Melt Instability *n* (melt-flow instability) A term applied to the early manifestations of ▶ Melt Fracture.

Melton \ˈmel-tᵊn\ *n* [*Melton* Mowbray, town in England] (1823) A heavily full, hard, plain coating fabric that was originally all wool but is now also seen in wool blends.

Melt Pressure *n* The gauge pressure exerted at any point in a processing apparatus that develops pressure. In extruders, melt pressure in the head is usually monitored. In injection machines the location is analogous but melt pressures have also been measured in mold cavities. Not to be confused with (though related to) ▶ Injection-Molding Pressure.

Melt Spinning See ▶ Spinning.

Melt Spinning Process *n* Molten polymer is pumped first through sand-bed filters, then through one to thousands of tiny orifices, called jets or spinnerets by small gear pumps operating at extremely high pressures. The fibers are then oriented to realize their optimal strength and modulus, four times or more that of the unoriented fibers. See ▶ Spinning.

Melt Strength *n* Denotes the viscous flow of a polymer melt under tensile stress.

Melt Strength *n* The strength of a plastic while in the molten state. This property is pertinent to extrusion of a parisons for blow molding, to drawing extrudates from dies, as in making monofilaments and cast film, and to sheet thermoforming. It is also important when a plastic film is reheated for shrink-packaging. This property is very difficult to measure because of the ease with which a filament stretches in elongational flow at the temperatures of interest.

Melt Temperature *n* The temperature of molten or softened plastic at any point within the material being processed. In extrusion and injection molding, melt temperature is an important indicator of the state of the material and the process. Many types of instruments, most of them based on thermocouples or resistance thermometers, have been employed in extruders, where melt temperature is usually measured in the head and sometimes in the die. In thermoforming, temperatures of softened sheets are measured with ▶ Infrared Pyrometers.

Melt Viscosity *n* The resistance to shear in a molten resin, quantified as the quotient of shear stress divided by shear rate at any point in the flowing material. Elongational viscosity, which comes into plan in the drawing of extrudates, is analogously defined. In polymers, the viscosity depends not only on temperature and, less strongly, on pressure, but also on the level of shear stress (or shear rate). See ▶ Viscosity, ▶ Power Law, and ▶ Pseudoplastic Fluid.

Melt Volume Index *n* The volume of plastic extruded in 10 min at a given load on a specified die.

Membrane Osmometry *n* The pressure difference between a solution and the pure solvent is measured for the case where the solvent is separated from the solution by a semipermeable membrane, isothermally; the measurement yields Δp (change in pressure) which corresponds to M_n number average molecular weight – a colligative property of polymer solutions:

$$dG = Vdp - SdT$$

where G = Gibbs energy (H-TS), p = pressure, S = entropy, H = enthalpy, T = temperature.

Memory Zone *n* (elastic memory, plastic memory) The tendency of a plastic article to revert in dimensions to a size previously existing at some stage in its manufacture. For example, a film that has been oriented by hot stretching and chilled while under tension, will, upon reheating, tend to revert to its original prestretched size due to its "memory". See also ▶ Orientation.

Menaccanite Old name for the titanium mineral, ilmenite.

Mending *n* A process in woven fabric manufacture in which weaving imperfections, tears, broken yarns, and similar defects are repaired after weaving; especially on woolen and worsted fabrics to prepare them for dyeing, finishing, or other processing.

Mer \mər\ *n* [ISV, fr. Gk *meros* part] Derived from the Greek *meros*, meaning a part or unit, the mer is the smallest repeating structural unit (mono + mer) of a polymer (poly + mer). In addition polymers such as polyethylene the mer weight is the same as the monomer's molecular weight. Saving a small correction for end groups, the molecular weight of a polymer chain equals the mer weight times the degree of polymerization. Dimers, trimers, tetramers, oligomers, and polymers contain two, three, four, several, and many mer units, respectively.

Meraklon *n* Poly(propylene). Manufactured by Montecatini, Italy.

Mercadium Orange See ▶ Cadmium-Mercury Sulfides.

Mercerization \ˈmər-sə-ˌrīz\ *vt* [John *Mercer* † 2866 English calico printer] (1859) A treatment of cotton yarn or fabric to increase its luster and affinity for dyes. The material is immersed under tension in a cold sodium hydroxide (caustic soda) solution in warp or skein form or in the piece, and is later neutralized in acid. The process causes a permanent swelling of the fiber and thus increases its luster.

Mercurials \(ˌ)mər-ˈkyur-ēl-əls\ *n* (1676) Fungicides and bactericides containing mercury.

Mercuric Chloride *n* (1874) (corrosive sublimate, mercury bichloride) $HgCl_2$. White crystals, used as a polymerization catalyst for PVC. Mercuric chloride is highly toxic, so must be handled with care and requires special disposal procedures.

$$Cl^-$$
$$Hg^{++} \quad Cl^-$$

Mercuric Sulfide *n* HgS. Pigment Red 106 (77766). (1) A naturally occurring mineral, cinnabar. (2) Synthetically produced by addition of Hg to alkali sulfides depending on temperature, the colors vary from red orange to bluish red. Has good alkali resistance. Density, 8.0 g/cm³ (66.6 lb/gal); O.A., 11–15. Syn: are ▶ Vermillion, ▶ Carmine Vermillion, ▶ Chinese Vermillion, ▶ English Vermillion, ▶ Patent Vermillion, ▶ Cinnabar, ▶ Cenobrium, ▶ Cinaper, ▶ Cinoper, ▶ Cynoper, ▶ Vermiculus, ▶ Zunsober and ▶ Red Cinnabar.

$$Hg^{++} \quad S^{--}$$

Mercury-Cadmium Lithopones *n* See ▶ Cadmium-Mercury Lithopones.

Merge *n* A group to which fiber production is assigned based on properties and dyeability. All fibers within a merge can be expected to behave uniformly, and for this reason, can be mixed or used interchangeably.

Meridional (mer-) Isomer *n* An isomer of an octahedral complex in which a plane contains three identical ligands and the central ion.

Merino \mə-ˈrē(ˌ)nō\ *n* [Spanish] (1810) (1) Wool from purebred Merino sheep. Merino wool usually has a mean fiber diameter of 24 microns or less. (2) A yarn of blended wool and cotton fibers.

Merinova *n* Casein fiber, manufactured by Snia Viscosa, Italy.

Mesh \ˈmesh\ *n* [ME, prob. fr. MD *maesche*; akin to OH Gr *masca* mesh, Lithuanian *mazgos* knot] (14c) (1) The square opening of a sieve. (2) The number of apertures per linear inch in a woven or electroformed metal screen or sieve, made especially for laboratory testing of high gravity dry powders or pigments for fineness and impurity content. Such screens are available to a mesh size of about 400.

Mesh Analysis See ▶ Sieve Analysis.

Mesh Fabrics *n* A broad term for fabric characterized by open spaces between the yarns. Mesh fabrics may be woven, knit, lace, net, crochet, etc.

Mesh Number *n* (1) The designation of size of an abrasive grain. Its name is derived from the openings per linear inch in the control sieving screen. Syn: ▶ Grit Number. (2) The deprecated, but still widely used (in the US) nomenclature for screen sizes, meaning the number of wires per inch of screen width. In standard square-mesh screens used in ▶ Sieve Analysis, the count and wire diameter are the same in both directions. thus, the widths of the standard-screen openings (inches) are in approximate inverse proportion to the mesh numbers, ≈ 0.6 (mesh number)$^{-1}$. Modern nomenclature, in accordance with SI, designates open-mesh screens by the minimum width of the openings in millimeters.

Mesitylene *n* $C_6H_3(CH_3)_3$. Powerful high-boiling hydrocarbon solvent. Bp, 165°C.

Mesityl Oxide *n* (4-methyl-3-pentene-2-one) $CH_3COCH=C(CH_3)_2$. An oily, colorless liquid used as a powerful solvent for cellulosic and vinyl resins, and as an intermediate in the production of plasticizers. Bp, 130°C; flp 25°C (78°F); vp, <10 mmHg/30°C.

Mesocolloid *n* Between hemicolloids and encolloids in size. Colloid particles limited to aggregates of from 100 to 1,000 molecules and from 25 to 250 nm long.

Mesomerism *n* Essentially synonymous with resonance. The term is particularly associated with the picture of pi electrons as less localized in an actual molecule than in a Lewis formula. The term is intended to imply that the correct representation of a structure is intermediate between two or more Lewis formulae.

Meson \ˈme-ˌzän, ˈmā-, ˈmē-, -ˌsän\ *n* [ISV *mes-* + ^{2}-*on*] (1939) Two types of particles of mass intermediate between that of the electron and proton have been discovered in cosmic radiation and in the laboratory. The one particle with mass about 215 m_e is called μ-meson, the other with about 280 m_e π-meson. Mesons of both positive and negative charge have been found and there is now reasonably good evidence for neutral mesons. Both types of mesons decay spontaneously. Some evidence exists for a meson of mass about 100 m_e.

Mesopic Vision *n* Vision at luminosities intermediate between luminosities required for completely photopic or completely scotopic vision; sometimes called twilight vision.

Meta- prefix [NL & ML, fr. L or Gk; L, fr. Gk, among, with, after, fr. *meta* among, with, after; akin to OE *mid*, *mith* with, OH Gr *mit*] (*m*-) A prefix used in naming aromatic organic compounds, ignored in alphabetization that designates the 3- and 5-positions relative to the substituted 1-position in a benzene ring. Compare *ORTHO-* and *PARA-* (3).

Metafiltration *n* Edge filtration through superimposed metallic strips with beveled edges, involving a change from coarse filtration (due to the strips) to fine filtration (due to the filter bed formed in their interstices).

Metal \ˈme-t°l\ *n* [ME, fr. OF, fr. L *metallum* mine, metal, fr. Gk. *metallon*] (14c) An element which has high electrical and thermal conductivities, a characteristic luster, and a low ionization energy, electron affinity, and electronegativity.

Metal Alloying *n* Combining two or more metals into an alloy, materials with different advantages exist. The number of available alloys increases factorially, each with its specific set of properties.

Metal Chelate Polymers *n* A polymer which contains metal atoms bonded to organic functional groups by coordinate bonds.

Metal Decorating *n* The process of lithographic printing on metal. The term most often applies to the can coating industries and includes the coating of can liners, outside basecoats and overprint varnishes as well as lithographic printing. The coatings are generally applied by direct roller coating or by spray onto tinplate or aluminum, on individual metal sheets, or by continuous coil coating.

Metallic Bond *n* Bonding present in metals. Metallic bonding can be described as the movement of valence electrons through-out the metal lattice.

Metallic Brown See Brown Iron Oxide Pigment.

Metallic Elements *n* Are distinguished from the non-metallic elements by their luster, malleability, conductivity and usual ability to form positive ions. Non-metallic elements are not malleable, have low conductivity and never form positive ions.

Metallic Fiber *n* A manufactured fiber composed of metal, plastic-coated metal, metal coated plastic, or a core completely covered by metal (FTC definition). They are available in "yarn" form as well as in staple form for spinning with other fibers. A core yarn with a metal surface is produced by twisting a strip of metal around yarn of natural or manufactured fibers. The most important characteristic of metallic fiber and the chief reason for its use in textiles is glitter. Metallic fibers are used as a decorative accent in fabrics for apparel, bedspreads, towels, draperies, and upholstery. A relatively new application for metallic fibers is in carpet pile, where they are being used in small percentages for control of static electricity.

Metallic Fillers *n* Powdered nickel, etc., to impart special properties, usually conductivity.

Metallic-Flake Pigment *n* Flat, thin particles of either aluminum, copper or copper alloy that reflect light specularly when incorporated into a plastic substance or coating vehicle with their reflecting surfaces approximately parallel. The aluminum pigments reflect very strongly throughout the visible spectrum, producing brilliant blue-white high-lights. The copper-based pigments, called gold bronzes but actually brasses, range from the characteristic red copper to progressively more yellow with rising zinc content.

Metallic Inks *n* Inks composed of aluminum or bronze powders in varnish to produce gold or silver color effects.

Metallic Paint *n* Paint which, on application, gives a film with a metallic appearance. This effect is normally produced by the incorporation of fine flakes of such metals as copper, bronze or aluminum. The aluminum used may be leafing or nonleafing, the former giving a far more brilliant metallic effect. These metals can be used in tinted or colored media to give polychromatic finishes. *Also known as Metallic Pigmented Paint.*

Metallic Pigment *n* Particles or flakes of nonoxidized metals or alloys used as pigments to modify the optical characteristics of a paint, to hide the substrate, modify the color or adjust other properties. The metals most commonly used as aluminum, bronze and zinc. See also ▶ Metallic Paint.

Metallic Pigmented Paint See ▶ Metallic Paint.

Metallic Pigments *n* A class of pigments consisting of thin opaque aluminum flakes or copper alloy flakes. Added to plastics, they produce unusual silvery and other metal-like attractive effects.

Metallic Soap *n* Salts derived from metals and organic acids, usually fatty acids. They include not only the sodium and potassium salts, which are popularly known as soaps, but compounds such as lead linoleate, calcium resinate, aluminum stearate, etc. See ▶ Soap, Metallic.

Metallic Solid *n* A solid in which positive ions are bonded together by delocalized electrons.

Metallized Glass *n* Glass spheres, flaks, or fibers that have been coated with silver or aluminum and, as fillers, provide increased electrical conductivity and light-reflecting pigmentation.

Metallizing \ˈme-tᵊl-ˌīz\ *vt* (1594) (1) Applying a thin coating of metal to a nonmetallic surface. May be done by chemical deposition or by exposing the surface to vaporized metal in a vacuum chamber. (2) See ▶ Flame Spray. (3) Adding metallic pigments, such as aluminum, to a paint. (4) A term covering all processes by which plastics (and some other base materials) are coated with metal. The most commonly used processes are described under *Electroless Plating, Silver-Spray Process, and Vacuum Metallizing*. Other methods include spraying with metallic pigments, chemical reduction, gas plating and vapor pyrolysis.

Metallizing Agents See ▶ Electroplating Chemicals.

Metalloid *n* An element which has properties which are intermediate between those of a typical metal and those of a typical nonmetal. *Also called a Semimetal.*

Metal Marking *n* In the coil coating industry, the degree or amount of a defect left on the surface of a film when the edge of a piece of metal is pulled across its surface. Pencil line marks on a painted surface caused by scratching with metal. *Also known as Coin Marking or Marring.*

Metal Oxide Catalysts *n* CrO_3, CrO_2Cl_2, etc.

Metal Primer *n* The first coat of paint on metal; a primer. One coat.

Metal Spraying *n* Application of a spray coat of metal (usually zinc or aluminum) onto a prepared surface (usually shot blasted mild steel). The metal to be sprayed is rendered molten by passing it, in wire or powder form, through a flame pistol which projects the semi-molten metal onto the surface by means of a jet of compressed air.

Metamer *n* (1) One of a pair of colors which matches the other color when viewed in a described way but which does not match under all viewing conditions. For example, if the light source, observer or angle of viewing is changed, the color will no longer match the other color of the pair. See ▶ Metameric Pair. (2) From the Greek *meta* (change, transposition, transfer) and *meros* (part or portion), the term metamer was formerly used in chemistry for a specific kind of isomer having to do with group-positional differences in molecules of the same composition and functionality. The term Isomer is now used in this limited sense (as well as in broader ones).

Metameric Color Match *n* A color match between two materials in which the colors are identical under some lighting conditions but not under others. Metameric color matches are common when different pigments or dyestuffs are used to color the two materials.

Metameric Match *n* A conditional identity of color exhibited by a pair of colors, each with different spectral distribution curves. *Also known as Conditional Match.* See ▶ Metamerism.

Metameric Pair *n* A pair of colors which match when viewed in a described way but which do not match if the viewing conditions are changed. Thus, a metameric pair of samples exhibit the same tristimulus values for a described set of viewing conditions (observer, light source, geometry of the illumination and viewing arrangement) but have different spectral distributions. Hence, they exhibit a match which is conditional only. See ▶ Conditional Match and ▶ Metamerism.

Metamerism *n* A term sometimes used in the color industry for the phenomenon exhibited by two surfaces that appear to be of the same color when viewed under one light source (e.g., sunlight), but that appear different when viewed under a different light source (e.g., incandescent lamp). The term *geometric metamerism* refers to a change in perceived color of a surface with a change in viewing angle. Metamerism should not be confused with "flair" or color constancy, which terms apply to the apparent color change exhibited by a single color when the spectral distribution of the light source is changed or when the angle of illumination or viewing is changed. See ▶ Metameric Pair.

Metamerism, Degree of *n* Metamerism exists in varying amounts, depending on the magnitude of the differences in the spectral distribution curves of the two colors. Thus, the degree of metamerism may be slight to moderate to severe and is generally described by one of a number of types of metamerism (or metameric) indices.

Metamerism Index See ▶ Index of Metamerism.

Metap Weave-Knit Process *n* A technique combining weaving and knitting in one operation with two independent yarn systems wound on warp beams. In the fabrics produced, woven strips are linked together with wales of stitches. Generally the fabrics have 75–85% woven and 25–15% knitted structure.

Metastable \-ˈstā-bəl\ *adj* [ISV] (1897) A temporary state of structure in a plastic, such as a crystalline plastic in which the final crystallinity is attained after passage of hours or days following molding. No physical or mechanical tests should be made while the test material is in a metastable condition (unless data regarding that condition are desired).

Meter \ˈmē-tər\ *n* [F *mètre*, fr. *Gk metron* measure] (1797) (1) (m, metre) The SI unit of length, one of the seven basic units of the system, defined as 1,680,763.73 wavelengths of the radiation in vacuum corresponding to the transition between the levels $2p_{10}$ and $5d_5$ of the krypton-86 atom (an orange spectral line). One foot equals (exactly) 0.3048 m. (2) Any device for measuring a physical or chemical quantity in which the measurement is indicated digitally, or analogically on a scale. In this sense, -meter is often used as a suffix, as in *thermometer*.

Metering Pump *n* A positive displacement device that pumps a measured amount of polymer solution to the spinnerets.

Metering Screw *n* An extruder screw whose final section, from four to ten flights, has a shallow channel of constant depth and lead. As its name suggests, the metering section of such a screw is intended to regulate the amount delivered per rotation of the screw. It also provides time for the equalization of melt temperature and helps to control the steadiness of the extrusion rate.

Metering Zone *n* (metering section) The final portion of a ▶ Metering Screw that builds pressure to force the melt through the screens and die. The metering section usually has a constant lead and a shallower channel than the preceding sections of the screw.

Methacrylate Ester \ˌme-ˈtha-krə-ˌlāt\ *n* [ISV] (1865) Any of the esters of methacrylic acid having the general formula $CH_2=C(CH_3)COOR$, wherein R is usually methyl, ethyl, isobutyl, or *n*-butyl to *n*-octyl. These esters are polymerizable to acrylic resins.

Methacrylate Plastic See ▶ Acrylic Resin.

Methacrylate Resins *n* A class of resins produced by the polymerization of methacrylate esters.

Methacrylic Acid \ˌme-thə-ˈkri-lik-\ *n* [ISV] (1865) (α-methacrylic acid, 2-methyl-2-propenoic acid) $CH_2=C(CH_3)COOH$. A colorless liquid prepared by the acid hydrolysis of acetone, from which are derived all of the methacrylate compounds. Most important of these are the esters, especially methyl methacrylate.

Methacrylonitrile *n* (MAN, α-methyl acrylonitrile) A vinyl monomer containing the nitrile group whose homopolymers are true thermoplastics with good mechanical strength and high resistance to solvents, acids, and alkalis. Modified properties can be obtained through blending, grafting, or copolymerization with other monomers such as styrene and methyl methacrylate. **MAN** is also used as a replacement for acrylonitrile in preparing nitrile elastomers.

γ-Methacryloxypropyltrimethoxy Silane *n* $CH_2=CHCOO(CH_2)_3SI(O-CH_3)_3$. A silane coupling agent used in reinforced polyesters, epoxies, and many thermoplastics to achieve improved adhesion between resin and glass fibers.

Methanol \ˌme-thə-ˌnól, -ˌnōl\ *n* [ISV] (1894) (carbinol, methyl alcohol, wood alcohol) CH_3OH. A colorless, toxic liquid usually obtained by synthesis from hydrogen and carbon monoxide. It is sometimes called *wood alcohol*, but the methanol obtained from the destructive distillation of wood also contains additional, contaminating compounds. Methanol is used as an intermediate in producing formaldehyde, phenolic, urea, melamine, and acetal resins, and as a solvent for cellulose nitrate, ethyl cellulose, polyvinyl acetate, and polyvinyl butyral. *Also known as Methyl Alcohol, Carbinol, Wood Alcohol, Colonial Spirits, and MeOH.* Syn: ▶ Formaldehyde.

Methenamine Pill Test See ▶ Flammability Tests.

Method of Least Squares See ▶ Regression Analysis.

Methoxybutyl Acetate *n* $CH_3OCH(CH_3)CH_2CH_2COOCH_3$. High-boiling solvent. Bp, 169°C; flp, 60°C (140°F).

Methoxyethylacetoxy Stearate *n* $C_{17}H_{34}(OCOCH_3)COOCH_2CH_2OCH_3$. A plasticizer for vinyl and cellulosic resins.

Methoxyethylacetyl Ricinoleate *n* A plasticizer for cellulosic and vinyl resins.

Methoxyethyl Stearate *n* (1,2-propylene glycol Monostearate) $C_{17}H_{35}-COOCH_2CH_2OCH_3$. A solvent and plasticizer for cellulosic plastics.

Methoxyl Group *n* The monovalent group, $-OCH_3$, characteristic of methyl alcohol and its esters or ethers.

Methyl Abietate *n* $C_{19}H_{29}COOCH_3$. A derivative of abietic acid (from rosin) used as a plasticizer for cellulosic, acrylic, and vinyl resins, polystyrene, and urea-formaldehyde resins.

***N*-Methyl Acetamide** *n* (NMA) $CH_3CONHCH_3$. A solvent useful in making aromatic-mer polymers, such as polyimides.

Methyl Acetate *n* (1885) CH_3COOCH_3. A colorless, volatile liquid with a fragrant odor, a solvent for acetyl cellulose and cellulose esters. A low-boiling ester solvent, exhibiting rapid evaporation; bp, 58°C, vp, 173 mmHg/20°C.

Methylacetyl Ricinoleate *n* $C_{17}H_{32}(OCOCH_3)COOCH_3$. A plasticizer for some vinyl resins and polystyrene.

Methyl Acrylate *n* $CH_2=CHCOOCH_3$. A colorless, volatile liquid, a monomer for acrylic resins used in the manufacture of synthetic resins. Sp gr, 0.953; bp, 80.5°C.

Methyl Alcohol *n* (ca. 1847) Syn: ▶ Methanol.

Methyl Amyl Carbinol See ▶ Heptanol-2.
Methyl Amyl Ketone *n* Boiling range, 147–153°C; flp, 106°F; vp, 4 mmHg/20°C. *Also known as MAK and 2-Heptanone.*

Methylated Spirit *n* A mixture of ethyl alcohol and a small amount of methyl alcohol; used industrially as a solvent for paints, lacquers, and varnishes.
Methyl Benzene *n* (methyl benzol) Syn: Toluene.

Methyl Butadiene *n* Syn: ▶ Isoprene.

Methyl Butyl Ketone *n* (MBK, propylacetone) $CH_3COC_4H_9$. A solvent for vinyl and many other resins, often used in conjunction with methyl ethyl ketone to control the drying rate of lacquers. A higher content of MBK slows the rate. Boiling range, 114–137°C; flp, 73°F; vp, 10 mmHg/20°C. *Also known as MBK and 2-Hexanone.*

Methyl Butynol *n* HC≡CCOH(CH$_3$)$_2$. A viscosity stabilizer and solvent for some nylons.

Methyl Butyrate *n* CH$_3$(CH$_2$)$_2$COOCH$_3$. A medium-boiling solvent for ethyl cellulose and cellulose nitrate. Bp, 102°C.

Methyl Cellosolve See ▶ Ethylene Glycol Monoethyl Ether.

Methyl Cellosolve®, Acetate Ether-ester solvent. Bp, 143°C; vp, 6 mmHg/30°C.

Methyl Cellosolve®, Methoxyethanol *n* CH^{3}OC$_2$H$_4$OH. Ether-alcohol solvent, Bp, 124°C.

Methyl Cellulose *n* A cellulose ether in which some of the cellulosic –OH groups have been replaced by –OCH$_3$. The degree of substitution determines properties and uses as thickeners and emulsifiers. Commercially, a granular, white, flakey material which acts as a water-soluble thickener and stabilizer; used in water-based paints.

Methyl Chavicol *n* Ether constituent found in some terpene solvents. Boiling range, 215–216°C.

Methyl Citrate (Tri) *n* Plasticizer with solvent properties, Bp, 176°C/16 mmHg.

Methyl-2-Cyanoacrylate *n* A fast setting adhesive used for bonding cellulosics, nylon, polyesters, acrylics, polystyrene, and polyurethanes to each other and to other materials such as woods, metals, and glass. Catalyzed by atmospheric moisture or lightly applied methanol, the adhesive polymerizes without loss of solvent. For best results, the surfaces to be bonded should mate closely.

Methyl Cyclohexane *n* CH$_3$C$_6$H$_{11}$. Hydrocarbon solvent. Bp, 101°C.

Methyl Cyclohexanol *n* CH$_3$C$_6$H$_{10}$OH. Alcoholic solvent. Because of its high boiling range, 160–180°C, it is used to improve flowing properties. Small quantities added to weaker solvents often exert a very pronounced effect on the viscosity of a given solution. Flp, 60°C (140°F).

Methyl Cyclohexanone *n* High-boiling solvent. Boiling range, 160–170°C; flp, 49°C (120°F); vp, 4 mmHg/30°C. It is often used in small amounts to improve flow or brushing properties.

Methyl Cyclohexyl Acetate n Boiling range, 175–190°C; flp, 68°C (155°F). *Known also as Methyl Hexaline Acetate and Sextate.*

Methyl Cyclohexyl Stearate n $CH_3(CH_2)_{16}COOC_6H_{10}CH_3$. Boiling range for the commercial product, 220–240°C/4 mmHg; flp, 170°C (338°F).

2,2′-Methylene-bis-(Cyclohexyl Isocyanate) n (H_{12}MDI) A diisocyanate used in making urethane elastomers and foams.

2,2′-Methylene-bis-(6-*tert*-Butyl-4-Ether Phenol) n An antioxidant for acrylonitrile-butadiene-styrene packaging, appliances, pipe, and automotive items.

2,2′-Methylene-bis-(6-*tert*-Butyl-4-Methyl Phenol) n A phenolic-type antioxidant for polyolefins and acrylonitrile-butadiene-styrene resins.

Methylene Blue n Blue 9 (52015). A thiazine dye used to make a type of lime blue.

Methylene Chloride n (dichloromethane, methylene dichloride) CH_2–Cl_2. A low-boiling chlorinated hydrocarbon which is a colorless, fairly dense, nonflammable liquid used as a solvent for cellulose triacetate and vinyl resins, a solvent in the polymerization of polycarbonate resins, and as a reactant for certain phenolic resins. It was widely used as a paint stripper and solvent for cured epoxy resins, but is less used now in the effort to keep chlorinated solvents out of the atmosphere. Bp, 40°C; vp. 230 mmHg/10°C; Sp gr. 1.34/4°C. *Also known as Dichloromethane.*

Methylene Group n The radical –CH_2– or =CH_2, existing only in combination.

Methyl Ethyl Ketone n (1876) (MEK, 2-butanone) $CH_3COC_2H_5$. A low-boiling colorless, flammable solvent, possessing all the properties of acetone without its extremely high volatility. One of the most widely used solvents for several thermoplastics including cellulosics, acrylics, polystyrene, and vinyl copolymers. Bp, 80°C; flp, 3°C (35°F); vp, 77 mmHg/20°C. *Also known as MEK and Butanone.*

Methyl Ethyl Ketone Peroxide n (MEKP, MED, peroxide) A complex peroxide mixture made by reacting hydrogen peroxide with MED, with the approximate formula $(CH_3COOC_2H_5)_3$. MEKP is an initiator for free-radical polymerization and a curing agent for polyester resins. In combination with an accelerator such as cobalt naphthenate, MEKP can bring about cure at room temperature. Because it is unstable, it is often handled in solution. MEKP should be kept only in small quantities and stored in a freezer when not in use.

Methylfluorosilicones n Silicone rubbers containing pendant fluorine and methyl groups. Have good chemical and heat resistance. Used in gasoline lines, gaskets, and seals. Also called FMQ.

Methyl Glucoside n $CH_2OHCH(CHOH)_3CHOOCH_3$. A plasticizer for alkyd, amino and phenolic resins. It is also used as a polyol for urethane-foam production.

Methyl Glycol Phthalate (Di) *n* Plasticizer. Boiling range, 210–260°C/20 mmHg; flp, 174°C (345°F).

Methyl Group *n* The radical –CH$_3$, existing only in combination.

Methyl Hexyl Ketone *n* (2-octanone) CH$_3$COC$_6$H$_{13}$. A colorless, high-boiling liquid with a pleasant odor, used as a solvent for epoxy coatings. Boiling range, 169–173°C; vp. 3 mmHg/30°C.

Methylimidazole *n* (EMI) An epoxy-resin curing agent with a heterocyclic structure. EMI is used with epoxies formed from epichlorohydrin and bisphenol A or –F, and for novolac epoxy resins. It provides ease of compounding, long pot life, low viscosity, and non-staining characteristics.

Methyl Isoamyl Ketone *n* (5-methyl-2-hexanone) CH$_3$COC$_2$H$_4$CH(CH$_3$)$_2$. A colorless liquid with a pleasant odor, used as a solvent for cellulose esters, acrylic resins, and certain vinyl polymers. It has a high solvent power and low evaporation rate, making it useful as a retarder solvent that promotes flow-out of coatings and reduces blushing.

Methyl Isobutyl Ketone *n* (MIBK, hexanone, 4-methyl-2-pentanone) (CH$_3$)CHCH$_2$COCH$_3$. A solvent with a moderate evaporation rate, used with cellulosic, vinyl, alkyd, acrylic, phenolic, and coumarone-indene resins, and polystyrene. Bp, 118°C; flp, 74°F; vp. 13 mmHg/20°C. *Known also as Hexone and MIBK.*

Methyl Isopropenyl Ketone *n* CH$_2$=C(CH$_3$)COCH$_3$. A flammable liquid used as a copolymerizable monomer.

Methyl Methacrylate *n* (1933) C$_5$H$_8$O$_2$. A volatile flammable liquid that polymerizes readily and is used as a monomer for resins.

Methylol Phenol \▎me-thəl-ōl ▎fē- ▎nól\ *n*. A phenol having one or more –CH$_2$OH groups in its ring, a first stage in the formation of phenolic resin by reaction of phenol with formaldehyde.

Methylol Urea *n* H$_2$NCONHCH$_2$OH. Colorless crystals derived from combination of urea with formaldehyde, the first stage in the production of urea-formaldehyde resins.

Methylolurea *n* H$_2$NCONHCH$_2$OH. First stage in the formation of urea formaldehyde resins. Reaction products of urea and formaldehyde. Monomethylol urea is obtained by reacting one molecule of formaldehyde and one molecule of urea.

Methylpentene Resin Syn: ▶ Poly(4-Methylpentene-1).

Methylphenylsilicones *n* Silicone rubbers containing pendant phenyl and methyl groups. Have good resistance to heat, oxidation, and radiation, and compatibility with plastics.

1-Methyl-2-Pyrrolidinone (NMP) *n* CH$_3$$\overline{\text{NCH}_2\text{CH}_2\text{C}}$ = O. A solvent with a low order of inhalation toxicity, good thermal and chemical stability, and a high flash point. It is capable of dissolving resistant resins such as polyamide-imides, epoxies, urethanes, nylon, and PVC. It is a solvent of choice for spinning PVC fibers from solution. Previously known an *N*-methyl-2-pyrrolidone.

Methyl Ricinoleate *n* CH$_3$(CH$_2$)$_5$CH(OH)CH$_2$CH= CH(CH$_2$)$_7$COOCH$_3$. A plasticizer for cellulosic resins, polyvinyl acetate, and polystyrene with the composition of methyl ester of ricinoleic acid. Bp. 245°C/10 mmHg; Sp gr. 0.9236.

Methyl Rubber *n* Poly(2,3-dimethylbutadiene). Manufactured by Bayer, Germany.

Methylsilicone *n* Silicone rubbers containing pendant methyl groups. Have good heat and oxidation resistance. Used in electrical insulation and coatings. Aslo called MQ.

α-Methylstyrene *n* C$_6$H$_5$C(CH$_3$)=CH$_2$. A colorless liquid, easily polymerizable by heat or with catalysts, and typically copolymerized with methyl methacrylate or styrene.

Methyl Tartrate (Di) *n* A plasticizer with a bp of 280°C.

Methylvinylfluorosilicone *n* Silicone rubbers containing pendant vinyl, methyl, and fluorine groups. Can be additionally crosslinked via vinyl groups. Have good resistance to petroleum products at elevated temperatures.

Methylvinylsilicone *n* Silicone rubbers containing pendant methyl and vinyl groups. Can be additionally crosslinked via vinyl groups. Vulcanized to high degrees of crosslinking. Used in sealants, adhesives, coatings, cables, gaskets, tubing, and electrical tape.

Methyl Violet *n* A class of strong purple dyes and pigments.

Metier \ˈme-ˌtyā, me-ˈ\ *n* [F, fr. (assumed) VL *misterium*, alter. of L *ministerium* work, ministry] (1792) A spinning machine for producing manufactured fibers. The bank of cells or compartments and associated equipment used in the dry spinning of fibers, such as cellulose acetate and cellulose triacetate.

Metier Twist *n* The amount of twist present in yarn wound at the métier.

Metre \ˈmē-tər\ *chiefly British variant of meter* The basic unit of length in the International System of Units (SI), equal to 39.37 in. SI spelling of METER.

Metrication \ˌme-tri-ˈkā-shən\ *n* (1965) Any act tending to increase the use of the International System of Units (SI).

Metric Count *n* The number of kilometers per kilogram of yarn.

Metrology *n* A study of measurement: the scientific study of units of measurement

Mev Abbreviation for Million Electron Volts, a measure of kinetic energy for subatomic particles. 1 Mev = 1.60219×10^{-13} J.

Meyer Bar *n* See ▶ Wire-Wound Rod and ▶ Equalizer Rod.

Mezzotints \ˈmet-sō-ˌtint, ˈmed-zō-\ *n* [mod. of It *mezzatinta*, fr. *mezzo* (feminine of *mezzo*) + *tinta* tint] (1738) Prints made from ground (e.g., crossed in several directions with knife-edge cuts) copper plates, the burred surface of which is scraped away to create the design.

MF See ▶ Melamine-Formaldehyde Resin.

MFC *n* Abbreviation for ▶ Multifunctional Concentrate. See ▶ Color Concentrate.

MFI *n* Abbreviation for Melt-Flow Index.

Mg *n* Chemical symbol for the element magnesium.

M-Glass *n* A high-modulus glass whose fibers are sometimes used for reinforcing plastics when high modulus at moderate cost is desired. Major constituents are SiO_2 54%, CaO 13%, MgO 9%, BeO and TiO_2 8% each, Li_2O and CeO_2 3% each, and ZrO_2 2%. Fiber density is 2.89 g/cm^3, modulus (E) is 110 GPa, and tensile strength is 3.5 GPa.

MHZ *n* Abbreviation for ▶ Megahertz.

MI *n* Abbreviation for ▶ Melt Index, a term replaced by Melt-Flow Index.

MIBK *n* Abbreviation for ▶ Methyl Isobutyl Ketone.

Mica \ˈmī-kə\ *n* [NL, fr. L, grain, crumb; perhaps akin to Gk. *mikros* small] (1777) Any of a family of crystalline silicate minerals characterized physically by a perfect basal cleavage, consisting essentially of orthosilicates of aluminum and potassium. They occur naturally, mainly as the minerals *muscovite* (white mica), *phlogopite* (amber mica), and *biotite*; and are also synthesized from potassium fluorilicate and alumina. Micas are used as fillers in thermosetting resins, imparting good electrical properties and heat resistance. A grade having high aspect ratios (HAR) with flakes 3–5 μm thick and aspect ratios as high as 200 can be processed, although the optimum aspect ratio appears to be about 70. The larger flakes increase flexural modulus and strength, have lower moisture content, and raise the deflection temperature of compounds containing them. See ▶ Aluminum Potassium Silicate.

Micaceous Iron Oxide *n* Naturally occurring iron ore which resembles mica only in appearance and not chemically. It is converted to a grayish pigment which has value as a constituent of anti-corrosive paints and protective coatings. Density, 4.90 g/cm^3 (40.8 lb/gal); O.A., 11 lb/200 lb; hardness (mohs), 6.0; ph, 7.7.

Mica Fillers *n* High-aspect-ratio minerals used traditionally in thermosets as phenolics and epoxies for mechanical improvements, as well as for electrical and heat insulation benefits.

Micelle \mī-ˈsel\ *n* [NL *micella*, fr. L *mica*] (1881) A colloidal particle formed by the reversible aggregation of dissolved molecules. Micelles may be in the shape of spheres, cylinders, or platelets. Soaps, detergents, and other emulsifying agents used in emulsion polymerization contain micelles generally composed of from 50 to 100 molecules of emulsifier, within which the polymerization reaction may be initiated.

Michel-Levy Chart *n* A chart relating thickness, birefringence, and retardation so that any one of these variables can be determined when the other two are known.

Micro- *adj* (μ) The SI prefix meaning × 10^{-6}.

Microballoons *n* (1) Tiny, hollow plastic spheres used to reduce evaporation of liquids such as oils by floating a layer of spheres on the surfaces of stored liquids. (2) Syn: ▶ Microspheres.

Microbial Degradation *n* See ▶ Biodegradation and ▶ Pink Staining.

Microcrystalline \ˈmī-krō-ˌkris-tᵊl\ *n* (1886) Pertaining to crystallinity that is visible only under a microscope, sometimes taken to mean that the crystals referred to are no larger than 1 μm.

Microcrystalline Silica See ▶ Silica, Microcrystalline.

Microcrystalline Silicate *n* A derivative of chrysotile asbestos, consisting of tiny rod-shaped particles of hydrated magnesium silicate. The particles have hydroxyl groups on their surfaces that bond with hydrogen-bonding sites on the molecules of a fluid in which they are incorporated. The material has also been used as a viscosity-building agent in unsaturated polyester and other resins.

Microcrystalline Wax *n* Any of a group of petroleum-derived waxes that differ from paraffin waxes in having finer crystal structure, higher melting points – between 60°C and 93°C, higher liquid viscosities, and greater ductility. They are used in fiberboard coatings, paper-container linings, and polishes.

Microdenier Refers to fibers having less than 1 denier per filament or 0.1 tex per filament.

Microemulsions *n* Transparent solutions of water and oil, which are thermodynamically stable and which spontaneously form when the components are brought in contact.

Microencapsulation \-in-ˌkap-sə-ˌlāt\ *vt* (1963) The process of encasing a small solid particle or a discrete amount of liquid or gas in a capsule. The term applies to capsules ranging in diameter from a few micrometers to about 500 μm. The capsule is usually made of a synthetic plastic, although waxes, glass, and metals are also used. Methods used for forming polymeric microcapsules fall into three broad classes: phase separation, interfacial reaction, and physical methods. Phase separation methods include ▶ Coacervation, applying meltable dispersions, and spray-drying of a suspension of the material in a vaporizable solvent, in-situ polymerization, and chemical-vapor deposition. The physical methods include fluidized-bed coating processes, spray coating, electrostatic coating methods, and extrusion. Typical examples of microencapsulation are "carbonless" carbon paper, timed-release drugs and fertilizers, and battery separators.

Microgel *n* A small particle of cross-linked polymer of very high molecular weight and containing closed loops. Microgels may be present in trace amounts due to impurities in monomers, and can influence polymer properties and molecular-weight studies.

Micrometer (μm) \mī-▮krä-mə-tər\ *n* [F *micromètre*, fr. *micr-* + *-mètre* –meter] (1670) (1) Instrument for measuring small lengths under the microscope. (2) Micrometer caliper; instrument used in measuring the dry film thickness of a coating or the thickness of metal or other materials. (3) The usual unit of length for light microscopical measurements (1 μm = 10^{-3} mm): it is still often referred to by its former name, "micron."

Micrometer *n* A unit of length measure used to describe the wavelength of radiant energy equal; to one thousandth part of a millimeter or one millionth of a meter, 1,000 nm. This term is replacing the term micron. Micrometer is designated as μm.

Micron \▮mī-▮krän\ *n* [NL, fr. Gk *mikron*, neuter of *mikros* small] (1885) This long deprecated but still used length unit and its abbreviation, the Greek letter m, were dropped by action of the General Conference on Weights and Measures on October 13, 1967. The symbol "μ" is to be used solely as the abbreviation for the prefix Micro-. The old micron should now be spoken as micrometer (μm) or 10^{-6} m.

Micronaire Method *n* A means of measuring fiber fineness by determining the resistance of a sample to a flow of air forced through it.

Micronized Pigment Pigment with a narrow, particle-size distribution.

Micronizer Mill Used for the dry grinding of pigments. Precrushed crudes or solids are fed into a relatively flat, cylindrical grinding chamber, the opening being in the bottom or top plate. The fluid medium, superheated compressed air or steam, enters the grinding chamber through nozzles located in the peripheral wall. The grinding action is done by the pigment particles hitting each other. The fines are carried out through an opening in the center of the mill, the coarser, larger particles return for further grinding.

Microorganisms \-▮ór-gə-▮ni-zəm\ *n* [ISV] (1880) In paint technology, bacteria and fungi which are harmful to liquid paint and dry paint films. Bactericides and fungicides are added to paints to inhibit the growth of these organisms.

Microphotograph \-▮fō-tə-▮graf\ *n* [ISV] (1858) A small, microscopic photograph, in which the image is minified; it requires enlarging or the use of a lens system in order to view it (cf., photomicrograph).

Microporous \▮mī-krə-▮pōr, -▮pór\ *n* [ISV] (1884) Having pores of microscopic dimensions. Some plastic films and fabric coatings are rendered microporous in order to permit the passage of water vapor ("breathing") while preventing the penetration of raindrops.

Microscopy \mī-▮kräs-kə-pē\ *n* (ca. 1665) The application of any tool or technique helpful in characterizing microscopic objects.

Microspheres \▮mī-krə-▮sfirs\ *n* (1894) Tiny, hollow spheres of glass or plastic used as fillers to impart low density to plastics, such plastics being known as *syntactic foams*. Plastics used to make microspheres include phenolic, epoxy and a copolymer of vinylidene chloride and acrylonitrile. The last contains a heat-activated blowing agent that expands the spheres either before their incorporation into a matrix polymer or afterward. The copolymer spheres impart better mechanical properties to the matrix than do the glass or epoxy microspheres. See also ▶ Glass Spheres.

Microstructure \▮mī-krō-▮strək-chər\ *n* [ISV] (1885) The detailed structure of plastics as seen through light and electron microscopes, approximately the magnification range of 100–100,000×, including such features as crystalline form, spherulites, voids, distribution of filler and pigment particles, discontinuous-phase particles in blends, and, in reinforced plastics, configuration, length distribution, and cross-section distribution of yarns and, within the yarns, the filament ends, etc.

Microtensile Specimen *n* A small specimen as specified in ASTM D1708 for determining tensile properties of plastics. It has maximum thickness 3.2 mm and minimum length 38.1 mm. Tensile properties determined with this specimen include yield strength, tensile strength, tensile strength at break and elongation at break.

Microvoids (1) A region in a polymer of lower electron density than its surroundings, of about 100 angstroms in size and amounting to about 1% of the volume. (2) Small voids or holes in a paint medium of such size that when filled with air (or some other material of different refractive index) they scatter light much as a pigment does because of the difference in refractive index between the material in the void and the vehicle. Some air-filled microvoids account for the white color of snow, for example. See ▶ Bubble Coating.

Microwaveable *vt* (1973) Said of plastics for kitchen use, and of the utensils made from them, that are heated little or not at all by the direct action of the high-frequency waves generated by microwave ovens, and that withstand many repeated heating by the foods contained in them without warping, shrinking, or staining.

Microwave Drying *n* Drying of printing inks by the use of microwave radiation. The presence of polar materials is a prerequisite.

Microwave Heating *n* A heating process similar to dielectric heating, but using frequencies in the 10^9–10^{10} Hz (radar) range. The Federal Communications commission has allocated the specific frequencies 915, 2450, and 5850 MHz for industrial use. Microwave ovens similar to those used in restaurants and households for rapidly cooking foods have been used experimentally for preheating molding powders, vacuum-bag curing, autoclave molding, and curing of nylon overwraps. Plastic films coated with water-containing materials such as polyvinylidene chloride can be dried rapidly and economically by microwave energy. Line speeds about 5 m/s have been attained with polyethylene film, by means of a microwave cabinet only 2.4 m long.

Middle Chrome See ▶ Chrome Yellow, Light and Primrose.

Mie Theory *n* Theory which relates the scattering of a single spherical particle in a medium to the diameter of the particle, the difference in refractive index between the particle and the medium, and the wavelength of radiant energy in the medium which is incident on the particle. This theory relates to the direct observation of the scattering of a single particle as compared to the Kubelka-Munk Theory, which relates to multiple scattering between particles. Mie Theory also takes into account the absorption which the particle may also exhibit.

Migration *n* (1) Movement of certain materials in a coating or plastic (e.g., plasticizers in vinyl) to the surface or into an adjacent material. (2) Movement of dye from one area of dyed fabric to another. Includes movement of color from the dyed area to the undyed area of cloth. (3) Movement of fibers which go from the center to the outside surface of yarn and back again periodically.

Migration of Plasticizer *n* In plasticized thermoplastics or elastomers, the movement of molecules of plasticizer from their interior locations when the article was originally formed to the surface layer of the article, where the plasticizer appears as a greasy or oily layer and may be rubbed off or dissolved away. The phenomenon occurs most often in vinyl compounds containing incompatible plasticizers.

Mil \❙mil\ *n* [L *mille* thousand] (1721) A unit of thickness equal to 0.001 inch, often used for specifying diameters of wires and glass fibers, and thicknesses of films. It is gradually being replaced by the SI units, the millimeter and micrometer. 1 mil = 0.0254 mm = 25.4 μm.

Milanese Knitting See ▶ Knitting.

Mildew \❙mil-❙dü, -❙dyü\ *n* [ME, fr. OE *meladēaw* honeydew; akin to OH Gr *militou* honeydew] (14c) Superficial growth produced by fungi on various surfaces, forms of organic matter and on living plants which are exposed to moisture: it results in discoloration and decomposition of the surface. Two types are common: (1) Spore type, which resembles caviar in appearance and (2) Mycelium or filament type.

Mildewcide *n* Chemical agent which destroys retards or prevents the growth of mildew.

Mildew (Fungus) Resistance The ability of a coating to resist fungus growth that can cause discoloration and ultimate decomposition of a coating's binding medium. See ▶ Mildew.

Mileage *n* The surface area covered by a given quantity of ink or coating material.

Milk Acid see ▶ Lactic Acid.

Milkiness *n* Whitish or translucent appearance in an unpigmented liquid coating or film which could normally be transparent.

Mill [ME *mille*, fr. OE *mylen*, fr. LL *molina*, molinum, fr. feminine and neuter of *molinus* of a mill, of a millstone, fr. L *mola* mill, millstone; akin to L *molere* to grind] (before 12c) (1, *n*) In the plastics industry, the term mill is generally taken to refer to a roll mill such as a two-roll mill used in compounding. More broadly, it includes all mechanical devices for converting raw materials into a conduction ready for use, as well as machine tools that cut materials with rotating bits and many types of size-reduction machines. (1552) (2, *v*) To process components of a plastic mixture in a two-roll mill.

Mill Base *n* The portion of the coating formulation which is charged in the dispersion mill.

Milled Fibers Small lengths of glass filaments produced by hammer-milling continuous glass strands. They are useful as anticrazing and reinforcing fillers for adhesives.

Mill End *n* A remnant or short length of finished fabric.

Miller Indices The notations usually used for naming crystal faces; they have the form hkl. These notations are based on the assignment of crystallographic axes and on an expression of the intercepts of the face on the three axes (hexagonal has four.)

Milli- *combining form* [F, fr. Latin *milli-* thousand, fr. *mille*] (m) The SI prefix meaning $\times\ 10^{-3}$.

Millimeter of Mercury (mmHg) *n* A unit of pressure: 1 mm Hg $= \frac{1}{760}$ atm. *Also known as Torr.*

Millimicron \mi-lə-ˈmī-ˌkrän\ *n* [ISV] (1904) (Deprecated) A unit of length used to describe the wavelength of electromagnetic radiation, particularly that in the visible region, equal to 10^{-9} meter. The use of this term is largely being replaced by the term nanometer.

Milling *n* (1) The process of treating fabric in a fulling mill, i.e., fulling. (2) In silk manufacturing the twisting of the filaments into yarn. (3) A grinding process, i.e., ball-milling of dyes and pigments.

Millipoise *n* One thousandth part of a poise (measure of viscosity) or 1/10 of a centipoise.

Milliliter *n* One thousandth of a liter.

Mill Run *n* A yarn, fabric, or other textile product that has not been inspected or that does not come up to the standard quality.

Mills, Ball & Pebble *n* Roll mills, such as two-roll mills used in compounding.

Mill Scale *n* The heavy oxide layer formed during hot fabrication or heat treatment of metals.

Milori Blue A green shade iron blue pigment. A pigment of the ferriferrocyanide family. *Also called Chinese Blue.* see ▶ Iron Blue.

Milori Green see ▶ Chrome Green.

Mineral \ˈmin-rəl, ˈmi-nə-\ *n* [ME, fr. ML *minerale*, fr. neuter of *mineralis*] (15c) Any naturally occurring, homogeneous inorganic substance having a definite chemical composition and characteristic crystalline structure, color and hardness.

Mineral Acids *n* Strong inorganic acids, e.g., nitric, sulfuric, hydrochloric, etc.

Mineral Black *n* Black pigment made by grinding and/or heating black slate, shale, or salty coal coke, or coal. It usually contains a high percentage of carbon, mixed with substantial amounts of mineral matter, which may include aluminum silicate, and oxides of silicon, iron, calcium, and magnesium. By comparison with true carbon blacks, its staining strength is inferior. See also ▶ Biddiblack.

Mineral Brown See Brown Iron Oxide Pigment.

Mineral Fibers A generic term for all non-metallic, inorganic fibers, which may be natural, such as asbestos, or manufactured from such sources as rock, ore, alloys, slag, or glass.

Mineral Fillers *n* Minerals compounded into plastics that provide one or more mechanical or thermal property improvements to the polymer matrix. They can be either extenders or reinforcing fillers.

Mineral Oil *n* (1805) Any liquid product of petroleum within the viscosity range of products commonly called oils and consisting of high molecular weight hydrocarbons.

Mineral Orange see ▶ Orange Mineral.

Mineral Pigments See ▶ Earth Pigment.

Mineral Spirits *n plural but singular or plural in construction* (1927) (naphtha) An aliphatic-hydrocarbon fraction of petroleum evolved in the distillation range of about 150–200°C. An example is "VM&P naphtha", used as a diluent in organosols. It is a petroleum fraction with boiling range between 300°F and 400°F. Due to having a low aromatic hydrocarbon content, with volatility, flash point, and other properties this makes it suitable as a thinner and solvent in paints, varnishes, and similar products. Syn: is ▶ Petroleum Spirits. See also ▶ Odorless Solvent.

Mineral Surfaced Roofing *n* Felt or fabric saturated with bitumen, coated on one or both sides with a bituminous coating and surfaced on its weather side with mineral granules.

Mineral Violet *n* (77742). A complex manganese ammonium phosphate. Its very low tinting strength, low color intensity and low hiding power largely restrict its use to toning white finishes, in which service it offers low cost and excellent bake and bleed resistance; however, tints fade severely in steam sterilizers. Because of its poor alkali resistance, it is not suitable for some household appliance finishes; in such cases, quinacridone violet can be used to advantage. Syn: ▶ Manganese Violet.

Mineral White See ▶ Gypsum.

Miniemulsion *n* A form of emulsion in which an organic liquid is dispersed into a continuous aqueous phase. Unlike a conventional emulsion where the droplets may be of the order of 10 microns or greater, the droplets sizes are submicron. This is accomplished through a combination of high shear, and surfactant/costabilizer combination. The surfactant prevents emulsion degradation via calescence (as in a conventional emulsion), while the costabilizer prevents diffusional degradation caused by Ostwald ripening.

Minimized Spangle *n* Galvanized sheet obtained by treating the regular galvanized sheet during the solidification of the zinc to restrict the normal spangle formation. This product usually has a dull appearance not characterized by a high degree of uniformity, and dissimilarity from coil to coil is not unusual. This minimizes the crystalline pattern from photographing through the applied coating resulting in a smoother appearing finish.

Minimum Care *n* A term describing home laundering methods. Minimum care fabrics, garments, and household textile articles can be washed satisfactorily by normal home laundering methods and can be used or worn after light ironing. Light ironing denotes ironing without starching or dampening and with a relatively small expenditure of physical effort.

Minimum Detectable Amount n (MDA) In chemical analysis, the least amount of a substance being sought that balances two risks, Type I, the risk of falsely finding the substance to be present when in fact it is not, and Type II, the risk of not detecting that least amount. Typically, the two risks are made equal and, if both are 5%, the MDA is very nearly four times the standard deviation of the method. Lowering the risk increases MDA.

Minimum Deviation n The deviation or change of direction of light passing through a prism in a minimum when the angle of incidence is equal to the angle of emergence. If D is the angle of minimum deviation and A the angle of the prism, the index of refraction of the prism for the wavelength used is

$$n = \frac{\sin\frac{1}{2}(A+D)}{\sin\frac{1}{2}A}$$

Minimum Perceptible Difference n The minimum color difference which can be observed between two colors. The magnitude is dependent on the conditions of viewing, i.e., size, illumination level and character, surrounding area, adaptation, etc.

Minium n Naturally occurring and synthetically prepared red lead oxide; used as a pigment.

Minute Value n Voltage with a unit thickness of insulator (e.g., varnish) will withstand for 1 min without breakdown.

Mipolam n Poly(vinyl chloride). Manufactured by Dynamit Nobel, Germany.

Mirbane Oil n Nitrobenzene. Liquid with an almond-like smell, used as a deodorant, chiefly in polishes. See ▶ Nitrobenzene and ▶ Oil of Mirbane.

Mired n A unit used to measure the reciprocal of color temperature, equal to the reciprocal of a color temperature of 10 K. Derived from micro-reciprocal-degree.

Mirlon n Polyamide, manufactured by Viscose-Suisse, Switzerland.

Miscibility \mi-sə-bəl\ *adj* [ML *miscibilis*, fr. L *miscēre* to mix] (1570) (solubility) The greatest percentage of one liquid or polymer that forms a true, homogeneous solution, i.e., a single phase, in another liquid of polymer. Few binary polymer systems are miscible over the entire range of composition, but many have limited miscibility at either end of the range. Miscibility usually increases with rising temperature. See also ▶ Compatibility.

Misclip See ▶ Scalloped Selvage.

Mispick n A weaving defect in which a pick is improperly interlaced, resulting in a break in the weave pattern. Mispicks can result from starting the loom on the incorrect pick after a pick-out.

Misses See ▶ Holidays.

Missing End See ▶ End Out.

Miss-Stitch n A knitting construction formed when the needle holds the old loop and does not receive new yarn. It connects two loops of the same course that are not in adjacent wales. *Also known as Float-Stitch*.

Mist Coat n (1) Very thin sprayed coat. (2) A thin coat of volatile thinners, with or without a small amount of lacquer which is sprayed over a dry lacquer film to improve the smoothness and luster.

Misting See ▶ Flying.

Mitt See ▶ Painter's Mitt.

Mittler's Green See ▶ Hydrated Chromium Oxide.

Mixed Aniline Point n Minimum equilibrium solution temperature of a mixture of two volumes of aniline, one volume of sample, and one volume of normal heptane of specified purity. Refer to ASTM D 1012.

Mixed End or Filling n Warp or filling yarn differing from that normally used in the fabric, e.g., yarn with the incorrect twist or number of plies, yarn of the wrong color, or yarn from the wrong lot.

Mixer \mik-sər\ n (ca. 1611) Any of a wide variety of devices used to intermingle two or more materials to some defined state of uniformity. Some equipment intended mainly to provide size reduction may also accomplish mixing. Types used in the plastics industry are:

Ball Mill	Internal Mixer
Banbury Mixer	Kneader
Centrifugal Impact Mixer	Mill
Change-Can Mixer	Propeller Mixer
Colloid mill	Ribbon Blender
Conical Dry-Blender	Rod Mill
Disk-and-Cone Agitator	Roll mill
Drum Tumbler	Sand Mill
High-Intensity Mixer	Static Mixer
Homogenizer	Tumbling Agitator
Intensive Mixer	Vibratory Mill

Mixing, Entropy, Free Energy n The change in free energy ΔG_{mix}, for mixing two substances at certain temperature, T, is given by the Gibbs function:

$$\Delta G_{mix} = \Delta H_{mix} - T\Delta S_{mix}$$

Because the solution process is an endothermic process for most solutions, the heat of mixing, ΔH_{mix}, is a positive quantity. The entropy of mixing, ΔS_{mix}, is normally positive because of the more random nature of solutions compared with that of the unmixed components. A necessary, although not sufficient, condition for a polymer to dissolve is that $\Delta G_{mix} \leq 0$, and if $\Delta G_{mix} \approx 0$ then mixing is marginal, and if $\Delta G_{mix} \geq 0$ then components will not mix.

Mixing Screw n Any extruder screw that incorporates some modification (from standard designs) intended to improve mixing, mainly ▶ Distributive Mixing but sometimes improving dispersion, too. One simple method is to insert one or more rings of closely spaced pegs arranged circumferentially in the screw channel and having nearly the same height as the flight. The pegs divide and re-divide the melt streaming in a complex but regular path down the channel, accomplishing a kind of braiding of substreams. See also ▶ Dulmadge Mixing Section, and ▶ Cavity-Transfer Mixer.

Mixing Varnish n General term for a variety of varnishes, or paint vehicles, that are used by mixing with: (1) aluminum pigment. See ▶ Aluminum Mixing Varnish. (2) Pigments paste in oil or colors in oil. (3) Other vehicle types, such as gloss oil (if the mixing varnish is long in oil), or a spar varnish (if the mixing varnish is short in oil), to produce a desired oil length for the complete vehicle. (4) The mill base discharged from pigment-dispersing equipment, as contrasted to the "grinding" varnish for vehicle in which the pigment has been dispersed. (5) Paint, to alter its properties.

Mixing White n A white ink, either transparent or opaque, used in making tints.

Mixture \ miks-chər\ n [ME, fr. MF, fr. OF *misture*, fr. L *mixtura*, fr. *mixtus*] (15c) A combination of two or more different substances intermingled with varying percentage composition (unlike a true solution), in which each component retains its chemical identity.

M.K.S. System n A system of units derived from the meter, kilogram, and second. Now superceded for scientific purposes by the SI units which are based on the m.k.s. system.

MMA n Abbreviation for ▶ Methyl Methacrylate.

M$_n$ n (1) Abbreviation for ▶ Number-Average Molecular Weight. (2) Chemical symbol for the element manganese.

Mo n Chemical symbol for the element molybdenum.

Mobility n The property of a material which allows it to flow when a shearing force larger than the yield value has been applied. The coefficient of mobility is the rate of shear induced by a shearing force per square cm of one dyne in excess of the yield value. Mobility pertains to plastic materials and is the analogue of fluidity. It is calculated from the slope of the straight-line portion of the flow curve. The coefficient of mobility is the reciprocal of the coefficient of plastic viscosity.

Mobilometer n Rheological instrument for measuring the consistency of paints and similar products. The "mobility" is determined by the time in seconds required for a loaded perforated disk to pass through a specified depth of the sample contained in the cylinder. See ▶ Gardner Mobilometer.

MOCA® n DuPont's trade name for methylene-bis-*o*-chloroaniline, much used until about 1980 as a curing agent for urethane rubbers and epoxy resins, prior to its being declared to be a carcinogen by OSHA.

Mock Dyeing n A heat stabilization process for yarns. The yarns are wound onto packages and subjected to package dyeing conditions (water, pressure, temperature) but without dye an chemicals in the bath.

Mock Leno n A combination of weaves having interlacings that tend to form the warp ends into groups (with empty spaces intervening) in the cloth, thereby giving an imitation of the open structure that is characteristic of leno fabrics. Mock leno fabrics are used for summer shirts, dresses, and other apparel, and as a shading medium in Jacquard designs.

MOD n Modacrylic fibers (EEC abbreviation).

Modacrylic Fibers \ mä-də- kri-lik-\ n [*modified acrylic*] (1960) A manufactured fiber in which the fiber-forming substance is any long-chain synthetic polymer composed of less than 85% but at least 35% by weight of acrylonitrile units (Federal Trade Commission). CHARACTERISTICS: Although modacrylics are similar to acrylics in properties and application, certain important differences exist. Modacrylics have superior resistance to chemicals and combustion, but they are more heat sensitive (lower safe ironing temperature) and have a higher specific gravity (less cover). END USES: The principal applications of modacrylic fibers are in pile fabrics, flame-retardant garments, draperies, and carpets.

Modacrylics n Generic name for a fiber containing between 35% and 85% acrylonitrile repeating units in the polymer chain (poly(acrylonitrile)), excluding rubbers, manufactured by Viscose-Suisse, Switzerland.

Modal Generic name for fibers from regenerated cellulose of modified structure.

Moderator \ˈmä-də-ˌrā-tər\ *n* (ca. 1560) A material used for slowing down neutrons in an atomic pile or reactor. Usually graphite or "heavy water" (deuterium oxide).

Modes of Appearance *n* Various manners in which colors can be perceived, depending on spatial distributions and temporal variations of the light causing the sensation. Five modes are generally recognized: surface mode, volume mode, film (or aperture) mode, illumination mode and illuminant mode. The first two of these, surface mode and volume mode, together make up the object mode. Color perceived as falling on an object, thus filling space around the object, describes the illumination mode. The illuminant mode is the appearance of the glow from a light source. The film or aperture mode describes color perceived in space, where no object is discernible.

Modified Phenolic Resins *n* Resins in which the basic phenolaldehyde product has been modified by the introduction of rosin or other natural resin. The products are often esterified.

Modified Polyphenylene Ether *n* Thermoplastic polyphenylene ether alloys with impact polystyrene. Have good impact strength, resistance to heat and fire, but poor resistance to solvents. Processed by injection and structural foam molding and extrusion. Used in auto parts, appliances, and telecommunication devices. Also called MPE, MPO, modified polyphenylene oxide.

Modified Polyphenylene Oxide See ▶ Modified Polyphenylene Ether.

Modified Resin *n* Any synthetic resin into which has been incorporated a natural resin, an elastomer, or an oil that alters the processing characteristics or physical properties of the material.

Modifier \ˈmä-də-ˌfī(-ə)r\ *n* (1583) Any chemically inert ingredient added to an adhesive formulation that changes its properties. See also ▶ Additive, ▶ Filler, ▶ Extender and ▶ Plasticizer.

Modulus \ˈmä-jə-ləs\ *n pl* –li [NL, fr. L, small measure] (1753) (1) A modulus is a measure of a mechanical property of a material, most frequently a stiffness property. See ▶ Flexural Modulus, ▶ Shear Modulus, ▶ Modulus of Elasticity and ▶ Modulus of Resilience. (2) The absolute value of a complex number or quantity, equal to the square root of the sum of the squares of the "real" and "imaginary" parts. (3) Modulus at 300% *n*: The tensile stress required to elongate a specimen to three times its original length (200% elongation) divided by 2. Although other elongations are used, 300% is the one most often employed for rubbers and flexible plastics.

Modulus in Compression See Compressive Modulus.

Modulus in Flexure See ▶ Flexural Modulus.

Modulus in Shear See ▶ Shear Modulus.

Modulus of Elasticity *n* (1) (elastic modulus, tensile modulus, Young's modulus) The ratio of nominal tensile stress to the corresponding elongation below the proportional limit of a material. Since elongation is dimensionless, modulus has the units of stress. The relevant ASTM test is D 638. In contrast to structural metals such as mild steel, the stress-strain graphs for many plastics exhibit some curvature, even at very low strains. Since there is then no significant linear region whose slope would give the modulus, a ▶ Secant Modulus at 1–3% elongation may be reported for stiff materials. (2) More generally, any of the several elastic moduli characterizing behavior in shear (torsion), flexure, or change in volume under pressure. See ▶ Bulk Modulus. In SI, all types of elastic moduli are reported in pascals, usually megapascals (MPa). 1,000 psi = 6.894,757 MPa. *Also known as Elastic Modulus or Young's Modulus.* See ▶ Hardness.

Modulus of Resilience *n* The energy that can be absorbed per unit volume of a stressed specimen without creating a permanent deformation. It is equal to the area under the stress–strain graph from zero to the elastic limit divided by the volume of specimen undergoing deformation.

Modulus of Rigidity See ▶ Shear Modulus.

Modulus of Rupture (MOR) *n* During a stress vs. strain test, tensile pull test, of a material a stress (force or load) is approached where the material ruptures, breaks and separates which is sometimes referred to as the modulus (stress/strain) at rutpure. Also referred to as the ultimate tensile strength and force at break.

Modulus of Rupture in Bending See ▶ Flexural Strength.

Mohair \ˈmō-ˌhar, -ˌher\ *n* [mod. of obs. It *mocaiarro*, fr. Arabic *mukhayyar*, literally, choice] (1619) See ▶ Angora (1).

Mohr Balance Balance used to measure the specific gravity of fluids.

Mohs Hardness \ˈmōz-, ˈmōs-, ˈmō-səz-\ [Friedrich Mohs † 1839 German mineralogist] A system of ranking materials according to their ability to scratch, and resist being scratched by, lower-ranking materials, diamond being the hardest material known and having the highest rank. Mohs' original scale ranked diamonds as 10, corundum as 9, etc., and talc as 1. The scale has been modified to recognize some newer hard materials ranking in the large gap between corundum and diamond. The modified scale is listed below, in order of decreasing a scratch hardness.

Modified

Mohs number	Material
15	Diamond
14	Boron carbide
13	Silicon carbide
12	Fused alumina
11	Fused zirconia
10	Garnet
9	Topaz
8	Quartz or Stellite®
7	Vitreous silica
6	Orthoclase
5	Apatite
4	Fluorite
3	Calcite
2	Gypsum
1	Talc

There is a strong positive correlation between rank on the Mohs scale and *Knoop Microhardness*. See also ▶ Scratch Hardness.

Mohs Scale *n* [Friedrich *Mohs* † 1839 German mineralogist] (1879) Scale for determining the relative hardness of a mineral, according to is resistance to scratching by one of the following minerals: (1) talc; (2) gypsum; (3) calcite; (4) fluorite; (5) apatite; (6) feldspar; (7) vitreous silica; (8) quartz; (9) topaz; (10) garnet; (11) fused zirconia; (12) fused alumina; (13) silicone carbide; (14) boron carbide; and (15) diamond. Other useful hardnesses are: fingernail, slightly more than (2); penny, about (3); pocket knife, slightly more than (5); window glass, (5.5) and a steel file, (6.5). See ▶ Hardness.

Moiety \ˈmói-ə-tē\ *n* [ME *moite*, fr. OF *moité*, fr. LL *medietat-*, *medietas*, fr. L *medius* middle] (15c) An indefinite amount of a constituent present in a material or compound.

Moil A rarely used Syn: molding ▶ Flash.

Moiré \móˈra, mwä\ *n* [F *moiré*, fr. *moiré* like moire, fr. *moire*] (1818) A wavy or watered effect on a textile fabric, especially a corded fabric of silk, rayon, or one of the manufactured fibers. Moiré is produced by passing the fabric between engraved cylinders which press the design into the material, causing the crushed and uncrushed parts to reflect light differently.

Moire Effect *n* An optical effect which results from light interference, exhibiting a pattern of light and dark areas. The effect of superimposing a repetitive design, such as a grid, on the same or a different design to produce a pattern distinct from its components.

Moire Papers *n* Wallpapers having a watered silk sheen effect. See ▶ Moire Effect.

Moisture Absorption *n* The pickup of water vapor by a material upon exposure for a definite time internal to an atmosphere of specified humidity and temperature. No ASTM test exists for this property. Moisture absorption should not be confused with ▶ Water Absorption, for which there *is* an ASTM test.

Moisture Barrier *n* Treated paper or metal that retards or bars water vapor, used to keep moisture from passing into walls or floors.

Moisture Content The amount of moisture in a material under prescribed conditions and ex-pressed as a percent of the mass of the moist specimen that is, the mass of the dry substance plus the moisture. Also, it can be described as the water in solid waste. Expressed as the percentage of weight lost when a sample is dried at more than 100°C until it reaches a constant weight.

Moisture Equilibrium *n* The condition reached by a sample when the net difference between the amount of moisture absorbed and the amount desorbed, as shown by a change in weight, shows no trend and becomes insignificant.

Moisture-Free Weight *n* (1) The constant weight of a specimen obtained by drying at a temperature of 105°C in a current of desiccated air. (2) The weight of a dry substance calculated from an independent determination of moisture content (e.g., by distillation with an immiscible solvent or by titration with Fischer reagent.

Moisture Properties *n* All fibers when exposed to the atmosphere pick up some moisture; the quantity varies with the fiber type, temperature, and relative humidity. Measurements are generally made at standard conditions, which are fixed at 65% RH and 70°F. Moisture content of a fiber or yarn is usually expressed in terms of percentage regain after partial drying.

Moisture Regain *n* The loss of weight on drying, expressed as percent of dry weight, of a predried material exposed for a specified time to a specified humidity and temperature, then over-dried at a temperature above 100°C. ASTM D 885 (Section 07.01) describes a procedure recommended for rayon yarns and tire cords.

Moisture Sensitivity *n* (1) The degree to which the performance of a plastic part or product is affected by changes in its moisture content or, for some persons, by changes in the relative humidity of the environment in which the product is situated. (2) The degree to which processing performance is affected by moisture pickup prior to processing.

Moisture-Set Inks *n* Inks that dry or set principally by precipitation. The vehicle consists of a water insoluble resin dissolved in a hygroscopic solvent. Drying occurs when the hygroscopic solvent has absorbed sufficient moisture either from the atmosphere, substrate or external application to precipitate the binder. An important characteristic of these inks is their low odor. See ▶ Steam-Set Ink.

Moisture Vapor Permeability See ▶ Specific Permeability (of a Film to Moisture).

Moisture-Vapor Transmission Rate of movement of moisture vapor through a membrane.

Moisture Vapor Transmission Rate (MVTR) See Water Vapor Transmission Rate.

Molality *n* (**Molal Concentration; *m***) A concentration unit: the number of moles of solute per kilogram of solvent.

Molal Solution *n* A solution that contains 1 mole of the solute per kilogram of the solvent.

Molar Heat of Fusion *n*, ΔH_{fus} The heat necessary to melt 1 mole of a substance.

Molar Heat of Solution *n*, ΔH_{sol} The heat liberated (if negative) or absorbed (if positive) when 1 mole of solute dissolves in a solvent.

Molar Heat of Vaporization *n*, ΔH_{vap} The heat necessary to vaporize 1 mole of a substance.

Molarity *n* (**Molar Concentration *M***) A concentration unit: number of moles of solute per liter of solution.

Molar Solution *n* (1 – M) A solution that contains 1 mole of solute per liter of solution.

Molar Volume *n*, V_m (1) The volume occupied by one mole of a substance. For an ideal gas, 22.4 Lmol^{-1}. (2) The molar volume, symbol V_m is the volume occupied by one mole of a substance (chemical element or chemical compound) at a given temperature and pressure. It is equal to the molar mass (M) divided by the mass density (ρ). It has the SI unit cubic meters per mole (m^3/mol), and often practical to use the units cubic decimeters per mole (dm^3/mol) for gases and cubic centimeters per mole (cm^3/mol) for liquids and solids.

$$V_m = mole/\rho = m^3/mole$$

For a mixture of components, the following expression exists,

$$V_m = \sum_{i=1}^{N} XiMi/\rho_{mix}$$

For gases the following expression is accepted,

$$V_m = V/n = \frac{RT}{P}$$

For a given temperature (K) and pressure (atmospheres), the molar volume is the same for all ideal gases and is known to the same precision as the gas constant: R = 8.314 472(15) J mol–1 K^{-1}. The molar volume of an ideal gas at 100 kPa (1 bar) is

22.710980(38)dm^3/mole at 0°C

22.789598(42)dm^3/mole at 25°C

The molar volume of an ideal gas at 1 atm (most commonly used) of pressure is in liters (L)

22.414 L/mole at 0°C

24.465 L/mole at 25°C

where moles are measured n = m/M, n = moles, m = mass (grams, g) and M = molar mass (grams per mole, g/mole). (International Union of Pure and Applied Chemistry (1993). Quantities, Units and Symbols in Physical Chemistry, 2nd edition, Oxford: Blackwell Science. ISBN 0-632-03583-8. p. 41; NIST. http://physics.nist.gov/cgi-bin/cuu/Value?r. Retrieved 2007-10-14)

Mold **mōld**\ [ME, mod. of OF *modle*, fr. L *modulus*, dim. of *modus* measure] (13c) n A hollow form or matrix into which a liquid or molten plastic material is placed and which imparts to the material, upon cooling or curing, its final shape as a finished article (14c) *v*. To impart shape to a plastic mass by means of a confining cavity or matrix, by a process usually involving high pressure and changes in temperature. The term *molding* is usually employed for processes using dry thermoplastic or thermosetting compounds, and in injection or transfer molding. The term *casting* is preferred for processes employing liquids – solutions or suspensions – that are sufficiently fluid to be poured into a mold and to fill it by gravity flow. (A glossary of plastics terminology in 5 languages, 5th edn. Glenz W (ed). Hanser Gardner, Cinicinnati, OH, 2001; Whittington's dictionary of plastics. Carley, James F (ed) Technomic, 1993)

Mold Base *n* An assembly of ground-flat steel plates, usually containing dowel pins, bushings and other components of injection or compression molds excepting the cavities and cores.

Mold-Clamping Force See ▶ Clamping Force.

Mold Efficiency *n* In a multimold blow-molding system, the percentage of the total turn-around time of the mold actually required for forming, cooling, and ejecting the part.

Molding *n* (1) A projecting or depressed surface, generally used at or around edges of walls, furniture, etc., either plain or decorated, employed to ornament a wall surface, cornice or capital. (2) In plastics processing, any process at some stage of which the plastic is softened or melted, usually by heating, and forced to flow into a shaped cavity or mold that essentially determines all the final dimensions of the product. (Whittington's dictionary of plastics. Carley James F (ed) Technomic, 1993; A glossary of plastics terminology in 5 languages, 5th edn. Glenz W (ed) Hanser Gardner, Cinicinnati, OH, 2001)

Molding Compound *n* Granules or pellets of a resin containing all desired additives such as plasticizes, stabilizers, colorants, processing aids, and fillers, prepared by blending these ingredients with the neat resin, then reducing the hot mix to pellets by extrusion or milling, cutting, and chilling, ready for further processing into finished products. See also ▶ Molding Powder.

Molding Cycle *n* (1) The sequence of operations necessary on a molding press to produce a set of moldings. (2) The period of time occupied by the complete sequence of operations required for the production of one set of moldings.

Molding Defects *n* Structural and other defects in material caused inadvertently during molding by using wrong tooling, process parameters or ingredients. Also called molding flaw.

Molding Index *n* A practical measure of the difficulty of molding of thermosetting compound. A calculated weight of the candidate molding powder, is placed into a flash-type cup mold that has been preheated to the temperature prescribed for the material. The mold is closed and the total minimum force required to close it is reported as the molding index of the compound.

Molding Flaw See ▶ Molding Defects.

Molding Powder *n* This term usually denotes pellets or granules of a neat resin or a ▶ Molding Compound. Also See ▶ Dry Blend.

Molding Pressure *n* The pressure applied to the ram of an injection machine or press to force the softened plastic to completely fill the mold cavities. It is expressed in force per unit of cross-sectional area of the ram surface acting upon the material.

Molding Shrinkage *n* (mold shrinkage, shrinkage, contraction) The fractional difference in corresponding dimensions between a mold cavity and the molding made in the cavity, both the cavity and the molding being at normal room temperature when measured. Shrinkage is often found to be different in different directions. It may be expressed as a percent, in mils/in., or mm/m.

Mold Pressure *n* The pressure measured inside a mold cavity, usually by a flush-mounted pressure transducer, at any time during a molding cycle, but in particular the highest pressure recorded during the cycle. Compare with *Melt Pressure*.

Mold Release See ▶ Parting Agent.

Mold Seam *n* A visible line on a molded or laminated piece, often very slightly raised above the general surface, impressed by the parting line of the mold.

Mole \ˈmōl\ *n* [Gr *Mol*, short for *Molekulargewicht* molecular weight, fr. *molekular* molecular + *Gewicht* weight] (1902) (mol) (1) In SI, the mole is defined as the amount of a substance of a system that contains as many elementary entities as there are atoms in 0.012 kg of carbon-12. The elementary entities must be specified and may be atoms, molecules, ions, electrons, other particles, or specified groups of such particles. (2) In practical usage, a mass of a substance equal to its molecular weight. To agree with the SI definition, the mass must be in grams. However, engineers often find it convenient to work with pound-moles, kilogram-moles, and even ton-moles of materials. If the mass unit isn't prefixed, the mole is always the gram-mole. (Whitten KW, Davis RE, Davis E, Peck LM, Stanley GG (2003) General chemistry. Brookes/Cole, New York; Perry's chemical engineer's handbook, 7th edn. Perry RH, Green DW (eds). McGraw-Hill, New York, 1997)

Molecular Formula *n* (ca. 1903) A formula expressing the actual number of atoms of each element in one molecule.

Molecularity *n* The number of molecules which collide to form the activated complex in an elementary process.

Molecular Orbital (MO) *n* (1932) An electronic energy level in a molecule and the corresponding charge-cloud distribution in space.

Molecular Orientation See ▶ Orientation.

Molecular Sieve *n* (1926) A porous mineral or synthetic inorganic material, such as a zeolite (hydrous silicate), usually in the form of porous pellets or fine granules, having the ability to strongly absorb molecules of other (fluid) materials.

Molecular Solid *n* A solid in which molecules are held together by dipole–dipole or London forces.

Molecular Volume *n* (molar volume) The volume occupied by 1 mole, numerically equal to the gram-molecular weight divided by the density at the prevailing pressure and temperature.

Molecular Weight *n* (1880) (formula weight, molecular mass) The sum of the atomic weights of all atoms in a molecule. (Goldberg DE (2003) Fundamentals of chemistry. McGraw-Hill Science/Engineering/Math, New York) In most nonpolymeric compounds the molecular weight is a known constant value. In high polymers, the individual molecules range widely in the number of atoms they contain, and therefore, in molecular weight. Hence an average must be used to characterize a particular sample of polymer. The two averages most commonly used are *Number-Average Molecular Weight* (Mn) and *Weight-Average Molecular Weight* (Mw). Methods for determining these averages include measurements of light scattering and osmotic pressure in solutions, sedimentation in an ultracentrifuge, and depression of freezing points and vapor pressures of solutions, dilute-solution viscosity, end-group titration, and spectroscopy. (Slade PE (2001) Polymer molecular weights, vol 4. Marcel Dekker, New York)

Molecular-Weight Distribution *n* The percentages by number (or weight) of molecules of various molecular weights that comprise a given specimen of a polymer. Two samples of a given polymer having the same number-average molecular weight may perform quite differently in processing because one has a broader distribution of molecular weights than the other. Two basic groups of methods are used for measuring molecular-weight distribution. Fractionation methods, which actually divide the specimen into portions having relatively narrow ranges of molecular weight, include: fraction precipitation and fraction solution (the two most widely used), chromatography, liquid–liquid partition, ultracentrifugation, zone refining, and thermo gravimetric diffusion. After fractionation by any of these methods, the weight percent of each fraction is plotted versus the average molecular weight for that fraction to obtain a histogram of the distribution, which may be smoothed into a curve. Non-fractionation methods include light-scattering studies, electron microscopy, dilute-solution viscosity, size-exclusion chromatography, ultracentrifugation, and diffusion. A popular measure of the breadth of a distribution is the ratio of weight-average to number-average molecular weight, M_w/M_n. (Goldberg DE (2003) Fundamentals of chemistry. McGraw-Hill Science/Engineering/Math, New York) See also ▶ Polydispersity.

Molecular-Weight Distribution Poisson *n* The ratio of the weight-average (M_w) to the number-average (M_n); gives an the of the distribution. MWD gives a general picture of the ratio of the large, medium, and small molecular chains in the plastic. (Goldberg DE (2003) Fundamentals of chemistry. McGraw-Hill Science/Engineering/Math, New York)

Molecular Weight Ratio See ▶ Molecular Weight Distribution.

Molecular Weights of Polymers *n* The sum of the atomic weights of all atoms in a molecule. In most nonpolymeric materials the molecular weight is a fixed constant value. In high polymers, the molecular weight weights of individual molecules vary widely so that must be expressed as averages, shown by a bar above the symbol, e.g., $\bar{M}_n$. Number-average molecular weight,

$$M_n = \sum n_i M_i / \sum n_i = \sum w_i / \sum (w_i/M_i)$$

where, n_i = number of molecules with molecular weight M_i, w_i = weigh fraction of material having molecular weight, M, Σn = total number of molecules and, Weight-average M_w

$$M_w = \sum w_i M_i / \sum w_i$$

Z-Average M_z

$$Mz = \sum w_i M_i^2 / \sum w_i M_i$$

Methods of determining molecular weight include osmotic pressure, light scattering, gel permeation chromatography, dilute solution viscosity, vapor pressure, freezing temperatures and others. (Goldberg DE (2003) Fundamentals of chemistry. McGraw-Hill Science/Engineering/Math, New York)

Molecule \ˈmä-li-ˌkyü(ə)l\ *n* [F *molécule*, fr. NL *molecula*, dim. of L *moles* mass] (1794) The smallest unit quantity of a compound that can exist by itself and still retain the chemical identity of the substances as a whole. (Goldberg DE (2003) Fundamentals of chemistry. McGraw-Hill Science/Engineering/Math, New York)

```
      O                    O⁻⁻
     / \                  / \
    H   H                H   H
                         +   +
The conventional view   The actual view
 of the water molecule   of the water molecule
```

Mole Fraction *n* (*X*) A concentration unit: the number of moles of one component in a solution divided by the total number of moles of all components. (Goldberg DE (2003) Fundamentals of chemistry. McGraw-Hill Science/Engineering/Math, New York)

Mole Percent *n* (**mol%**) Mole fraction multiplied by 100. (Goldberg DE (2003) Fundamentals of chemistry. McGraw-Hill Science/Engineering/Math, New York)

Moleskin \\▮skin\\ *n* (1668) A heavy sateen-weave fabric made with heavy, soft-spun filling yarns. The fabric is sheared and napped to produce a suede effect.

Moltopren *n* Polyester or polyether + diisocyanate + water (foam). Manufactured by Bayer, Germany.

Mol Volume *n* The volume occupied by a mol or a gram molecular weight of any gas measured at standard conditions is 22.414 L. A **molal solution** contains 1 mole/1,000 g of solvent. A **molar solution** contains 1 mole or gram molecular weight of the solute in 1 L of solution.

Molybdate Orange *n* \\mə-▮lib-▮dāt\\ $25PbCrO_4 \cdot 4Pb\text{-}MoO_4 \cdot 1PbSO_4$. Pigment Red 104 (77605). Bright orange, inorganic pigment which is a solid solution of lead chromate, lead molybdate and lead sulfate. Exhibits excellent hiding and tinting strength, brilliant hue, good permanency, but only fair chemical resistance.

Molybdate Orange Pigment *n* Any of a range of solid solutions of lead chromate, lead molybdate, and lead sulfate, used as dark-orange to light-red pigments for plastics. Their advantages are high opacity, bright color, light-fastness, good heat stability, and freedom from bleeding.

Molybdenum Disulfide *n* (ca. 1931) (molybdic sulfide, molybdenum sulfide) MoS_2. A black, shiny, flaky-crystalline material used as a filler in nylons, fluorocarbons, and polystyrene to improve stiffness and strength, and, principally, to provide lubricity. The compound occurs naturally as the ore *molybdenite*.

$$S^{--}$$
$$Mo^{+4} \quad S^{--}$$

Molybdenum FR Any of several molybdenum compounds, such as the oxide or ammonium dimolybdate, $(NH_2)_2Mo_2O_7$, added to plastics to improve their fire retardancy (FR) and smoke suppression.

Mombassa Gum See ▶ Animi.

Moment of Force or Torque *n* The effectiveness of a force to produce rotation about an axis, measured by the product of the force and the perpendicular distance from the line of action of the force to the axis. Cgs unit – the dyne-centimeter. If a force F acts to produce rotation about a center at a distance d from the line in which the force acts, and force has a torque, $L = Fd$.

Moment of Inertia *n* A measure of the effectiveness of mass in rotation. In the rotation of a rigid body not only the body's mass, but the distribution of the mass about the axis of rotation determines the change in the angular velocity resulting from the action of a given torque for a given time. Moment of inertia in rotation is analogous to mass (inertia) in simple translation. The cgs unit is g-cm². Dimensions – $[m\ l^2]$. If m_1, m_2, m_3, etc., represent the masses of infinitely small particles of a body; r_1, r_2, r_3, etc., their respective distances from an axis of rotation, the moment of inertia about this axis will be

$$I = (m_1 r_1^2 + m_2 r_2^2 + m_3 r_3^2 + \cdots)$$

or

$$I = \sum (mr^2).$$

(Giambattista A, Richardson R, Richardson RC, Richardson B (2003) College physics. McGraw Hill Science, New York)

Momentum \\mō-▮men-təm, mə-\\ *n* [NL, fr. L, movement] (1610) Quality of motion measured by the product of mass and velocity. Cgs unit – 1 g-cm/s. A mass m moving with velocity v has a momentum, $M = mv$. If a mass m has its velocity changed from v_1 to v_2 by the action of a force F for a time t_1, $mv_2 - mv_1 = Ft$.

Momentum Flux *n* In hydrodynamics, an interpretation of shear stress τ in which each of the six shear components of the stress tensor is viewed as the rate of flow, per unit of shear area, and perpendicular to that area, of momentum directed along a principal axis in the surface of shear. In laminar flow through a circular orifice, with radial coordinate r and axial coordinate z, the only nonzero shear component is t_{rz}, the flux of z-directed momentum in the r-direction. Newton's law of viscosity for this situation becomes

$$t_{rz} = -\mu (dv_z/dr)$$

where v_z is the fluid velocity at any radius r and the viscosity μ has the dimensions: (momentum/area-time) shear rate, and for which the SI unit is 1 [kg m/s]/(m² s)]/s^{-1} = 1 kg/m s = 1 Pa s. (Goodwin JW, Goodwin J, Hughes RW (2000) Rheology for chemists. Royal Society of Chemistry, UK, August 2000; Kamide K, Dobashi T (2000) Physical chemistry of polymer solutions. Elsevier, New York; Handbook of chemistry and physics, 52nd edn. Weast RC (ed) The Chemical Rubber, Boca Raton, FL)

Monkey *n* Trade term for a batch of resin which has become thermoset and unusable during processing.

Monk's Cloth *n* Wallcoverings that simulate a basket-weave material of cotton, jute, or flax.

Mono- A prefix designating the entity that follows it as the only one or as containing only one of that kind, e.g., monomer, or monohydric alcohol.

Monobasic \\▮mä-nə-▮bā-sik\\ *adj* [ISV] (1842) (1) Pertaining to acids having one active hydrogen per molecule, e.g., hydrochloric acid, HCl. (2) Designating an acid salt in which one hydrogen (of two or three) has been replaced by a metal, e.g., potassium monophosphate, KH_2PO_4.

Monocarboxylic Acid \-ˌkär-(ˌ)bäk-ˈsi-lik\ *adj* (ca. 1909) Any organic acid containing a single –COOH group in the molecule. Many of the larger acids of this type are derived from natural fats and oils and are used in the production of alkyd resins and polyesters. Esters of oleic, stearic, pelargonic and richinoleic acid, all monocarboxylic, are widely used as plasticizers.

Monochloroethylene *n* Syn: ▶ Vinyl Chloride.

Monochromatic Color Scheme *n* Combining of colors in a room based on one color used in various values and saturation or chromas.

Monochromatic Emissive Power *n* The ratio of the energy of certain defined wavelengths radiated at definite temperatures to the energy of the same wavelengths radiated by a black body at the same temperature and under the same conditions.

Monochromatic Light *n* Light of a single wavelength. It may be obtained by the use of a laser or by gaseous discharge tubes in combination with proper filters. An approximation is obtained by interference filters or monochromators. The degree of monochromaticness depends upon the dispersion device and accompanying optical scheme used to separate the different wavelengths from a continuous or "white" light source. Emission lines from elements which, when excited, emit light, such as mercury, sodium, etc., are most nearly true monochromatic light. (Moller KD (2003) Optics. Springer, New York)

Monochromator \ˌmä-nə-ˈkrō-ˌmā-tər\ *n* [*mono-chrom*atic + illumin*ator*] (1909) A device for dispersing white light into individual wavelengths, generally a prism or diffraction grating. (Moller KD (2003) Optics. Springer, New York)

Monochrome \ˈmä-nə-ˌkrōm\ *n* [ML *monochroma*, fr. L, feminine of *monochromos* of one color, fr. Gk *monochrōmos*, fr. *mono-* + *-chrōmos* –chrome] (1662) Painting in different values and chromas of one hue. (Moller KD (2003) Optics. Springer, New York)

Monoclinic \ˌmä-nə-ˈkli-nik\ *adj* [ISV] (ca. 1864) Of a crystal (or crystal system), having two axes that are mutually perpendicular to the third one, but not to each other. An example is the β form of elemental sulfur, which is stable between 112°C and 119°C, but slowly converts to the rhombic form below 112°C. (Rhodes G (1999) Crystallography made crystal clear: A guide for users of macromolecular models. Elsevier Science and Technology Books,

New York; Hibbard MJ (2001) Mineralogy. McGraw-Hill, New York)

Monocyclic Terpenes *n* Heterogeneous mixture of monocyclic, bicyclic and other related terpene hydrocarbons recovered or removed in the fractionation of certain terpenes or other essential oils, or as a by-product in the chemical conversion of pinenes; generally sold under trade names.

Monodisperse \ˌmä-nō-dis-ˈpərs\ *adj* [*mon-* + *disperse*, adj, fr. *disperse*, v.] (1925) Of a polymer, all the molecules having the same molecular weight. It has long been possible to make (nearly) monodisperse polystyrene of various molecular weights and a few other monodisperse polymers are now available as laboratory chemicals. (Slade PE (2001) Polymer molecular weights, vol 4. Marcel Dekker, New York) Compare *Polydisperse*.

Monodispersity *n* Refers to a polymer system which is homogeneous in molecular weight, i.e., lacks molecular weight distribution. (Slade PE (2001) Polymer molecular weights, vol 4. Marcel Dekker, New York)

Monofil See ▶ Monofilament.

Monofilament \ˌmä-nə-ˈfi-lə-mənt\ *n* (1940) A single filament of indefinite length, strong enough to function as a yarn in textile operations or as an entity in other applications. Monofilaments are generally produced by extrusion. Their outstanding uses are in the fabrication of brush bristles, surgical sutures, fishing lines, racquet strings, screen materials, ropes, and nets. The finer monofilaments are woven and knitted on textile machinery. (Kadolph SJJ, Langford AL (2001) Textiles. Pearson Education, New York)

Monoglyceride \ˌmä-nə-ˈgli-sə-ˌrīd\ *n* (1860) Partial ester of glycerol in which one of the three hydroxyl groups is reacted with an organic acid.

Monolithic Flooring See ▶ Seamless Flooring.

Monomer \ˈmä-nə-mər\ *n* [ISV] (1914) A relatively simple compound that can react with itself or other compounds to form long-chain compounds by either (1) utilizing its C=C bonds for addition or (2) by having two or more functional groups that can react with receptive groups in other molecules. Monomers are the basic building blocks of polymers. Monomer is sometimes referred to as *mer*. (Odian GC (2004) Principles of polymerization. Wiley, New York) See ▶ Polymer.

Monomeric Pertaining to a ▶ Monomer.

Monomeric Cement See ▶ Adhesive.

Monotropic *n* Of a material or element, having two forms, one of which is metastable toward the other. The metastable form tends to change spontaneously to the stable form but the reverse change does not occur.

Monotype Printing *n* Process of printing from mechanically assembled pieces of type, each of which bears one character.

Montan Wax \ ▌män-t°n-\ *n* [L *montanus* of a mountain] (1908) (lignite wax) A hard, white wax derived from *lignite*, (a lower-grade hydrocarbon fossil mineral between peat and bituminous coal). It is sometimes described as a bitumen wax. The crude product is very dark in color, almost black, but after refining it becomes pale yellow. Its mp varies from 72°C to 82°C; its acid value, from 25 to 99; and its saponification value, from 58 to 104. The wax is used as a mold lubricant.

Monthier's Blue See ▶ Iron Blue.

Mooney Equation *n* (M. Mooney) An empirical modification of the ▶ Einstein Equation applicable to higher solids concentrations, and relating the viscosity of a suspension of monodisperse spheres η_f to that of the pure liquid η_o.

$$\ln(\eta_f/\eta_o) = 2.5\ f/(1-Sf)$$

where f = the volume fraction of solids and $S \cong 1.4$ for spheres. (ASTM, www.astm.org; Goodwin JW, Goodwin J, Hughes RW (2000) Rheology for chemists. Royal Society of Chemistry, UK, August 2000) See also ▶ Eilers Equation.

Mooney Scorch Time *n* For a rubber specimen tested in a ▶ Mooney Viscometer, the time elapsing after the minimum torque has been reached for torque to increase by five "Mooney units".

Mooney Viscosimeter *n* An instrument invented by M. Mooney in 1924, used to measure the effects of time of shearing and temperature on the comparative viscosities of rubber compounds. It consists of a motor driven disk, tooth-surfaced on the sides, enclosed within a die cavity formed by two halves maintained at controlled temperature and closing force. The specimen is a double disk, joined at the edges and trapped between the die halves and the rotor. (ASTM, www.astm.org)

Mop Board See ▶ Baseboard.

Moplen *n* Poly(ethylene) and poly(propylene), manufactured by Montedison, Italy.

Mop Polishing *n* (1) Application of a modified French polish with a camel hair "mop" to carved work. (2) Polishing by friction with a rotary mop.

Mordant Dyes *n* Dyes which develop their characteristic colors when precipitated on suitable bases to form lakes. It is often possible to obtain considerable variation in the color of the final lake by altering the metallic radical of the base. Mordant dyes in their original state are colorless.

Mordants \ ▌mór-d°nt\ *n* [ME, fr. MF, pp of *mordre* to bite, fr. L *mordēre*; perhaps akin to Sanskrit *mṛdnāti* he presses, rubs] (1791) Substances capable of uniting with both dyes and textile fibers so as to improve the bond between dye and textile and give improved textures and alter the colors.

Moresque *n* A multicolored yarn formed by twisting or plying single strands of different colors.

Morphological Analysis *n* The identification of particles based on easily observed microscopical characteristics.

Morphology \mór- ▌fä-lə-jē\ *n* [Gr *Morphologie*, fr. *morph-* + *-logie* –logy] (1830) The study of the physical form and structure of a material. This includes a wide range of characteristics, extending from the external size and shape of large articles to dimensions of crystal lattices but, with polymers, it most often refers to microstructure. (Kamide K, Dobashi T (2000) Physical chemistry of polymer solutions. Elsevier, New York; Physical properties of polymers handbook. Mark JE (ed) Springer, New York, 1996)

Mortar \ ▌mór-tər\ *n* [ME *morter*, fr. OF *mortier*, fr. L *mortarium*] (14c) (1) Material used in a plastic state which can be troweled, and becomes hard in place, to bond units of masonry structures. NOTE — The word "mortar" is used without regard to the composition of the material, and is defined only with reference to its use as a bonding material as contrasted with the words "stucco" and "plaster"). (2) A mixture of gypsum plaster with aggregate or hydrated lime, or both, and water is to produce a trowelable fluidity.

Mosaic \mō- ▌zā-ik\ *n* [ME *musycke*, fr. MF *mosaique*, fr. OIt *mosaico*, fr. ML *musaicum*, alter. of LL *musivum*, fr. L *museum, musaeum*] (15c) (1) Picture or surface decoration made by embedding small pieces of colored stone, marble, pottery or glass in mortar.

Mosley's Law *n* The frequencies of the characteristic X rays of the elements show a strict linear relationship with the square of the atomic number.

Mote *n* A small piece of seed or vegetable matter in cotton. Motes are removed by boiling the fiber or fabric in sodium hydroxide, then bleaching. When not removed, they can leave a dark spot in the fabric.

Motion, Laws of See ▶ Newton's Law of Motion.

Motionless Mixer See ▶ Static Mixer.

Mottle \ ▌mä-t°l\ *n* [prob. back-formation fr. *motley*] (1676) (1) An irregular distribution or mixture of colorants or colored materials giving a more or less distinct appearance of specks, spots, or streaks of color. Mottling is often deliberate although it may occur accidentally due to inadequate mixing. (2) Presence of irregularly shaped and randomly distributed areas of nonuniform appearance in color, gloss or sheen. Syn: ▶ Blotching.

Mouldrite *n* Urea, phenol, and melamine/formaldehyde resins manufactured by ICI, Great Britain.

Mountain Blue See ▶ Azurite.

Mounting Plate *n* In blow molding, the plate to which the mold is attached. See also ▶ Clamping Plate.

Movable Platen *n* The large back plate of an injection molding machine to which the back half of the mold is secured during operation. This platen is moved either by a hydraulic ram or a toggle mechanism.

Movil, Mowil *n* Poly(vinyl chloride), manufactured by Polymer Ind., Italy.

Moviol *n* Poly(vinyl alcohol), manufactured by Hoechst, Germany.

Mowilith *n* Poly(vinyl acetate), manufactured by Hoechst, Germany.

MPE See ▶ Modified Polyphenylene Ether.

MPO See ▶ Modified Polyphenylene Ether.

MQ See ▶ Methylsilicone.

MS Abbreviation for Mass Spectrometry.

Mucilage \ˈmyü-s(ə)lij\ *n* [ME *muscilage*, fr. LL *mucilago* mucus, musty juice, fr. L *mucus*] (15c) (1) Flocculant or slimy deposit which separates from unrefined vegetable oils on heating. (2) An adhesive prepared from a gum and water. Also in a more general sense, a liquid adhesive which has a low order of bonding strength.

Mudcracking *n* Paint film defect characterized by a broken network of cracks in the film.

Muff *n* A loose skein of textured yarn prepared for dyeing or bulking. In the bulking operation, the yarn contracts and the resulting skein resembles a muff.

Muff Dyeing See ▶ Dyeing.

Mullen Bursting Strength *n* An instrumental test method that measures the ability of a fabric to resist rupture by pressure exerted by an inflated diaphragm.

Muller \ˈmə-lər\ *n* [alter. of ME *molour*, prob. fr. *mullen* to grind] (1612) An instrument usually of glass used for dispersing pigments in varnish for test purposes. The two glass grinding surfaces are generally ground glass. Pigment and vehicle are placed between them and the two rubbed together, either by hand or by making power. Power driven automatic mullers have two circular glass grinding surfaces.

Mullion \ˈməl-yən\ *n* [prob. alter. of *monial* mullion] (1567) Narrow dividers between windows or glass panels.

Multicavity Mold *n* (multiple-cavity mold, multiple-impression mold) A mold having several to hundreds of cavities so that many parts may be molded with each shot. In many cases, the parts are identical, but that need not be so. In a type of multicavity mold known as a ▶ Family Mold, some of the cavities may be identical while others are different, or they may all be different.

Multicolor Finish *n* A speckled coating containing flecks of small, individual colored particles different from the base color. Also called *Speckled Finish*.

Multicomponent Polymerization *n* A polymerization which in general involves the use of more than one comonomer. The term is usually reserved for copolymerization of three or more comonomers. (Odian GC (2004) Principles of polymerization. Wiley, New York)

Multifilament *n* A yarn consisting of many continuous filaments or strands, as opposed to monofilament which is one strand. Most textile filament yarns are multifilament.

Multifilament Yarn *n* A manufactured yarn composed of many fine continuous filaments or strands.

Multifunctional Concentrate *n* A plastic compound that contains high percentages of at least two of such additives as colorants, stabilizers, flame retardants, lubricants, antistatic agents, antiblocking agents, blowing agents, fillers, etc, that will be diluted in base resin to provide a tailored compound with the desired final concentrations of the additives in the extruded or molded product.

Multigated *n* Of an injection-mold cavity, having two or more gates. Multigating is common in molding large or complex parts that would be difficult to fill through one gate. Also, shrewd placement of the several gates can direct the weld lines to areas of the part where the stresses expected in service will be low.

Multilayer Fabric *n* A fabric for reinforced-plastic structures formed by braiding to and fro or overlapping in one direction. Layers may be biaxial or triaxial, fibers mixed, and braid angles varied.

Multilayer Film *n* A film comprised of layers of two or more different materials, all polymeric. The goal is to make a film that has a combination of properties not achievable with a single polymer. Most multilayer films are made by ▶ Coextrusion. The principal layers are usually joined by an adhesive resin compatible with both the adjacent principals and coextruded with them.

Multiple Proportions, Law of *n* If two elements form more than one compound, the weights of the first element which combine with a fixed weight of the second element are in the ratio of integers to each other.

Multiple-Regression Analysis See ▶ Regression Analysis.

Multiple Scattering See ▶ Scattering Multiple.

Munsell Book Notation *n* The Munsell color notation of a specimen obtained by visual or computational comparison with the Munsell hue, value, and chroma scales of the Munsell Book of Color.

Munsell Book of Color *n* A collection of color chips which illustrate the Munsell Color System, manufactured by the Munsell Color Co., in both matte and glossy finishes. It consists of pages of constant hue at 2.5 hue step intervals, with darkest colors (of lowest value at the bottom) increasing in value upwards to the top and colors of lowest chroma in the center fold increasing to highest chroma at the outside edge of the page.

Munsell Chroma *n* The color term applied to the psychological sensation dimension of saturation. In Munsell notation, neutrals, white, graphs, etc., have a chroma of zero. Increasing deviation from gray to purest color in equally perceptible steps are given increasing numerical notations, with intermediate steps designated by decimals proportional to the perceptability differences.

Munsell Value *n* (1) That portion of Munsell notation that corresponds to the sensation of lightness. It is described by a number from zero for a perfect black to ten for a perfect white and is calculated from the equation

$$Y_o = 1.2219V - 0.23111V^2 + 0.23951V^3 - 0.021009V^4 + 0.0008404V^5$$

(2) The daylight reflectance of a specimen expressed on a scale extending from 0 for ideal black to 10 for ideal white by steps of approximately equal visual importance. Achromatic or neutral colors are designated N followed by the value notation, thus: N 5.0. (Colour physics for industry, 2nd edn. McDonald, Roderick (eds) Society of Dyers and Colourists, West Yorkshire, England, 1997)

Mu Oil *n* Name sometimes given to tung oil obtained from *Aleurites Montana*.

Mural Painting *n* Usually, large picture which is painted on a wall, or fastened to a wall surface. (Gair A (1996) Artist's manual. Chronicle Books LLC, San Francisco)

Muscovite \ˈməs-kə-ˌvīt\ *n* [ML or NL *Muscovia, Moscovia* Moscow] (1535) See ▶ Aluminum Potassium Silicate.

Muslin \ˈməz-lən\ *n* [F *mousseline*, fr. It *mussolina*, fr. Arabic *mawsilīy* of Mosul, fr. al-*Mawsil* Mosul, Iraq] (1609) A broad term describing a wide variety of plain-weave cotton or polyester/cotton fabrics ranging from lightweight seers to heavier shirting and sheeting.

Muted Colors Colors whose chroma or saturation has been lessened or moderated by use of their complementary color or a neutral. Also called *Grayed Colors or Tones* which, on losing most of their character as a color, approach *Neutrals*.

M_v *n* Symbol for Viscosity-Average Molecular Weight.

MVP *n* Abbreviation for ▶ Moisture Vapor Permeability.

MVT *n* Abbreviation for Moisture Vapor Transmission. See ▶ Specific Permeability (of a Film to Moisture).

MVTR *n* Abbreviation for Moisture-Vapor-Transmission Rate.

M_w *n* Symbol for Weight-Average Molecular Weight.

MW Abbreviation for Molecular Weight.

MWD See ▶ Molecular Weigh Distribution.

Mycology \mī-ˈkä-lə-jē\ *n* [NL *mycologia*, fr. *myc-* + Latin *-logia* -logy] (1836) The science dealing with fungi.

Mylar \ˈmī-ˌlär\ *tradename* Polyester (film) DuPont's registered trade name for biaxially oriented film composed of polyethylene glycol terephthalate.

Myristoyl Peroxide $(C_{13}H_{27}CO)O_2$. A soft, granular powder, used as a polymerization catalyst for vinyl monomers.

N

n \ˈen\ *n* {*often capitalized, often attributive*}(before 12c) (1) SI abbreviation for ▶ Nano-. (2) *n*-: In organic chemistry, abbreviation for Normal, signifying a straight (unbranched) aliphatic chain. (3) A subscript denoting the last of a series of *n* ordered numbers or data.

N (1) Chemical symbol for the element nitrogen. (2) SI abbreviation for ▶ Newton. (3) -N: In solution chemistry, following a numeral, abbreviation for Normal or Normality. See ▶ Normal Solution.

Na Chemical symbol for the element sodium.

NACE Abbreviation for ▶ National Association of Corrosion Engineers.

N-Acetyl Ethanolamine \-ˌe-thə-ˈnä-lə-ˌmēn, ˈnō-, British also ˌē-\ *n* (hydroxyethyl acetamide) CH$_3$CONHC$_2$H$_4$–OH. A plasticizer for polyvinyl alcohol and cellulosic plastics.

Nacreous \ˈnā-kər\ *n* [MF, fr. OIt *naccara* drum, nacre, fr. Arabic *naqqārah* drum] (1718) Pertaining to, or having the appearance of, mother-of-pearl. (Solomon DH, Hawthorne DG (1991) Chemistry of Pigments and Fillers, Krieger Publishing Co., New York) See ▶ Pearlescent Pigment.

Nacreous Pigment n Translucent pigment which, when added to transparent coating material, creates a deep lustrous and pearlescent finish. See ▶ Pearlescent Pigment. (*Kirk-Othmer Encyclopedia of Chemical Technology: Pigments-Powders*, John Wiley and Sons, New York, 1996)

NAD See ▶ Nonaqueous Dispersion.

Nailhead Rusting *n* Rust from iron nails that penetrates or bleeds through the coating and stains the surrounding area.

Nainsook \ˈnān-ˌsuk\ *n* [Hindi *nainsukh*, fr. *nain* eye + *sukh* delight] (1790) A fine, lightweight, plain-weave fabric, usually of combed cotton. The fabric is often mercerized to produce luster and is finished soft. Nainsook is chiefly used for infants' wear, lingerie, and blouses.

N-β(Aminoethyl)-γ-Aminopropyltrimethoxy Silane *n* A silane coupling agent used in reinforced epoxy, phenolic, melamine and polypropylene resins.

Nano- {*combining form*} [ISV, fr/ Gk nanos dwarf] The SI prefix meaning $\times\ 10^{-9}$ (one billionth).

Nanometer (nm) \ˈna-nə-ˌmē-tər\ *n* [ISV] (1963) The usual unit of linear measurement with the electron microscope and of measuring ultraviolet and visible light wavelengths. (1 nm = 10^{-9} m or about 4×10^{-8} in.) replaces the former name, "millimicron" (mμ). An SI unit of length equal to 10^{-9} meter, convenient for stating light wavelengths and superseding the older angstrom unit and millimicron, both now deprecated.

Nanotechnology *n* The ultra-miniaturization of nanomaterial, nanodevices and nanoanalysis methods including nanorheology and molecular tribology. (*Springer Handbook of Nanotechnology*, Bhushan, ed., Springer-Verlag, New York, 2004)

Nap *n* [ME *noppe*, fr. MD, flock of wool, nap] (15c) (1) The dense, soft or fuzzy fabric attached to a cylindrical cover used on paint rollers. (2) A downy surface given to a cloth when part of the fiber is raised from the basic structure.

Naphtha \ˈnaf-thə\ *n* [L, fr Gk, of Iranian origin, akin to Per *neft* naphtha] (1572) (solvent naphtha) Any of a family of petroleum and coal-tar distillates with 30°C boiling ranges within the interval 125 to 200°C. They are useful as solvents for natural resins and rubber, and as paint thinners. Naphtha's contain substantial portions of paraffin and naphthalenes.

Naphthalene \ˈnaf-thə-ˌlēn\ *n* [alter. of earlier *naphthaline*, irreg. fr. *naphtha*] (1821) A crystalline aromatic hydrocarbon usually obtained by distillation of coal tar (napththalin, tar camphor). An aromatic hydrocarbon, C$_{10}$H$_8$, derived from coal-tar oils or petroleum fractions and having the structure shown below. Once used as a moth repellent, it is now important as a reactant in the production of phthalic anhydride, which in turn is used for making plasticizers, alkyd resins, and polyester resins. See ▶ Magdala Red.

Naphthas \ˈnaf-thəs\ *n* [L, fr. Gk, of Iranian origin, akin to Persian *neft* naphtha] (1572) Aromatic hydrocarbons derived from coal tar, although the term "petroleum naphtha" is sometimes used for petroleum spirits of substantially aliphatic type. The coal tar naphthas are the distillation fractions which contain a complex mixture of aromatic hydrocarbons, including xylenes. Generally refers to hydrocarbon solvents, both aromatic and aliphatic. Hi-flash naphtha is an example of the former and VM & P naphtha an example of the latter.

Naphthenate *n* Salt of naphthenic acid. A drier used in paints; made with naphthenic acid and lead, cobalt, calcium, iron, zinc or manganese salts. See ▶ Naphthenic Acid.

Naphthenate Driers *n* Compounds of naphthenic acid with metals, usually lead, cobalt, or manganese used to accelerate the oxidation of the ink film.

Naphthenes \ naf- thēns\ *n* (1884) C_nH_{2n}. Found in certain types of crude petroleum. *Known also as Cycloparaffins or Hydrogenated Benzenes.*

Naphthenic Acid *n* A carboxylic acid derived from petroleum refining and usually one of a mixture of similar compounds. The mixed acids and some of their soaps, e.g., cobalt naphthenate and calcium naphthenate, are useful as catalysts or accelerators in curing polyester resins and as drying agents in paints and varnishes. Commercial naphthenic acids are not pure compounds, but consist of a mixture of acids based on cyclopentane rings.

Naphthenic Solvent *n* Hydrocarbon solvents comprised wholly or partially of cycloparaffinic (naphthenic) hydrocarbon compounds. The only common commercial naphthenic solvent is cyclohexane.

Naphthol *n* \ naf- thól\ *n* [ISV] (1849) $C_{10}H_8O$
(1) Either of two isomeric derivatives of naphthalene used as an antiseptics and in the manufacture of dyes. (2) Any of various hydroxy derivatives of naphthalene that resemble the simpler phenols.

NAPIM The National Association of Printing Ink Manufacturers. NAPIM is a trade association whose purpose is to provide information and assistance to its members to better manage their businesses, and to represent the printing industry in the United States. NAPIM (ww.napim.org) is located 581 Main Street, Woodbridge, NJ 07095.

Naples Yellow *n* A light yellow pigment; the true pigment is a basic antimonite of lead, but is imitated by mixtures.

Napping *n* A finishing process that raises the surface fibers of a fabric by means of passage over rapidly revolving cylinders covered with metal points or teasel burrs. Outing, flannel, and wool broadcloth derive their downy appearance from this finishing process. Napping is also used for certain knit goods, blankets, and other fabrics with a raised surface.

1,5-Napththalene Diisocyanate *n* (NDI) $OCNC_{10}H_6NCO$, An isocyanate used in the production of urethane elastomers and foams.

Narrow Fabric *n* Any nonelastic woven fabric, 12 inches or less in width, having a selvage on either side, except ribbon and seam binding.

National Association of Corrosion Engineers *n* (NACE) Address: 1440 S Creek Dr., Houston, TX 77084. A leading publisher of technical information on protection and performance of materials in corrosion environments. NACE also sponsors educational seminars on anticorrosion properties and applications of plastics in the chemical-process industries.

National Association of Printing Ink Manufacturers (NAPIM) 581 Main Street, Woodbridge, NJ 07095. The NAPIM is a trade association whose purpose is to provide information and assistance to its members to better manage their businesses, and to represent the printing ink industry in the United States.

Natta Catalyst \ nät-( )tä-\ *n* Any of several catalysts used in the stereospecific polymerization of olefins, e.g., ethylene and propylene, particularly a catalyst made from titanium chloride and aluminum alkyl or similar materials by a special process including grinding the materials together to produce an active catalytic surface. (Odian GC, (2004) Principles of Polymerization. John Wiley and Sons, Inc., New York)

Natural Fibers *n* A fiber of plant or animal origin such as cotton (nearly pure cellulose), flax, sisal, abaca, hemp, jute, etc., the wood of sheet and other animals, horsehair, and swine bristle. Cellulose from cotton linters and wood pump is the starting material for cellulosic plastics and is by far the most important natural fiber for the plastics industry.

Natural Finish *n* Any finish resulting from the application of a transparent substance (such as a varnish, water-repellent preservative, sealer, or oil) which does not affect significantly the original color or grain.

Natural Gums *n* Fossilized or animal gum resins. (Whistler JN, BeMiller JN (eds) (1992) Industrial Gums, Polysaccharides and Their Derivatives, Elsevier Science and Technology Books)

Natural Iron Oxides *n* Natural mineral deposits consisting chiefly of ferric oxide. Color range of yellow, red, brown, and black, consisting chiefly of ferric oxide. See ▶ Iron Oxides.

Natural Pigments See ▶ Earth Pigment.

Natural Polymers *n* A polymer that is produced by biosynthesis in nature, as opposed to human-controlled polymerized synthetic polymer.

Natural Red 4 (75470) See ▶ Carmine Lake.

Natural Resin *n* A resin produced by nature, mostly by exudation from certain trees from cuts or tears in the bark. Lac resin is secreted by the lac insect and is refined to make shellac. Some of the tree resins are copal, rosin, and sandarac, at one time widely used in wood finishes. (Langenheim JH (2003) Plant Resins: Chemistry, Evolution Ecology and Ethnobotany, Timber Press, Portland, OR) See ▶ Resin, Natural.

Natural Rubber *n* The rubber material obtained from the latex produced by certain plants and trees. See ▶ Rubber, Natural.

Natural Yellow Oxide see ▶ Iron Oxides, Natural.

Naval Stores *n* Chemically reactive oils, resins, tars, and pitches derived from the oleoresin contained in, exuded by, or extracted from, trees chiefly of the pine species *Genus Pinus*, or from the wood of such trees.

Navy Pitch See ▶ Pitch, Navy.

NBR *n* Elastomers from acrylonitrile and butadiene. See ▶ Acrylonitrile-Butadiene Copolymer.

Color Difference Equation *n* This expression was originally the National Bureau of Standards equation, now National Institute of Standards and Technology. Color difference equation devised by Judd and modified by Hunter. The total color difference, ΔE, was called an NIST unit (*Paint/Coatings Dictionary*, Compiled by Definitions Committee of the Federation of Societies for Coatings Technology, 1978).

$$\Delta E = f_g \left\{ \left[221 Y_m^{1/4} \left((\Delta \alpha)^2 + \Delta \beta \right)^2 \right)^{1/2} \right]^2 + \left[k(\Delta Y^{1/2}) \right]^2 \right\}^{1/2}$$

where

$$\alpha = \frac{2.4266x - 1.3631y - 0.3214}{1.0000x + 2.2633y + 1.1054}$$

$$\beta = \frac{0.5710x + 1.2447y - 0.5708}{1.000x + 2.2633y + 1.1054}$$

$$Y_m = \frac{Y_1 + Y_2}{2} \quad \Delta \alpha = \alpha_1 - \alpha_2, \Delta \beta = \beta_1 - \beta_2,$$
$$\Delta Y = Y_1 - Y_2$$

k = constant expressing the relative importance of lightness and chromaticness in a particular viewing arrangement; k = 12 for comparison made across a very narrow dividing line. With increasing separation of the two colors, the value K decreases; k = 10 is frequently used. fg = factor which takes into account the masking effect of gloss on the detection of color difference: for normal observation

$$f_g = \frac{Y_m}{Y_m + 2.5}$$

(Y_m is measured with specular reflectance excluded). If f_g is calculated in this way f_g decreases very rapidly when the Y_m becomes less an 10%, where $f_g = 0.8$. At 1% $f_g = 0.287$, for example. For ease in calculation k is frequently used as 10.000, especially where colors are not going to be used adjacently. (McDonald, Roderick (1997) Colour Physics for Industry, 2nd Ed., Society of Dyers and Colourists, West Yorkshire, England; Syszecki, G. and Stiles, W. S., *Color Science: Concepts and Methods, Quantitative Data and Formulas*, John Wiley and Sons. Inc., New York, 1967; Billmeyer, F. W. and Saltzman, M., *Principles of Color Technology*, John Wiley and Sons, Inc., New York 1966)

NIST Total Color Difference *n* The total color difference, ΔE, calculated by means of the NIST Color Difference Equation, sometimes referred to as a Judd or a Judd Unit. The term is frequently erroneously used to refer to the color difference calculated by other equations normalized to agree in magnitude on the average to an NBS unit, or to the ΔE measured on tristimulus colorimeters with electrical circuits designed to give an approximation of the NIST color difference equation, frequently one of the Hunter L, a, b equations. Such incorrect use of the term is to be discouraged. Description of the exact equation and method of measurement used will avoid confusion. (NIST, www.nist.gov).

n-Butanol \ ˈen- ˈbyü-tᵊn- ˌól\ *n* See n-▶ Butyl Alcohol. *n* $CH_2=CHCOOC_4H_9$. A colorless liquid that polymerizes readily on heating.

n-Butyl Alcohol *n* (ca. 1871) (1-butanol) $CH_3(CH_2)_2CH_2OH$. A medium-boiling alcohol, liquid above 35°C, used as a solvent for cellulosic, phenolic, and urea-formaldehyde resins. It is also used as a diluent/reactant in the manufacture of urea-formaldehyde and phenol-formaldehyde resins, and as an intermediate in the production of butyl acetate, dibutyl phthalate, and dibutyl sebacate.

n-Butyl Aldehyde See Butyaldehyde.

n-Butyllithium *n* (nBu-Li) An initiator for anionic polymerization; lithium alkyls are known to be highly associated in many solvents, and especially in nonpolar solvents such as benzene and heptane.

n-Butyl Methacrylate \-ǀme-ǀtha-krə-ǀlāt\ *n* $H_2C=CHCH_3COOC_4H_9$. A polymerizable monomer used in the production of acrylic resins and potting compounds.

n-Butyl Myristate *n* $CH_3(CH_2)_{12}COOC_4H_9$. The butyl ester of myristic acid, an oily liquid used as a plasticizer for cellulosic plastics.

n-Butyl Palmitate \-ǀpal-mə-ǀtāt\ *n* $C_{15}H_{31}COOC_4H_9$. A plasticizer for polystyrene and cellulosic plastics.

n-Butylphosphoric Acid *n* $C_4H_9H_2PO_4$. A reddish amber liquid used as a catalyst, e.g., in urea-resin production.

n-Butyl Propionate \-ǀprō-pē-ə-ǀnāt\ *n* $C_2H_5COOC_4H_9$. A colorless liquid with an apple-like odor, used as a solvent for nitrocellulose.

NC *n* (1) Abbreviation for Numerical Control. (2) See ▶ Cellulose Nitrate.

N-Carboxyanhydrides *n* Alternative name for N-carboxy-α-amino acid anhydride.

NCCA Abbreviation for National Coil Coaters Association.

NCNS See ▶ Triazine Resin.

n-Decyl-n-Octyl Phthalate *n* (NDOP) See *n*-▶ Octyl-*N*-Decyl Phthalate.

N,N′-Dinitroso-N,N′-Dimethylterephthalamide *n* A blowing agent that is a weak explosive in powder form and thus unsafe to handle, but is available in desensitized form by treatment with mineral oil (DuPont Nitrosan®). This blowing agent is unique in that its low decomposition temperature permits the expansion of vinyl plastisol prior to gelation (93°C). Subsequent fusion at 177°C produces open-cell vinyl foam. Closed-cell foam can be produced by heating to fusion in a closed mold, releasing the pressure and subsequently heating in an oven at 100°C.

NDOP *n* Abbreviation for *n*-Decyl-*N*-Octyl phthalate. See *n*-▶ Octyl-*N*-Decyl Phthalate.

NDPA Abbreviation for National Decorating Products Association.

Near-Net-Shape Configuration *n* In reinforced plastics molding, designating a fibrous-preform shape very close to the shape the perform will take after resin impregnation and curing in the mold, such that no layup and little or no trimming after molding are required.

Neat Resin *n* Strictly, a resin containing nothing but the main identified polymer(s). Usually, the term means that, while there may be fractional percentages of stabilizers and other additives present, there are no fillers, reinforcements, or pigments. Sometimes called a "pure" resin, though, since all commercial polymers are mixtures of homologs of various molecular weights, "pure" has a looser meaning here.

Neatsfoot Oil See ▶ Animal Oil.

Neck-In *n* In extrusion of film, sheet and coatings, the difference between the width of the extrude web as it leaves the die and the final width of the chilled film, etc., (before any edge trimming is done).

Necking *n* The localized reduction in cross section that may occur in a ductile material under tensile stress (ASTM D 883). A similar phenomenon can occur during extrusion under certain conditions as the extrudate leaves the die. Necking can also occur during drawing of fibers at temperatures below their melting ranges. Fibers of crystalline and some noncrystalline thermoplastics, e.g., polyethylene, exhibit necking at a critical stress near the yield point. (ShahV (1998) Handbook of Plastics Testing Technology, John Wiley and Sons, New York)

Neck Insert *n* (finish insert) In blow molding of bottles, a removable part of the mold that forms a specific neck finish of the bottle.

Needle **nē-dəl**\ *n* [ME *nedle*, fr. OE *nîdl*; akin to OHGr *nādala* needle, *nājan* to sew, L *nere* to spin, Gk *nēn*] (before 12c) (1) A thin, metal device, usually with an eye at one end for inserting the thread, used in sewing to transport the thread. (2) The portion of a knitting machine used for intermeshing the loops. Several types of knitting needles are available. Also see ▶ Spring Needle and ▶ Latch Needle. (3) In nonwovens manufacture, a barbed metal device used for punching the web's own fibers vertically through the web. (Kadolph SJJ and Langford AL (2001) Textiles, Pearson Education, New York)

Needled *n* The product of the needle loom. Needled fabrics are used for rug pads, papermaker's felts, padding, linings, etc.

Needle Loom *n* A machine for bonding a Nonwoven web by mechanically orienting fibers through the web. The process is called needling, or needlepunching. Barbed needles set into a board punch fiber into the batt and withdraw, leaving the fibers entangled. The needles are spaced in a nonaligned arrangement. By varying the strokes per minute, the advance rate of the batt, the degree of penetration of the needles, and the weight of the batt, a wide range of fabric densities can be made. For additional strength, the fiber web can be needled to a woven, knit, or bonded fabric. Bonding agents may also be used. (Kadolph SJJ and Langford AL (2001) Textiles, Pearson Education, New York)

Negative Catalyst *n* (inhibitor, retarder) An agent that slows a chemical reaction. See ▶ Catalyst.

Negative Crystals *n* A uniaxial crystal is optically negative if $\varepsilon < \omega$. A biaxial crystal is said to be optically negative is $\gamma - \beta < \beta - \alpha$.

Negatron *n* (1) A term used for electron when it is necessary to distinguish between (negative) electrons and positrons. (2) A four element vacuum tube which displays a negative resistance characteristic.

NEMA *n* Acronymical abbreviation for National Electrical Manufacturers Association, an organization that has strongly influenced electrical applications of plastics.

Nematic \ni-**ˈ**ma-tik\ *adj* [ISV *nemat*- + ^{1}-*ic*] (1923) See ▶ Liquid-Crystal Polymer.

Neo- {*combining form*} [Gl, fr. *neos* new] (1) Prefix meaning *new* and denoting a compound isomerically related to an older one whose name follows the prefix. (2) A prefix denoting a hydrocarbon in which at least one carbon atom is connected directly to four other carbon atoms, e.g., neopentane, $C(CH_3)_4$.

Neopentyl Glycol *n* (NPG, 2,2'-dimethyl-1,3-propanediol An important intermediate for the production of alkyd and polyester resins, urethane foam and elastomers, and polyester plasticizers. Gel coats based on NPG for reinforced polyesters have improved flexibility, hardness, and resistance to abrasion and weathering.

Neopentyl Glycol Diacrylate *n* (NPGDA) A highly reactive crosslinking monomer used in photocurable coatings. It provides solvent and strain resistance as well as improved response to light.

Neopentyl Glycol Dibenzoate *n* A solid plasticizer for rigid PVC.

Neoprene \ˈnē-ə-ˌprēn\ *n* [ne- + chlor*oprene*] (1937) (Polychloroprene, poly-2-chloro-1-butadiene) Elastomers made from the monomer $CH_2=CHCCl=CH_2$. They are available as dry solids and lattices, and are vulcanizable to tough products with excellent resistance to oils, gasoline, solvents, heat, and weathering. The original neoprene, produced by DuPont under the trade name "Duprene," was America's first successful synthetic rubbe, manufactured by DuPont.

Neoprene Paint *n* Paint based upon pigmented solutions of the proprietary synthetic rubber neoprene (Polychloroprene rubber), a vulcanizing agent being added before use.

Neoprene Rubber *n* Polychloroprene rubbers with good resistance to petroleum products, heat, and ozone, weatherability, and toughness.

Neou Oil *n* Seed oil of the tree, *Parinarium macrophyllum*, which grows in several parts of West Africa. The main constituent acids of the oil are an isomer of α - elaeostearic acid and linoleic acid. The oil has an iodine value of 135 and a saponification value of 190.

Nep *n* A small knot of entangled fibers that usually will not straighten to a parallel position during carding or drafting.

Nepheline \ˈne-fə-ˌlēn\ *n* [F *néphéline*, fr. Gk *nephelē* cloud] (ca. 1814) A naturally occurring mineral composed mainly of feldspar and nephelite. As a filler in PVC compounds, it has the unique property of contributing almost no opacity, so that it can be used in nearly transparent compounds. It is also used as a filler in epoxy and polyester resins.

Nepheline Syenite \-ˈsī-ə-ˌnīt\ *n* A mineral aggregate consisting chiefly of albite, microline, and nephelite, each in significant amount.

Nephelite \ˈne-fə-ˌlit\ *n* [F *néphéline*, fr. Gk *nephelē* cloud] (ca. 1814) See ▶ Nepheline.

Nernst Effect \ˈnern(t)st-\ *n* [Walther Hermann *Nernst* 1864–1941] When heat flows across the lines of magnetic force, there is observed an electromotive force in the mutually perpendicular direction. (Giambattista A, Richardson R, Richardson RC, Richardson B (2003) College Physics, McGraw Hill Science, New York)

Nerve \ˈnərv\ *n* [L *nervus* sinew, nerve; akin to Gk *neuron* sinew, nerve, *nēn* to spin] (14c) A condition difficult to define fully, but commonly used to denote the qualities of firmness, strength and elasticity in crude rubber. In crude rubber, nerve is reduced or destroyed by milling or breakdown.

Nesting *n* In reinforced plastics, the placing of plies of fabric so that the yarns of one ply lie in the valleys between the yarns of the adjacent ply.

Nest Plate A retainer plate with depressed area for cavity blocks used in injection molding.

Net \ˈnet\ *n* [ME *nett*, fr. OE, akin to OHGr *nezzi* net] (before 12c) An open fabric made by knotting the intersections of thread, cord, or wires to form meshes. Net can be made by hand or machine in a variety of mesh sizes and weights matched to varying end uses, i.e., veils, curtains, fish nets, and heavy cargo nets.

Net Equation *n* An equation which shows only actual reactants at the left and only actual products at the right of the arrow.

Net Flow *n* The output of an extruder's metering section, being, to a first approximation, the algebraic sum of the drag flow, pressure flow and leakae flow. In most plastics extruders with solid feeds and screws having the conventional three sections – feed, transition, and metering – the net flow may be conservatively estimated from the equation

$$\dot{m} = 2.0 D^2 N h \rho_o$$

where $\dot{m}$ is in lb/h, D is the nominal screw diameter, in., N is the screw speed in rpm, h is the channel depth in the metering section, and ρ_o is the plastic's room-temperature density, g/cm^3. If the rate is given in kg/h and the diameter and channel depth in cm, the equation becomes

$$\dot{m} = 0.055 D^2 N h \rho_o$$

(Strong AB (2000) Plastics Materials and Processing, Prentice Hall, Columbus, Ohio; *Handbook of Plastics, Elastomers and Composites*, 4th Ed., Harper, C. A., ed., McGraw-Hill, New York, 2002; Carley, James F (ed.) (1993) Whittington's Dictionary of Plastics Technomic, Publishing Co., Inc.)

Net Rate *n* In a fiber production process the total throughput less waste and inferior or off-grade material.

Netting *n* A crossing-strand sheet or tubular plastic structure, fused at the crossing points, produced by extrusion through (patented) oscillating dies. Plastic netting, in small pieces and baskets, is also made by injection molding.

Netting Analysis *n* The stress analysis of filament-wound structures that neglects the strength of the resin and assumes that the filaments carry only axial tensile loads and possess no bending or shearing stiffness.

Network Copolymers *n* Three-dimensional polymers, polymer gels, multifunctional reactive polymers, and crosslinked polymers. (Zaccaria VK and Utracki L (2003) Polymer Blends, Springer-Verlag, New York; Galina H, Spiegel S, Meisel I, Kniep CS and Grieve K (2001) Polymer Networks, John Wiley and Sons, New York)

Network Polymer *n* A polymer obtained by the polymerization of a monomer having two or more functional groups that become interconnected with sufficient interchain bonds to form a large three-dimensional network. The network can be formed during polymerization, or may be created by crosslinking the polymers after they have been formed. The vulcanization of rubber is an example of the formation of a network polymer from a preformed polymer. Copolymers of ethylene and propylene can be made into network polymers by crosslinking with ionizing radiation after reactive sites have been prepared by treating with heat or peroxides. (Galina H, Spiegel S, Meisel I, Kniep CS and Grieve K (2001) Polymer Networks, John Wiley and Sons, New York)

Network Structure *n* An atomic or molecular arrangement in which primary bonds form a three-dimensional network. (Zaccaria VK and Utracki L (2003) Polymer Blends, Springer-Verlag, New York; Galina H, Spiegel S, Meisel I, Kniep CS, Grieve K (2001) Polymer Networks, John Wiley and Sons, New York; Elias, Hans-Georg (2003) An Introduction to Plastics, John Wiley and Sons, New York; Allcock HR, Mark J and Lampe F (2003) Contemporary Polymer Chemistry, Prentice Hall, New York; Elias HG (1977) Macromolecules Vol. 1–2, Plenum Press, New York)

Neuberg Chalk *n* Aluminum silicate used to some extent as a filler or extender, but it has special properties as a constituent of metal polishes.

Neutral \ˈnü-trəl, ˈnyü-\ *n* (15c) (1) Having no distinguishable hue; achromatic. See ▸ Gray Scale. (2) Not decided in color; nearly achromatic; of low saturation.

Neutral Axis *n* In a beam or column subject to bending moments, the surface near the center of the beam and perpendicular to the applied loads upon which neither tensile nor compressive stress is acting. In homogeneous beams with depth-symmetrical cross sections, the neutral axis is exactly at the center.

Neutral Gray *n* An achromatic gray representing a portion of the gray scale.

Neutralization \ˌnü-trə-lə-ˈzā-shən, ˌnyü-\ *n* (1808) A chemical reaction in which the hydrogen ion of an acid and the hydroxyl ion of a base unite to form water and a salt.

Neutralization Equivalent or Number See ▸ Acid Number.

Neutralization Reaction *n* An acid-base reaction (Arrhenius).

Neutralizing *n* (British) Spray finish deliberately applied to produce an uneven or mottled appearance.

Neutral Oil *n* A light gravity mineral oil derived from petroleum and used as a lubricant in rubbing finished work.

Neutral Solution *n* An aqueous solution in which the concentrations of hydrogen and hydroxide ions are equal. A solution of pH = 7.0 at 25°C.

Neutral Toner *n* A stain of yellow-orange color used to blend colored wood streaks in wood finishing.

Neutrino \nü-ˈtrē-(ˌ)nō, nyü-\ *n* [It, fr. *neutro* neutral, neuter, fr. L *neutr-*, neuter] (1935) An electrically neutral particle of very small (probably zero) rest mass and of spin quantum number $\frac{1}{2}$. When the spin is oriented parallel to the linear momentum the particle is the antineutrino. When the spin is oriented antiparallel to the linear momentum the particle is the neutrino. Postulated by Pauli in explaining the beta decay process. Whenever a beta (positron) particle is created in a radioactive decay so is an antineutrino (neutrino).

The two particles and the parent nucleus share between them the available energy and momentum. Neutrinos and antineutrinos can penetrate amounts of matter measured in light years without appreciable attenuation. Detected by Reines and Cowan using antineutrinos from fission reactors and large scintillation detectors.(Galina H, Spiegel S, Meisel I, Kniep CS, Grieve K (2001) Polymer Networks, John Wiley and Sons, New York)

Neutron \ nü-,trän, nyü-\ *n* [prob. fr. *neutral*] (1932) A neutral elementary particle of mass number 1. It is believed to be a constituent particle of all nuclei of mass number greater than 1. It is unstable with respect to beta-decay, with half life of about 12 minutes. It produces no detectable primary ionization in its passage through matter, but interacts with matter predominantly by collisions and, to a lesser extent, magnetically. Some properties of the neutron are: rest mass, 1.00894 atomic mass unit; charge, 0; spin quantum number, $\frac{1}{2}$; magnetic moment, -1.9125 nuclear Bohr *magnetrons*. (Galina H, Spiegel S, Meisel I, Kniep CS and Grieve R (2001) Polymer Networks, John Wiley and Sons, New York; Weast RC (ed.), Handbook of Chemistry and Physics, The Chemical Rubber Co., Boca Raton, FL, 52nd Ed.)

Neutron-Absorbing Fiber *n* Polyethylene fiber modified with boron used in the nuclear industry for reducing neutron transmission.

Neutron Cross Section See ▶ Cross Section.

Newel \ nü-əl, nyü\ *n* [ME *nowell*, fr. MF *nouel* stone of a fruit, fr. LL *nucalis* like a nut, fr. L *nuc-, nux* nut] (14c) Large post at the termination of a stair rail used in coatings jargon.

News Inks *n* Printing inks designed to run on newsprint, consisting basically of carbon black or colored pigments dispersed in mineral oil vehicles, which dry by absorption. Recent developments utilize emulsion, oxidation and heat set systems.

Newsprint *n* \- print\ *n* (1909) A generic term used to describe paper of the type generally used in the publication of newspapers.

Newton \ nü-tən, nyü-\ *n* [for Sir Isaac *Newton*] The SI unit of force that, when applied to a body having a mass of one kilogram and free to move, gives it an acceleration of one meter per second per second (1 m/s^2). See also ▶ Force. (Giambattista A, Richardson R, Richardson RC and Richardson B (2003) College Physics, McGraw Hill Science/Engineering/Math, New York; Weast R C (ed.), Handbook of Chemistry and Physics, The Chemical Rubber Co., Boca Raton, FL, 52nd Ed.)

Newtonian Flow *n* An isothermal, laminar flow characterized by a viscosity that is independent of the level of shear, so that the shear rate at all points in the flowing liquid is directly proportional to the shear stress and vice versa. Simple liquids such as water and mineral oil usually exhibit Newtonian flow, whereas polymer melts and solutions usually do not, but are *pseudoplastic*. See ▶ Pseudoplastic Fluid. Newtonian flow can occur, at least ideally, under the influence of an infinitesimally small force. It is said to be distinguished from plastic flow which occurs only when a finite minimum force is exceeded. Oils, at sufficiently low rates of shear, exhibit Newtonian flow. (Munson BR, Young DF and Okiishi TH (2005) Fundamentals of Fluid Mechanics, John Wiley and Sons, New York; Kamide K and dobashi T (2000) Physical Chemistry of Polymer Solutions, Elsevier, New York; Van Wazer, Lyons, Kim and Colwell (1963) Viscosity and Flow Measurement, Lyons Kim, Colwell, Interscience Publishers, Inc., New York) See ▶ *Nonnewtonian Liquid*.

Newtonian Liquid *n* (Newtonian fluid) One is which the rate of shear is proportional to the shearing stress. The constant ratio of the shearing stress to the rate of shear is the viscosity of the liquid. If the ratio is not constant, the liquid is nonnewtonian. (Munson BR, Young DF and Okiishi TH (2005) Fundamentals of Fluid Mechanics, John Wiley and Sons, New York) See ▶ Viscosity.

Newtonian Viscosity *n* Another name for ▶ Viscosity. Of polymer melts, the viscosity at very low shear rates (< 0.01 s^{-1}) in low-density polyethylene, for example) where viscosity is independent of shear rate and the melt is essentially Newtonian. In some models of pseudoplastic flow, a second limiting viscosity (μ_∞) approached as shear rate rises to extreme values and observed in some polymer solutions at high (but not infinite) shear rates. (Munson BR, Young DF and Okiishi TH (2005) Fundamentals of Fluid Mechanics, John Wiley and Sons, New York; Patton TC (1964) Paint Flow and Pigment Dispersion, Interscience Publishers, Inc., New York)

Newton's Law of Motion *n* Every body continues in its state of rest or of uniform motion in a straight line except in so far as it may be compelled to change that state by the action of some outside force. Change of motion is proportional to force applied and takes place in the direction of the line of action of the force. Also, to every action there is always an equal and opposite reaction.

Newton's Second Law of Motion *n* (Newton's law of momentum change) See ▶ Force.

Newton's Series *n* The sequence of interference colors observed when a quartz wedge is turned to the 45° position and observed between crossed polars. It is divided into "orders" by the red bands which occur periodically as the thickness increases. It is identical with the interference colors from a thin isotropic film of gradually increasing thickness.

Nextel® *n* 3M Corporation's trade name for their high-performance fiber containing 62% Al_2O_3, 24% SiO_2, and 14% B_2O_3. Grades range in properties: density 2.71 to 3.10 g/cm^3; modulus, 150 to 240 GPa; and tensile strength, 1.3 to 2.0 GPa.

N'Gart Oil *n* Oil obtained from *Plukenetia conophora*, which grows in the Cameroons. It polymerizes readily at normal stand-oil-making temperatures (300°C). Average reported constants for this oil are: iodine value, 200, sp gr, 0.937/15°C; and saponification value, 191.

N'Gore Oil See ▶ Isano Oil.

NGR Grain See ▶ Nongrain-Raising Stain.

NHDP *n* Abbreviation for *N*-Hexylethyl-*N*-Decyl Phthalate.

Ni *n* Chemical symbol for the element nickel.

Niax *n* Polyether from propylene and glycerine or 1,2,6-hexantriol, manufactured by Union Carbide, U.S.

Nibs \ˈnib\ *n* [prob. alter. of *neb*] (1585) Small piece of foreign material, pieces of skin, coagulated medium, etc., which project above the surface of an applied film, usually varnish or lacquer. See ▶ Bitty.

Nickel \ˈni-kəl\ *n* [prob. fr. S *nickel*, fr. Gr *Kupfernickel* niccolite, prob. fr. *Kupfer* copper + *Nickel* goblin; fr. the deceptive copper color of niccolite] (1755) A silver-white hard malleable ductile metallic element capable of a high polish and resistant to corrosion that is used chiefly in alloys and as a catalyst.

Nicol Prism *n* A polarizing element made of two pieces of calcite specially cut, ground, polished, and cemented. A transmitted beam splits into two polarized components, one of which is refracted into and absorbed by the asphalt mount. The remaining polarized beam is transmitted.

Night Vision See ▶ Scotopic Vision.

Nigrosines *n* Deep blue or black aniline dyes.

Ninon \ˈnē-ˌnän\ *n* [prob. fr. F *Ninon*, nickname for *Anne*] (1911) A lightweight fabric of silk or manufactured fibers made in a plain weave with an open mesh. Used for curtains and evening wear. NIP: (1) The line or area of contact between two contiguous rollers. (2) A defect in yarn consisting of a thin place.

NIOSH *n* Acronymical abbreviation for National Institute for Occupational Safety And Health, a division of the Center for Disease Control (Public Health Service, under the Department of Health, Education, and Welfare). This agency conducts investigations and research projects on industrial safety and makes recommendations for the guidance of OSHA. However, it does not enforce its own findings or OSHA regulations.

Nip \ˈnip\ *n* [ME *nippen*; akin to ON *hnippa* to prod] (14c) (1) The curved, **V**-shaped gap between a pair of counter-rotating calender rolls, chills rolls, or rubber pull rolls where incoming material is "nipped" and drawn between the rolls. (2) In safety-management parlance, "nip" is broadened to include any convergent approach of two machine elements, such as meshing spur gears or a V-belt approaching its pulley.

Nip Creases *n* Creases occurring at regular intervals along a fabric selvage subsequent to a nipping operation such as calendering or padding. Such creases are caused by a loosely wound selvage or improper let-off tension which allows the fabric to fold over or gather at the selvage prior to entering the nip of the rolls.

Nip Rolls *n* (pinch rolls) In film blowing, a pair of rolls situated at the top of the tower that pinch shut the blown-film tube, seal air inside it, and regulate the rate at which the film is pulled away from the extrusion die. One roll is usually covered with a resilient material, the other being of metal and internally cooled.

NIST *n* National Institute of Standards and Technology, Gaithersburg, MD, www.nist.gov. NIST was established (formerly NBS or National Bureau of Standards founded in 1901) is a non-regulatory federal agency with the U.S. Commerce Department's Technology Administration. The NIST (www,nist.gov) is to promote U.S. innovation and industrial commmpetiveness by advancing measurement science, standards, and technology in ways that enhance economic security and improve our quality of life. NIST carries out its mission in four cooperative programs: the NIST Laboratories, the Baldridge National Quality Program, the Manufacturing Extension Partnership and the Advanced Technology Program.

Nitrate Green \ˈnī-ˌtrāt-\ *n* Green pigment made from nitrate lead chrome and ferriferro cyanide.

Nitration \nī-ˈtrā-shən\ *n* (1887) Process of chemically adding nitrogen in combination with oxygen to any material.

Nitriding *vt* (1928) A type of ▶ Case Hardening in which a steel surface is reacted for hours or days with gaseous ammonia at temperatures from 480 and 540° to produce a surface hardness of about Rockwell C70 to a depth from 0.2 to 2 mm. The process is used with injection-mold cavities and extrusion screws, being less expensive and less durable with the latter than inlaid hard facing.

Nitril \ˈnī-trəl, ˌtrīl\ *n* [ISV *nitr-* + *-il*, *-ile* (fr. L *–ilis¹-ile*) (1848) An organic cyanide containing the group CN which on hydrolysis yields an acid with elimination of ammonia.

Nitrile Barrier Resins *n* (high-nitrile polymer) One of a family of polymers generally containing greater than 60% acrylonitrile, along with comonomers such as acrylates, methacrylates, butadiene, and styrene. Both straight copolymers and copolymers grafted onto elastomeric spines are available. Their unique property is outstanding resistance to passage of gases and water vapor, making them useful in packaging applications.

Nitrile Resins *n* Used principally in the packaging of foods other than beverages, and in nonfood packaging in the U.S. and abroad. The monomer used primarily is acrylonitrile, which provides a good gas barrier, chemical resistance, and taste- and odor-retention properties. (Harper CA (2000) Modern Plastics Encyclopedia, McGraw Hill Professional, New York)

Nitrile Rubber *n* A synthetic rubber obtained by the copolymerization of acrylonitrile and butadiene, noted for its oil resistance. See ▶ Acrylonitrile-Butadiene Copolymer.

Nitrobenzene \ˌnī-trō-ˈben-ˌzēn, –ben-ˈ\ *n* [ISV] (1868) $C_6H_5NO_2$. Mol wt, 123; bp, 210.85°C; mp, 5.7°C; sp gr, 1.1987; flp, 88°C (190°F). Syn: ▶ Mirbane Oil.

o-**Nitrobiphenyl** \-(ˌ)bī-ˈfe-nᵊl\ *n* (2-nitrobiphenyl, ONB) $O_2NC_6H_4$–C_6H_5. An involatile plasticizer for cellulose-ester polymers, and compatible with many others.

Nitrocellulose \-ˈsel-yə-ˌlōs, -ˌlōz\ *n* [ISV] (1882) Another name for cellulose nitrate. The product obtained by treating cellulose with a mixture of nitric and sulfuric acids. It is primarily used in the coatings industry as a base for lacquers and as a film-forming material widely used in flexographic and gravure inks. See ▶ Dope Cotton, and ▶ Pyroxylin. See also ▶ Lacquer.

Nitroethane *n* $CH_3CH_2NO_2$. A colorless liquid used as a solvent for cellulosic, vinyl, alkyd, and other resins. Bp, 114°C; flp, 41°C (106°F); mp, 16 mmHg/20°C; sp gr, 1.052/20°C.

Nitrogen \ˈnī-trə-jən\ *n* [F *nitrogène*, fr. *niter* + *-gène*, gen] (1794) Nitrogen gas is the most widely used blowing agent for injection-molded, structural foams. It is less expensive than most chemical blowing agents, leaves no residue, is environmentally harmless, and is easy to handle. Nitrogen is added to the polymer melt by pumping it directly into the barrel of the injection machine. Nitrogen is also used in the same way in foam extrusion.

Nitromethane \-ˈme-ˌthän, *British usually* -ˈmē-\ *n* (1872) CH_3NO_2. A colorless liquid made by reacting methane with oxides of nitrogen or nitric acid under pressure, and used as a solvent for cellulosic, vinyl, alkyd, and other resins. Bp, 101°C; flp, 45°C (112°F); vp, 28 mmHg/20°C.

Nitron *n* Cellulose nitrate, manufactured by Monsanto, U.S.

Nitroparaffins \-ˈpar-ə-fən\ *n* [ISV] (1892) Nitrated hydrocarbons, e.g., nitroethane, nitromethane, etc.

Nitropropane \-ˈprō-ˌpān\ *n* (1-nitropropane) $CH_3CH_2CH_2NO_2$. A colorless liquid boiling at 130°C, a good solvent for vinyl resins.
Bp, 132°C; flp, 50°C; vp, 7.5 mmHg/20°C; sp gr, 1.003/20°C.

2-Nitropropane *n* $CH_3CHNO_2CH_3$. Bp, 120°C; vp, 13 mmHg/20°C; flp, 40°C; sp gr, 0.992/20°C.

Nitroso Rubber *n* A copolymer of tetrafluoroethylene and trifluoronitrosomethane.

nm *n* Abbreviation for ▶ Nanometer. See ▶ Angstrom Unit.

NMA *n* (1) Abbreviation for Methyl Nadic Anhydride. (2) Abbreviation for N-Methyl Acetamide.

NMR *n* A technique for characterization of organic molecules. Abbreviation for ▶ Nuclear Magnetic Resonance.

Noble Gas *n* (1902) A member of group 0 in the periodic table. Any of a group of rare gases that include helium, neon, argon, krypton, xenon, and sometimes radon and that exhibit great stability and extremely low reaction rates. *Also known as Inert Gas*.

NODA *n* Abbreviation for *N*-Octyl-*N*-Decyl Adipate.

Nodal Points *n* Two points on the axis of a lens such that a ray entering the lens in the direction of one, leaves as if from the other and parallel to the original direction. A *normal salt* is an ionic compound containing neither replaceable hydrogen nor hydroxyl ions. A *normal solution* contains one gram molecular weight of the dissolved substance divided by the hydrogen equivalent of the substance (that is, one gram equivalent) per liter of solution. (Moller KD (2003) Optics, Springer-Verlag, New York)

Node \nōd\ *n* [ME, fr. L *nodus* know, node; akin to MI *naidm* bond] (15c) (1) A characteristic of flax and hemp fibers, giving them a bamboo-like appearance under the microscope, seen in no other fiber. In flax, the nodes are fairly regularly spaced at intervals of about 0.5 mm. (2) In a vibrating body, a point, line, or surface which is wholly or mostly free of vibration and appears to be at rest.

NODP *n* Abbreviation for *N*-Octyl-*N*-Decyl Phthalate.

NODTM Abbreviation for Tri(*n*-Octyl-*n*-Decyl) Trimellitate.

No- Flow Point *n* The temperature at which gelation (crosslinking) of a plastic material reaches a degree of no flow in a capillary rheometer.

Noil \nói(ə)l\ *n* (ca. 1624) A short fiber that is rejected in the combing process of yarn manufacture.

Nomenclature \nō-mən-klā-chər\ *n* [L *nomenclatura* assigning of names, from *nomen* + calatus, pp of *calare*] (1610) The names of chemical substances and the systems used for assigning them. (Progress in Polymer Science and Technology (2002) IUPAC World Polymer Congress, Beijing, China, July 7–12, 2002, John Wiley and Sons, New York; IUPAC (2000) Handbook, 2000–2001, International Union of Pure and Applied Chemistry)

Nomex Polyamide from iosphthalic acid + *m*-phenylene diamine. Manufactured by DuPont, U.S.

Nonaqueous Dispersion (NAD) The solvent analog of a latex: the polymer is dispersed in a volatile organic liquid which is not a solvent for the polymer. NAD's have much higher solids than conventional high molecular weight solvent coatings. Like latices, the viscosity is independent of the molecular weight as opposed to solvent soluble resins.

Nonbreak Oil See ▶ Nonbreak Oil.

Noncombustible Incapable of being ignited and burned in air; fire-resistance; preferred to incombustible by fire authorities. See ▶ Flammability.

Nonconvertible Coating Film-former which, after being deposited from a solution, dries to give a film which can be redissolved in a solvent from which it was originally deposited. See also ▶ Convertible Coating.

Nondestructive Test (1) A test that yields information about failure under mechanical stress without actually stressing to failure. (2) More broadly, any test to evaluate a property of a material, part, or structure that does not significantly damage the part. Techniques used include ultrasound, magnetic inspection of metals and welds, X-ray inspection, infrared, nuclear magnetic resonance, and sonic analysis. Although an indentation-hardness test leaves a permanent mark, in many tests of parts it is nondestructive.

Nondrying Coating Coating that does not dry in the course of a regular schedule. See ▶ Grease Paint.

Nondrying Oils Oil which does not of itself possess to a perceptible degree the power to take up oxygen from the air and lose its liquid characteristics. *Also known as Fixed Oil*.

Nonelastic Woven Tape A woven narrow fabric, weighing less than 15 ounces per square yard, made principally of natural and/or manufactured fibers, including monofilaments, but not containing rubber or other similar elastic stands.

Nonelastomeric Thermoplastic Polyurethanes *n* See ▶ Rigid Thermoplastic Polyurethanes.

Nonelastomeric Thermosetting Polyerethane *n* Curable mixtures of isocyanate prepolymers or monomers. Have good abrasion resistance and low-temperature stability, but poor heat, fire, and solvent resistance and weatherability. Processed by reaction injection and structural foam molding, casting, potting, encapsulation, and coating. Used in heat insulation, auto panels and trim, and housings for electronic devices.

Nonelectrolyte \ˌnän-ə-ˈlek-trə-ˌlīt\ *n* (1891) A solute which does not dissociate into ions in solution. (Goldberg DE (2003) Fundamentals of Chemistry, McGraw-Hill Science/Engineering/Math, New York)

Nonflammable \-ˈfla-mə-bəl\ *adj* (1915) If combustible, burning without flame. Practically, whether or not a plastic material or part is "flammable" is a matter of its performance in a test – of which there are many – of ▶ Flammability. Note that the word "inflammable" has been deprecated by fire-safety authorities because of the ambiguity of the prefix "in-," and has long been superseded by "flammable" and "nonflammable." (*Tests for Comparative Flammability of Liquids*, UI 340, Laboratories Incorporated Underwriters, New York, 1997)

Nongrain-Rising Stain One of many liquid wood stains, based on alcohol or other solvent; almost totally free of water. Abbreviation is NGR.

Nonionic \-(ˌ)ī-ˈä-nik\ *adj* (1929) Pertaining to the material, atom, radical, or molecule that is incapable of being electrically charged. See also ▶ Ionic.

Nonionic Surfactants Those which contain hydrophilic groups which do not ionize appreciably in aqueous solutions. (Gooch, J. W., *Emulsification and Polymerization of Alkyd Resins*, Kluwer Academic/Plenum Publishers, New York, 2002; *McCutcheon's Emulsifiers and Detergents (2000): North American Edition*, Vol. 1, McCutcheon Division of McCutcheon Publishing Co., 2000; Ash, M, and Ash, I, *Industrial Surfactants*, Synapse Information Resourses, 2000)

Nonisothermal See ▶ Isothermal.

Nonmetal \-ˈme-tᵊl\ *n* (ca. 1864) An element with generally low electrical and thermal conductivities, dull luster, and a high ionization energy, electron affinity and electronegativity.

Nonmetameric Match \-ˈme-tə-ˌmer-ik-\ A pair of colors which appear to be identical to all observers under all conditions of illumination and viewing; an unconditional match. Popularly, the term is frequently used to apply to a pair of colors which has identical spectrophotomeric reflectance or transmittance curves, as measured in a prescribed geometric configuration, thus ignoring the effects of any differences in geometric characteristics between the samples in the pair. In practice, the term may be used to describe a pair of samples which does not appear to change in comparison to one another when viewed under several different light sources. See ▶ Metamerism. Also known as a *Nonconditional Match*. (McDonald, Roderick (ed.) Colour Physics for Industry, 2nd (1997) Society of Dyers and Colourists, West Yorkshire, England)

NonNewtonian Designating liquids whose viscosities are dependent on the rate of shear (as well as temperature and pressure), or the glow of such a liquid. See ▶ Newtonian Liquid, Pseudoplastic Fluid, and Viscosity.

NonNewtonian Liquid Any liquid that exhibits a viscosity which varies with changing shear stress or shear rate; thus, any liquid which does not satisfy the requirements for a Newtonian liquid, i.e., that displays plastic, pseudoplastic, or dilatant flow characteristics. Most paints are nonnewtonian liquids.

Nonnewtonian Oil or Fluid A fluid that exhibits a viscosity which varies with changing shear stress or shear rate.

Nonpenetrating Stain See ▶ Oil Stain.

Nonpolar \ˌnän-ˈpō-lər\ *adj* (1892) Having no concentrations of electrical charge on a molecular scale, thus incapable of significant dielectric loss. Examples of nonpolar resins are polystyrene and polyethylene.

Nonpolar Covalent Bond A covalent bond in which the bonding electron pair is shared equally between atoms of identical electronegativity.

Nonpolar Molecule A molecule in which the centers of positive and negative charge coincide.

Nonpolar Solvents The aromatic and petroleum hydrocarbon groups characterized by low dielectric constants are referred to as nonpolar solvents.

Nonrigid Plastic (as given by ASTM D 883) For the purpose of general classification, a plastic that has a modulus of elasticity either in flexure or tension of not over 70 MPa (10,000 psi) at 23°C and 50% relative humidity when tested in accordance with ASTM D 638, D 747, D 790, or D 882. See also ▶ Rigid Plastic.

Nonscratch Inks Inks which have high abrasion and mar-resistance when dry.

Nonskid Paint See ▶ Anti-Slip Paint.

Nonspecular Reflectance Reflectance other than the "mirror" reflectance that occurs at the angle equal and opposite to the incident angle; diffused reflectance. See ▶ Specular Reflectance Excluded and ▶ Fresnel Reflection.

Nontoxic Materials The toxicity status of resins and additives used for food contact and packaging changes frequently, and in many cases percentages and conditions of end use are stipulated. Therefore, we do not attempt to list such materials here. The current statuses of resins, plasticizers, stabilizers, and other additives with respect to their permissible use in contact with food ("indirect additives") are spelled out in detail in the Code of Federal Regulations (CFR), Title 21 (Food and Drugs), Parts 173 through 184. The Code is reprinted annually in book form and updated weekly by the Federal Register, where one should check for the latest word on any particular substance.

Nonvolatile Matter Ingredients of a coating composition which, after drying, are left behind on the material to which it has been applied, and which constitute the dry film. The term also applies to coatings components such as varnishes, resins, solvents, thinners, and diluents, driers and additives, etc. Abbreviation is NVM. *Known also as Solids and Total Solids.*

Nonwoven Fabric An assembly of textile fibers held together by mechanical interlocking in a random web or mat, by fusing of the fibers (in the case of thermoplastic fibers), or by bonding with a cementing medium such as starch, glue, casein, rubber, latex, or one of the cellulose derivatives or synthetic resins. Initially, the fibers may be oriented in one direction or may be deposited in a random manner. This web or sheet of fibers is bonded together by one of the methods described above. Normally, crimped fibers that range in length from 0.75 to 4.5 inches are used. Nonwoven fabrics are used for expendable items such as hospitable sheets, napkins, diapers, wiping cloths, as the base material for coated fabrics, and in a variety of other applications. They can also be used for semi-disposable items and for permanent items such as interlinings. (Tortora PG, Merkel RS (2001) Fairchild's Dictionary of Textiles, 7th ed., Fairchield Publications, New York)

Nonwoven Mat A mat of glass fibers in random arrangement, lightly bonded so as to be able to be handled and cut, used in making reinforced-plastics structures.

Nonwoven Scrim An open-mesh glass fabric in which two or more layers of parallel yarns are bonded to each other by chemical or mechanical means, the yarns in adjacent layers lying at an angle to each other.

Nor- A prefix for organic compounds indicating the parent compound from which the subject compound may be derived, usually by removal of one or more carbon atoms and their attaché hydrogens. Example: *norcamphor* is camphor from which three –CH$_3$ groups have been removed.

Norbornene-Spiro-Orthocarbonate A compound that, when added in small concentrations to matrix resin of carbon composites, apparently strengthens the fiber-matrix interfacial bond, as evidenced by improved strength properties and lower water absorption of the composites.

Norelac Resins Particular type of nylon resin in which the polymeric acids are those derived from drying oil fatty acids. These products differ from the true nylon resins in that they are much more soluble in solvents, and can be used as constituents of varnishes which possess air-drying properties.

Normal Boiling Point The boiling point of a substance at 1 atm pressure.

Normal Color Vision Vision of a normal observer who exhibits no symptoms of anomalous or defective color response.

Normal Covalent Bond A covalent bond in which one of the shared electrons appears to have been contributed by the first bonded atom, the other by the second. (Whitten KW, Davis RE, Davis E, Peck LM Stanley GG (2003) General Chemistry, Brookes/Cole, New York)

Normal Freezing Point The freezing point of a substance at 1 atm pressure.

Normal Hydrocarbon A hydrocarbon whose skeletal carbon chain is unbranched.

Normality (N) *n* [L *normalis*, fr. *norma*] (ca. 1696) A concentration unit: the number of equivalents of solute per liter of solution.

Normal Lead Silico Chromate Normal lead chromate (medium shade) formed as a strongly adherent coating on a core of silica at a 50/50 weight ratio. Its major use is in traffic paints in which its lower density and lower cost are advantageous compared to pigmentations of pure lead chromate and extender. Density, 3.6 g/cm^3 (29.2 lb/gal); O.A., 14; particle size, 5 μm. Syn: ▶ Coated. (Kirk-Othmer Encyclopedia of Chemical Technology: Pigments-Powders (1996) John Wiley and Sons, New York; Gooch JW (1993) Lead Based Paint Handbook, Plenum Press, New York)

Normal Salt *n* An ionic compound containing neither replaceable hydrogen nor hydroxyl groups.

Normal Solution *n* (1-N solution) A solution containing a mass of the dissolved substance per liter equal to one mole divided by the hydrogen equivalence of the substance, i.e., one gram-equivalent per liter.

Normal Stress *n* (1) A stress directed at right angles to the area upon which it acts. (2) In a flowing viscoelastic liquid, tensile/compressive stresses at any point in the fluid in the principal coordinate directions, one of which will be the main direction of flow. Rheologists are usually concerned with *differences* between normal stresses acting in the flow direction and directions perpendicular to the flow.

North Light *n* (14c) Light from the north sky, generally through a window so located as to allow only light from the north to enter the room. Because such light is the more constant throughout the day than light from other directions, it was and is preferred by artists, colorists, etc. for viewing color. Artificial north light has been simulated with modern lamps and filters for the same purpose. See ▶ Daylight.

North Standard See N. S.

Noryl® *n* Poly(phenylene oxide) Trade name for a family of blends of polyphenylene oxide (PPO) with much less costly styrenic polymers. These blends have the processability, low water absorption, and good dielectric properties associated with polystyrene, while the PPO contributes heat resistance. Glass-reinforced grades are available. Manufactured by General Electric, U.S.

Noryl Resins *n* Blends of poly-(2,6-dimethyl-1,4-phenyl oxide) with polystyrene or with high impact polystyrene.

Notch Effect *n* The effect of the presence of specimen notch or its geometry on the outcome of a test such as an impact strength test of plastics. Notching results in local stresses and accelerates failure in both static and cycling testing (mechanical, ozone cracking, etc.).

Notch Sensitivity *n* The extent to which a material's tendency to fracture under load, particularly an impact load, is increased by the presence of a surface in homogeneity such as a notch or sharp inside corner, a sudden change in section thickness, a crack, or a scratch. Low notch sensitivity is usually associated with ductility, while brittle materials exhibit higher notch sensitivity. Most engineers and physical testers consider the notched ▶ Izod and ▶ Charpy Impact Tests to be as much measures of notch sensitivity and they are of pure impact strength.

Notched Izod See Noticed Izod Impact Energy.

Notched Izod Impact See ▶ Notched Izod Impact Energy.

Notched Izod Impact Energy The energy required to break a notched specimen equal to the difference between the energy in the striking member of the Izod-type impact apparatus at the instant of impact and the energy remaining after complete fracture of the specimen. Note: Energy depends on geometry (e.g., width, depth, shape) of the notch, on the cross-sectional area of the specimen and on the place of impact (on the side of the notch or on the opposite side). In some tests notch is made on both sides of the specimen. Also called notched Izod impact strength, notched Izod impact, notched Izod.

Notched Izod Impact Strength See ▶ Notched Izod Impact Energy.

Novacite See ▶ Silica, Microcrystalline.

Novaculite \nō-ˈva-kyə-ˌlīt\ *n* [L *novacula* razor] (1796) A very fine-grained type of quartz found in Arkansas, Georgia, Massachusetts, North Carolina, Oklahoma, and Tennessee. A variety known as *altered novaculite*, typically about 99.5% quartz, is a solid crystalline substance with the basic hardness of quartz but more easily reduced to very fine particles. At their surfaces, these particles have high concentrations of ruptured Si-O bonds that readily combine with water to create surface hydroxides called *silanols*. Such novaculites are useful as semi-reinforcing filers in silicone rubber, epoxy resins, urethane foams, and PVC.

Novelty Siding See ▶ Drop Siding.

Novelty Yarn A yarn produced for a special effect. Novelty yarns are usually uneven in size, varied in color, or modified in appearance by the presence of irregularities deliberately produced during their formation. In singles yarns, the irregularities may be caused by inclusion of knots, loops, curls, slubs, and the like. In plied yarns, the irregularities may be affected by variable delivery of one or more yarn components or by twisting together dissimilar singles yarns. Nub and slub are examples of novelty yarns.

Novodur ABS polymer. Manufactured by Bayer, Germany.

Novolac (novolak) According to ASTM D 883, a novolac is a phenolic-aldehyde resin which, unless a source of methylene groups is added, remains permanently thermoplastic. For a preferred definition, see ▶ Phenolic Novolac. However, the term is also used in connection with epoxies. (Lenz RW (1967) Organic Chemistry of Synthetic High Polymers, Interscience Publishers Inc., New York) See ▶ Epoxy-Novolac Resin.

Novolak Phenol/formaldehyde condensate. Phenolic-aldehydic resin which, unless a source of methylene groups is added, remains permanently thermoplastic. Manufactured by Dynamit Nobel, Germany.

Novoloid Fiber A phenolic fiber made by crosslinking a melt-spun novolac resin with formaldehyde. Novoloid fibers have good flame resistance, can serve at temperatures to about 220°C, and are used as reinforcement in a range of thermosetting matrices.

Nozzle \ˈnä-zəl\ n [dimin. of *nose*] (1683) In injection or transfer molding, the orifice-containing fitment at the delivery end of the injection cylinder or transfer chamber that contacts the mold's sprue bushing and conducts the softened resin into the mold. The nozzle is shaped to form a seal under pressure against the sprue bushing. Its orifice is tapered and sometimes contains a check valve to prevent flow reversal, or an on/off valve to interrupt the flow at any desired point in the molding cycle.

Nozzle Manifold A series of injection nozzles mounted on a common feed tube, each nozzle positioned so as to feed a single cavity in the mold. Such manifolds have been used to eliminate runners in molds when molding articles such as cups and when it is desired to gate the cavities at the centers of the cup bottoms.

Nozzle, Mold-Gating In injection molding, a nozzle whose tip is part of the mold cavity, thus feeding material directly into the cavity, eliminating the sprue and runner.

NPCA Abbreviation for National Paint And Coatings Association.

NPIRI Abbreviation for National Printing Ink Research Institute.

NR Abbreviation for ▶ Natural Rubber.

N_Re Symbol for Reynolds Number.

N.S. Abbreviation for North Standard, used as a dimension in the measurement of the fineness of dispersion (fineness of grind is deprecated). See ▶ Fineness of Dispersion.

Nsa-Sana Oil See Essang Oil.

n-Type Semiconductor A semiconductor in which the charge carries are weakly bound electrons.

Nubs \ˈnəb(s)\ n [alter. of E dialect *knub*, prob. fr. LG *knubbe*] (1727) Size grading of natural resins, about the size of the end of a finger.

Nub Yarn A novelty yarn containing slubs, beads, or lumps introduced intentionally.

Nuclear Atom The atom of each element consists of a small dense nucleus which includes most of the mass of the atom. The nucleus is made up of roughly equal numbers of neutrons and protons. The positive charges of the protons, enables the nucleus to surround itself with a set of negatively charged electrons which move around the nucleus in complicated orbits with well defined energies. The outermost electrons which are least tightly bound to the nucleus play the dominant part in determining the physical and chemical properties of the atom. There are as many electrons in orbits as there are protons in the nucleus.

Nuclear Fusion See ▶ Fusion.

Nuclear Isomers Isotopes of elements having the same mass number and atomic number but differing in radioactive properties such as half-life period.

Nuclear Magnetic Resonance n (1942) (NMR) The spinning motion of atomic nuclei imparted by an alternating, high-frequency magnetic field. The phenomenon is the basis for an analytical method enabling identification and quantification of isotopes. NMR spectroscopy has been used to study the distribution of hydrogen in substituent groups, and the molecular structure of polymers, such as tacticity and occurrence of infrequent branches. Proton and ^{13}C - NMR technologies have been developed for different purposes. An example of an ^1H - NMR spectrum of toluene is shown. (Bovey FAA Mirau PA (1996) NMR of Polymers, Elsevier Science and Technololgy Books, New York)

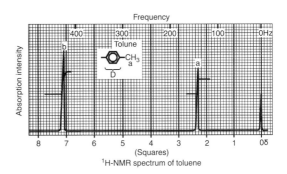

^{1}H-NMR spectrum of toluene

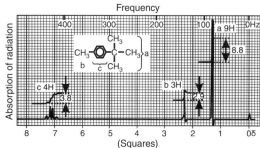

1H-NMR spectrum of p-tert-butyltoluene, proton counting (Source: Morrison and Boyd, 1973)

Nucleating Agent A chemical substance which, when incorporated in crystal-forming plastics, provide active centers (nuclei) for the growth of crystals as the melt is cooled through the melting range. In polypropylene, for example, a higher degree of crystallinity and more uniform crystalline structure is obtained by adding a nucleating agent such as adipic or benzoic acid or certain of their metal salts. Colloidal silicas are used as nucleating agents in nylon, seeding the polymer to produce more uniform growth of spherulites.

Nucleation \ˈnu-klē-ˌāt, ˈnyü-\ v [LL *nucleates*, pp of *nucleare* to become stony, fr. L *nucleus*] (ca, 1864) The formation of short-range ordered polymer aggregates in a melt or solution which acts as growth centers for crystallization.

Nucleon \ˈnu-klē-ˌän\ n [ISV] (1923) A particle in the nucleus of an atom; a proton or a neutron.

Nucleophile \ˈnu-klē-ə-ˌfīl\ n (1943) An electron-rich atom or group of atoms which seeks to share its electrons with a relatively positive atom.

Nucleic Acid \nü-ˈklē-ik-\ n [fr. their occurrence in cell nuclei] (1892) A family of macromolecules, of molecular masses ranging upward from 25,000, found in the chromosomes, nucleoli, mitochondria, and cytoplasm of all cells, and in viruses; in complexes with proteins, they are called nucleoproteins. On hydrolysis they yield purines, pyrimidines, phosphoric acid, and a pentose, either D-ribose or D-deoxyribose; from the last, the nucleic acid derive their more specific names, ribronucleic acid and deoxyribonucleic acid. Nuclear acids are liner (i.e., unbranched) chains of nucleotides in which the 5'-phosphoric group of each one is esterified with the 3'-hydroxyl of the adjoining nucleotide. (Black JG (2002) Microbiology, 5th ed., John Wiley and Sons, Inc., New York)

Nucleus \ˈnü-klē-əs\ n. pl **nulcei** \-klē-ˌī\ [NL fr. L, kernal, dim. of *nuc-*, *nux* nut] The dense central core of the atom, in which most of the mass and all of the positive charge is concentrated. The charge on the nucleus, an integral multiple of Z of the electronic charge, is the essential factor which distinguishes one element from another. Z is called the atomic number and gives the number of protons in the nucleus, which includes a roughly equal number of neutrons. The mass number A gives the total number of neutrons plus protons.

Nuclide \ˈnü-ˌklīd\ n [*nucleus* + Gk *eidos* form, species] (1947) A species of atom distinguished by the constitution of its nucleus. The nuclear constitution is specified by the number of protons, Z; number of neutrons, N; and energy content. (Or, by the atomic number, Z; mass number A (= N + Z) and atomic mass.)

Number-Average Molecular Weight n (M_n) The sum of the molecular weights of all the individual molecules in a given polymer sample divided by the total number of molecules. The defining equation is

$$M_n = \sum n_i M_i / \sum n_i = \sum w_i / \sum (w_i/M_i)$$

where,
n_i = number of molecules with molecular weight M_i
w_i = weight fraction of material having molecular weight M_i; $\Sigma\, n$ = total number of molecules and,
Weight-average M_w

$$M_w = \sum w_i M_i / \sum w_i$$

Z-Average M_z

$$M_z = \sum w_i M_i^2 / \sum w_i M_i$$

Methods of determining molecular weight include osmotic pressure, light scattering, gel permeation chromatography, dilute solution viscosity, vapor pressure, freezing temperatures, and others. (Odian GC (2004) Principles of Polymerization, John Wiley and Sons, Inc., New York; Elias and Hans-Georg (2003) An Introduction to Plastics, John Wiley and Sons, New York; Slade PE (2001) Polymer Molecular Weights, Vol. 4, Marcel Dekker, New York) See also ▶ Molecular Weight, ▶ Molecular-Weight Distribution and ▶ Weight-Average Molecular Weight.

Numerical Aperature n The numerical aperture of a lens system (objective or condenser) is the sine of one-half the angular aperture times the refractive index of the medium (1.0 for air, 1.515 for Cargille immersion oil, etc.) between objective and specimen. The numerical aperture is a measure of the light gathering capacity of the lens system and determines its resolving power and depth of field.

Numerical Prefixes Commonly Used in Forming Chemical Names

Numeral	Prefix	Numeral	Prefix
½	hemi-	21	heneicosa-
1	mono-	22	docosa-
1 ½	sesqui-	23	tricosa-
2	di-, bi-	24	tetracosa-
2 ½	hemipenta-	25	pentacosa-
3	tri-	26	hexacosa-
4	tetra-	27	heptacosa-
5	penta-	28	octacosa-
6	hexa-	29	nonacosa-
7	hepta-	30	triconta-
8	octa-	40	tetraconta-
9	ennea, nona-	50	pentaconta-
10	deca-	60	hexaconta-
11	hendeca-, undeca-	70	heptaconta-
12	dodeca-	80	octaconta-
13	trideca-	90	nonaconta-
14	tetradeca-	100	hecta-
15	pentadeca-	101	henhecta-
16	hexadeca-	102	dohecta-
17	heptadeca-	110	decahecta-
18	octadeca-	120	eicosahecta-
19	nonadeca-	200	dicta-
20	eicosa-		

Nun's Veiling *n* A soft, lightweight, plain-weave fabric that usually comes in black and white, nun's veiling is a rather flimsy, open fabric but always of high quality. It may be made from fine woolen yarn or yarns spun from manufactured fibers such as nylon, acrylic, or polyester.

Nurnbert See ▶ Mineral Violet.

Nutshell Flour *n* Ground peanut or walnut shells, dried by heating or solvent extraction, have been used as low-cost fillers in olyethylene. Physical properties are comparable to those of PE filled with wood flour.

NV *n* Abbreviation for ▶ Nonvolatile Matter.

NVM *n* Abbreviation for ▶ Nonvolatile Matter.

Nylon \ˈnī-ˌlän\ *n* [coined word] (1938) (polyamide) Generic name for all long-chain polyamides that have recurring amide groups (–CONH–) as an integral part of the main polymer chain. Nylons are synthesized from intermediates such as dicarboxylic acids, diamins, amino acids and lactams, and are identified by dual numbers denoting the number of carbon atoms in the polymer chain derived from specific constituents, that of the diamine being given first. Use of a single numeral signifies that the monomer was a lactam, as in *nylon 6*. The second number, if used, denotes the number of carbons derived from a diacid. For example, in nylon 6/6 the two numbers are the numbers of carbon atoms in hexamethylene diamine and adipic acid, respectively. However, in the literature these numbers may otherwise appear as 66, 6.6, 6,6, or 6–6, and sometimes precede, rather than follow, the word "nylon." The convention used here – numbers following, divided by slash mark – is almost universally used today. Nylon molding powders can be converted to useful products by injection molding, extrusion, and blow molding. Nylons are crystalline polymers. In nylon 6/6, a wide range of crystallinity is possible, depending on how quickly the melt is chilled and the presence or absence of nucleating agents. Injection-molded items normally have low-crystallinity skins with higher-crystallinity interiors. Very thin sections may have as little as 10% crystalline material. Finely powdered forms of nylon are available for fluidized-bed coating, rotational molding, and other powder processes. A casting process employs molten caprolactam monomer to which catalysts are added, polymerization occurring in the mold after pouring without additional heat or pressure. Large solid castings and rotationally cast parts have been made by this method. In 1992 U.S. production of nylon plastics was about 270 Gg (300,000 tons.) Polyamides are not commercially available in so many filled and reinforced varieties that *Modern Plastics Encyclopedia's* "Resins and Compounds" table for 1993 contained eight pages of polyamide listings, far more than for any other plastics family (*Modern Plastics Encyclopedia*, McGraw-Hill/Modern Plastics, New York, 1986; – 1990, – 1992, and – 1993 editions). Various types of nylons are described in the immediately following listings. Manufactured by DuPont, U.S.

Nylon *n* "Zytel ST" A DuPont rubber-toughened nylon, identified by the initials standing for *super-tough*, claimed to be the most rugged engineering resin then (1977) available. It is superior to polycarbonate in impact strength, with notched Izod = 9 J/cm (17 ft-lb$_f$/in.).

Nylon 3 *n* (polypropiolactam) A type of nylon that has been prepared and explored experimentally, but has not become commercial.

Nylon 4 *n* (polypyrrolidinone) A polymer of 2-pyrrolidinone. Early attempts to commercialize nylon 4 failed because much of the material was of low molecular weight and decomposed at a relatively low temperature, making it unusable for melt spinning. Improved catalyst systems resulted in a polymer with a molecular weight (M_n) of about 400,000 and a melting point of 256°C. Today's nylon 4 has better heat stability than other nylons. Its moisture absorption is higher than that of nylon 6 and 6/6. It can be molded and extruded. Artificial leathers have been made from slurries of nylon-4 fibers.

Nylon 4/6 *n* (polytetramethylenediamineadipamide) A condensation polymer of diaminobutane and adipic acid that melts higher than ▶ Nylon 6/6 so it can be used at somewhat higher temperatures than the latter.

Nylon 5 *n* A blend of aliphatic-, cycloaliphatic-, and aromatic-based polyamides. The material is clear in thick cross sections, has low water absorption, good dimensional stability and solvent resistance, and can be processed economically in injection molding or extrusion.

Nylon 6 *n* (polycaprolactam) A type of nylon made by the polycondensation of ▶ Caprolactam, the second-most widely used polyamide in the U.S. Melting at about 228°C, it is used for fibers, including tire cord, and as a thermoplastic molding powder. Nylon 6 is as structurally sound as type 6/6 at room temperature, but it picks up moisture more rapidly and loses strength more rapidly as humidity and temperature increase. It is available in many grades, including glass-fiber-filled, and in a broad range of molecular weights, suitable for injection molding, extrusion, blow molding, and rotational molding. Parts can be machined, welded, and adhesive-bonded.

Nylon 6/6 *n* (polyhexamthyleneadipamide) A type of nylon made by condensing hexamethylenediamine with adipic acid, first prepared by W. H. Carothers of DuPont in 1936. It is the leading commercial polyamide, being used extensively for staple fibers and monofilaments, and is the most widely used type in other applications. The bulk polymer is a tough, white, translucent, crystalline material that melts rather sharply near 269°C. Nylon 6/6 is the strongest of the nylons over the widest range of temperature and humidity, but absorbs up to 2% water from air in the normal range of climatic humidity. Water acts as a plasticizer, reducing moduli but improving impact resistance and flex life in humid environments, with opposite effects in arid ones. During molding and extrusion, moisture content must be less than 0.1%.

Nylon 6/6 Salt *n* (hexamethylenediammonium adipate) An intermediate in the manufacture of nylon 6/6, formed from one molecule each of adipic acid and hexamethylene diamine.

Nylon 6/T *n* Terephthalic acid + hexamethylene diamine. (polyhexamethyleneterephthalamide) A major member of an aliphatic-aromatic family of polyamides, none of which have gained commercial importance because they are difficult to prepare and to process. Manufactured by Celanese, U.S.

Nylon 6 *n*, **Thermal Degradation of** When exposed to elevated temperatures, unmodified nylons undergo molecular weight degradation, which results in loss of mechanical properties. The degradation process is highly time-temperature dependent.

Nylon 6/10 *n* (polyhexamethylenesebacamide) The product of condensation of Hexamethylenediamine with sebacic acid, used for brush bristles and monofilaments. It has lower water absorption and lower melting point than nylon 6 or 6/6. When a small amount of an alkyl-substituted hexamethylenediamine is added to the condensation mixture, a more elastic polymer known as *elastic nylon* is obtained.

Nylon 6/12 *n* (polyhexamethylenedodecanamide)
A nylon introduced by DuPont in 1970, made from Hexamethylenediamine and a 12-carbon dibasic acid. Nylon 6/12 is characterized by retention of physical and electrical properties over a wide humidity range, good dimensional stability, and low moisture absorption.

Nylon 7 *n* (polyenantholactam, polyheptanamide)
A type of nylon known commercially in Russia as "Enant," but not yet commercial in the U.S. Its properties are similar to those of NYLON 6.

Nylon 8 *n* (polycapryllactam, polyoctanamide) A nylon made by condensation polymerization from capryllactam. Its low melting temperature (200°C) and high cost of starting materials have limited the utilization of this polymer. It should not be confused with a type of nylon long marketed as "Type 8," which is actually a chemically modified nylon 6/6.

Nylon 9 *n* (polypelargonamide, polynonanamide)
A type of nylon made by melt condensation of aminopelargonic acid (9-aminononanoic acid). Nylon 9 has tensile yield and flexural strengths approaching those of nylon 6, having low water absorption like those of nylons 11 and 12. It is, however, in limited use.

Nylon 10 *n* An experimental polymer prepared, with difficulty, from aminodecanoic acid.

Nylon 11 *n* (polyundecanamide) A type of nylon produced by polycondensation of the monomer 11-aminoundecanoic acid, a derivative of castor oil. It is available in the form of fine powders for rotational molding and other powder processes; and in pellet form for extrusion or molding. Like nylon 12, nylon 11 has properties intermediate between those of nylon 6 and polyethylene: good impact strength, hardness, and abrasion resistance, but other mechanical properties are lower than those of most other nylons. However, due to its exceptionally low water absorption, the dimensional stability of nylon 11 is high. A modified nylon 11 trade named Rilsan N is flexible, transparent, and self-extinguishing.

Nylon 12 *n* (polylauryllactam, polydodecanamide)
A nylon made by the polymerization of lauric lactam (dodecanoic lactam) or cyclododecalactam, with 11 methylene groups between the linking –CONH– groups in the polymer chain. Its mechanical properties are intermediate between those of conventional nylons and polyethylene, and it is the lowest in water absorption (1.5%) and density (1.01 g/cm^3) of all the nylons.

Nylon Fiber *n* Generic name for a manufactured fiber in which the fiber-forming substance is any long-chain synthetic polyamide having recurring amide groups (–CONH–) as an integral part of the polymer chain, with less than 85% of the amide groups bonded directly to aromatic rings. (If more than 85%, it's considered to be a ▶ Polyimide. Nylon was the first fiber of major commercial importance to be made of wholly synthetic material. Carothers' pioneering research in 1929 culminated in DuPont's introduction of nylon hosiery in 1940. CHARACTERISTICS: Although the properties of the nylons described above vary in some respects, they all exhibit excellent strength, flexibility, toughness, elasticity, abrasion resistance, washability, ease of drying, and resistance to attack by insects and microorganisms. END USES: Nylon is used for apparel such as stockings, lingerie, dresses, bathing suits, foundation garments, and wash-and-wear linings; for floor coverings; for tire cord and industrial fabrics; and in-home furnishings such as upholstery fabrics.

Nylon Monofilament *n* Single strands much larger in diameter than those of staple fiber, used for fishing leaders and lines, brush bristles, racket strings, surgical sutures, and ropes.

Nylon MXD/6 *n* (poly-*m*-xyleneadipamide) A type of nylon with lower elongation at break than nylon 6 or 6/6, but capable of attaining their properties by reinforcement with glass fibers. The resin has low melt viscosity, good flexural strength and modulus, and resists alkalies and hydrolytic degradation.

Nylon, Nucleated *n* A nylon polymerized in the presence of a nucleating agent, e.g., about 0.1% of finely dispersed silica, which promotes the growth of spherulites and controls their number, type, and size. Nucleated nylons have higher tensile strength, flexural modulus, abrasion, resistance, and hardness, but lower impact strength and elongation than their unnucleated counterparts.

Nylon Resins *n* Polyamide resins made from the interaction of diamines and dicarboxylic acids. Hexamethylene diamine and adipic acid are typical reactants. These resins are composed principally of a long-chain synthetic polymeric amide which has recurring amide groups as an integral part of the main polymer chain.

Nylons *n* Nylons are one of the most common polymers used as a fiber. Nylons are also called polyamides, because of the characteristic amide groups in the backbone chain. Nylon is a DuPont trade mark.

Nylon Salt *n* Any of the intermediates in nylon synthesis formed by the combination of one molecule of diamine and one of diacid, such as ▶ Nylon 6/6 Salt.

Nylon, Transparent *n* Any of several nylon polymers based on aromatic ring units. The first such nylon, introduced in the early 1970s by Dynamit Nobel under the trade name Trogamid T, was made by polycondensation of terephthalic acid with 2,4-

trimethylhexamethylenediamine. This crystal-clear, amorphous polyamide has excellent resistance to stress cracking and a glass-transition temperature comparable to those of polycarbonates.

Nylsuisse *n* Adipic acid + hexamethylene diamine, manufactured by Viscose-Suisse, Switzerland.

Nytril Fiber *n* A manufactured fiber containing at least 85% by weight of a long chain polymer of vinylidene dinitrile $[-CH_2-C(CN)_2-]$ and having the vinylidene dinitrile group in no less than every other unit in the polymer chain (FTC definition). Nytril fibers have a low softening point so they are most commonly used in articles that do not require pressing such as sweaters and pile fabrics. They are also blended with wool to improve shrink resistance and shape retention.

O

o- \ō\ *n* Abbreviation for prefix *ORTHO-* and ignored in alphabetizing compound names.

O Chemical symbol for the element oxygen.

O.A Abbreviation for ▶ Oil Absorption.

Oak Varnish Short oil type of varnish for interiors use, normally based on natural copals.

Oatmeal \ōt-ˌmēl\ *n* A heavy, soft linen fabric with a pebbled or crepe effect.

Object Color *n* Color of light reflected by an object as normally observed, as contrasted to other types of observed color. (Saleh BEA, Teich MC (1991) Fundamentals of photonics. Wiley, New York) See ▶ Object Mode of Appearance.

Object Mode of Appearance *n* Mode of perceiving the appearance of tangible objects as located at a particular place; object mode includes two of the five recognized modes of appearance, that of surfaces and of bulk volume. See ▶ Modes of Appearance.

Objet d'Art \ˌób-ˌzhä-ˈdär\ *n* [F, literally, art object] (1865) Any object of artistic merit. A curio. (Gair A (1996) Artist's manual. Chronicle Books LLC, San Francisco, CA)

Obliterating Power See ▶ Opacity.

OBP Abbreviation for ▶ Octyl Benzyl Phthalate.

Observation Angle *n* (1) The angle between the incident ray measured relative to the perpendicular or normal and the viewing angle, also measured relative to the normal; therefore, the total subtended angle between incident and viewing. It is the same as "bend angle" in the literature relating to projection screens, and as "divergence angle" in some federal and military specifications. (2) It is synonymous with viewing angle.

Observer \əb-ˈzər-vər\ *n* (ca. 1550) The human viewer who receives a stimulus and experiences a sensation from it. In vision, the stimulus is a visual one and the sensation is an appearance.

Observer, Standard *n* The spectral response characteristics of the average observer defined by the CIE. Two such sets of data are defined, the 1931 data for the 2° visual field (distance viewing) and the 1964 data for the annular 10° visual field (approximately arm's length viewing). By custom, the assumption is made that if the observer is not specified the tristimulus data has been calculated for the 1931, or 2° field observer. The use of the 1964 data should be specified.

OBSH *n* Abbreviation for 4,4'-Oxybis(Benzenesulfonyl-Hydr-Azide).

Obtuse Bisectrix *n* (**Bx$_o$**) The angle supplementary to the acute optic axial angle of a biaxial crystal is obtuse and is bisected by the obtuse bisectrix, which may be either α or γ. (Handbook of chemistry and physics, 52nd edn. Weast RC (ed). The Chemical Rubber, Boca Raton, FL)

OCCA *n* Abbreviation for Oil and Colour Chemists' Association (British).

Occupational Safety and Health Act *n* (**OSHA**) According to this Act of 1970, inspectors may at any time or when requested by an employee, examine any company for violations of occupational safety and health standards set by the act.

Ocher, Ochre (Yellow, Golden, Red) \ˈō-kər\ *n* [ME *oker* fr. MF *ocre*, fr. L *ochra*, fr. Gk *ōchra*, fr. feminine of *ōchros* yellow] (14c) There are two types of ocher: (1) synthetic and (2) natural (a naturally occurring yellow-brown hydrated iron oxide). Synthetic ocher possesses little advantage over the natural earth pigments, the difference being the method of manufacture. Natural ocher is found in France, Italy, Spain, Africa, Great Britain, and the United States. See ▶ Iron Oxide.

OCR Inks *n* See ▶ Optical Character Recognition Inks.

Octabis(2-Hydroxypropyl) Sucrose A viscous, straw-colored liquid used as a crosslinking agent for urethane foams and as a plasticizer for cellulosics.

Octahedral Hole \ˌäk-tə-ˈhē-drəl-\ *n* A space in a close-packed structure which is bounded by six spheres located at the corners of an octahedron.

Octahedrite A term sometimes used for the Anatase form of ▶ Titanium Dioxide.

1-Octene *n* $C_6H_{13}CH=CH_2$. A comonomer, made from ethylene, and polymerized with ethylene to make linear, low-density polyethylene.

Octet \äk-ˈtet\ *n* (1879) *n* A filled shell of eight (8) electrons surrounding an element. (Morrison RT, Boyd RN (1992) Organic chemistry, 6th edn. Prentice Hall, Englewood Cliffs, NJ)

Octet Rule *n* A rule which states that a (valence-shell) ns^2np^6 configuration in an atom is an especially stable one.

Octoates *n* (1) Salts of octoic, or 2-ethyl hexoic acid. (2) Driers used in paints; made with the lead, cobalt, manganese, calcium, zinc or iron salts of 2-ethyl hexoic acid.

Octobromodiphenyl \oct-ˈbrō-(ˌ)mō-(ˌ)dī-ˈfe-nᵊl\ *n* [ISV] (octabromodi-phenyl) $(C_6H_2Br_4)_2$. A very dense involatile liquid useful as a fire-retardant additive.

Octyl- *nj* The general term for all saturated, 8-carbon aliphatic radicals having the formula, C_8H_{17}—, often used imprecisely for the actual radical 2-Ethylhexyl–.

Octyl Acetate *n* High-boiling solvent. Bp, 199°C; sp gr, 0.873/20°C; flp, 83°C (180°F); vp, 0.3 mmHg/20°C.

Octyl Alcohol *n* High-boiling solvent used for controlling flow and for coupling with hydrocarbon solvents to reduce high viscosities. Bp, 194°C; sp gr, 0.834/20°C; flp, 81°C (178°F); vp, 0.3 mmHg/20°C. It is also used as an etherifying agent for urea-formaldehyde resins to confer drying oil and aliphatic hydrocarbon compatibility.

Octyl Benzyl Phthalate *n* (OBP) A plasticizer for PVC, cellulosics, polystyrene, and polyvinyl butyral. It is similar to butyl benzyl phthalate but has lower volatility. It resists oil extraction.

Octyl Biphenyl Phosphate *n* A plasticizer for vinyl and other resins, with good permanence and low-temperature properties. It imparts flame resistance and is approved by FDA and use in food packaging.

Octyl Epoxy Tallate *n* A monomeric plasticizer and heat and light stabilizer for vinyls and cellulosics. It imparts good low-temperature flexibility, has low volatility, and is used primarily in coated fabrics, garden hose, film and sheeting, and slush-molded parts.

Octyl Isodecyl Phthalate See Ethylhexyl Isodecyl Phthalate.

n-Octyl Methacrylate A comonomer for acrylic resins.

n-Octyl-n-Decyl Adipate *n* (NODA, octyl decyl adipate, isooctyl isodecyl adipate) A plasticizer for cellulosics, synthetic rubbers, and vinyl resins. It imparts good low-temperature flexibility and resistance to extraction by water. NODA is also useful at low concentrations in polypropylene as a processing aid.

n-Octyl-n-Decyl Phthalate *n* (NODP, Ethylhexyl decyl phthalate, octyl decyl phthalate) One of an important family of phthalate-ester plasticizers derived from C_6 to C_{10} alcohols. These plasticizers may be used interchangeably in PVC compositions, to which they impart somewhat better drape, flexibility, and low-temperature resistance than does dioctyl phthalate. They are also compatible with vinyl chloride-acetate copolymers, cellulose nitrate, ethyl cellulose, cellulose acetate butyrate, polystyrene, acrylic and butadiene-acrylonitrile resins, neoprene and chlorinated rubber.

n-Octyl-n-Didecyl Trimellitate n $C_8H_{17}OOCC_6H_3(COOC_{10}H_{21})_2$. An ester of trimellitic acid (1,2,4-benzene tricarboxylic acid), used as a plasticizer for vinyl chloride polymers and copolymers. The trimellitate plasticizers are used especially for nonfogging applications, and in adhesives or laminates where low migration is important.

Octyl Phenol (Tertiary) Phenol used as an alternative to tertiary butyl phenol in the manufacture of pure oil-reactive phenolic resins.

p-Octylphenyl Salicylate n A white, crystalline powder, used as an ultraviolet absorber in polyolefins and cellulosics. It is reported to have increased the outdoor weathering life of polyethylene by 400%.

Octyl Phosphate See ▶ Trioctyl Phosphate.
OD Abbreviation for Outside Diameter (of an annulus or spherical shell).
Odor \ˈō-dər\ n [ME *odour*, fr. MF, fr. L *odor*; akin to L *olēre* to smell, Gk *ozein* to smell, *osmē* smell, odor] (13c) A reaction that is manifested by a physiological sensation with due to contact of their molecules with the human olfactory nervous system.
Odor Intensity Level n (Odor Threshold Number) A test to determine the intensity of an odorant, or the number of dilutions required for an odorant (gas, vapor or liquid) in order to become odorless or barely detectable as evaluated by a panel of humans sniffing samples. Also the character of the odorant sample may be evaluated (e.g., sweet, sour, ethereal, putrid). (ASTM D1292–86, Standard Test Methods for Odor in Water; and ASTM Publication DS 61, Atlas of Odor Character Profiles.)
Odorless Mineral Spirits See ▶ Odorless Solvent.
Odorless Paint n A paint such as a water-base latex paint or an oil- or alkyd-base paint which contains an odorless mineral spirit as a thinner; produces a minimum amount of odor during application.
Odorless Solvent n Solvents generally of the mineral spirits type, that are synthesized by the aklylation process and refined to remove odorous aromatics and sulfur compounds.
Oenanthal n Aldehyde decomposition product derived during the dehydration of castor oil. Also known as *Enanthaldehyde*.

Oersted \ˈər-stəd\ n [Hans Christian *Oerstead*] (1930) cgs emu of magnetic intensity exists at a point where a force of 1 dyne acts upon a unit magnetic pole at that point, i.e., the intensity of 1 cm from a unit magnetic pole.
Oeschle' n A method of measuring density of liquids named for its Swiss inventor. The Oeschle' scale is related to specific gravity(SG) by: °Oeschle' = (°SG−1) × 1,000. (See www.monashscientific.com, Oeschle')
Off-Clip See ▶ Scalloped Selvage.
Off Color n Unacceptable color difference. Nonmatching at different shade.
Offset \ˈȯf-ˌset\ n (ca. 1555) An indirect form of printing in which the ink is transferred from the printing plate to a rubber blanket and subsequently to the sheet. See ▶ Lithography.
Offset Adapter n In extrusion, a short, angled connector between extruder and die that orients the die axis on an axis different from, but sometimes parallel to that of the extruder axis.
Offset, Custom Decorating n A printing process in which the image to be printed is first applied to an intermediate carrier such as a rubber roll or plate then is transferred (in reverse) to the surface to be printed.
Offset Molding See ▶ Jet Molding.
Offset Paper n A variety of printing paper especially sized for offset lithography; it is noncuring and absorptive.
Offset Printing n A printing process in which the image to be printed is first applied to an intermediate carrier such as a rubber roll or plate, then is transferred (in reverse) to the surface to be printed.
Offsetting of Inks See ▶ Set Off of Ink.

Offset Yield Strength *n* The stress at which the strain exceeds by a specified amount (the offset) an extension of the initial proportional portion of the stress-strain curve. It is expressed in force per unit area, usually psi or MPa (ASTM D 638).

Off-Square *n* (1) A term to describe the difference between the percentage of warp crimp and the percentage of filling crimp. (2) A term referring to a fabric in which the number of ends and the number of picks per inch are not equal.

Off-White *n* (1927) Color which is obviously not white, but which is not sufficiently far removed from white to enable it to be called by a definite color name.

-OH *n* The hydroxyl radical, the characterizing group of inorganic basses, aliphatic alcohols, and phenols.

Ohm \ōm,\ *n* [Georg Simon *Ohn*] (1867) (Ω) The SI unit of electrical resistance, equal to 1 V divided by 1 A. The SI definition for one ohm is the resistance between two points of a conductor when a constant difference of potential of 1 V, applied between these two points, produce in this conductor a current of 1 A, this conductor not being the source of any electromotive force.

Ohm's Law \ōmz-\ *n* (1863) Current in terms of electromotive force *E* and resistance *R*.

$$I = \frac{E}{R}$$

The current is given in amperes when *E* is in volts and *R* in ohms. (Handbook of chemistry and physics, 52nd edn. Weast RC (ed). The Chemical Rubber, Boca Raton, FL)

OIDP *n* Abbreviation for ▶ Octyl Isodecyl Phthalate.

Oil \ói(ə)l\ *n* [ME *oile*, fr. OF, fr. L *oleum* olive oil, fr. Gk *elaion*, fr. *elaia* olive] (13c) Any of numerous mineral, vegetable and synthetic substances, and animal or vegetable fats that are generally unctuous, slippery, combustible, viscous, liquid or liquefiable at room temperatures, soluble in various organic solvents, but not in water. See ▶ Drying Oil, ▶ Nondrying Oil and ▶ Semidrying Oil.

Oil Absorption *n* The percentage increase in weight of a specimen after immersion in oil for a specified time. This property is important with respect to fillers used in plasticized thermoplastics, which can absorb plasticizer from a resin in a compound.

Oil Absorption (Value) *n* Quantity of oil required to wet completely a definite weight of pigment to form a stiff paste when mechanically mixed. The oil absorption number or value of a pigment is the number of milliliters or grams of oil, usually acid refined linseed oil, used to bind together 100 g of pigment under specified conditions of test. The unit used should be stated. The figure is not absolute, but depends on the operator as well as on the method of determination. NOTE – The term "oil absorption" is frequently used as a measure of plastic viscosity. As such it can lead to unpredictable results, since oil absorption as measured bears little relation to finished mill-ground pigment dispersion. Oil absorption has meaning of a qualitative sort, however, in relation to a preliminary mixer operation. See also ▶ Binder Demand.

Oil, Blown See ▶ Blown Oil.

Oil-Bound *n* Description of a water paint, the medium of which contains a proportion of drying oil in the binder.

Oil-Bound Distemper *n* A distemper which contains a drying oil.

Oil Cake *n* (1743) A cake or mass of linseed, cottonseed, etc., from which the oil has been expressed.

Oil Can *n* In the coil coating industry, metal irregularities of an otherwise flat looking metal sheet when viewed at near grazing angle.

Oilcanning *n* (1) A problem of distortion or buckling encountered in extrusion of rigid-vinyl sheet and shapes, usually attributed to uneven cooling of the extrudate. (2) A similar problem with dark-colored, rigid-vinyl house siding caused by solar heating and cured by adding the extrusion compound a heat-resistant resin that elevates the glass-transition and heat-deflection temperatures of the siding.

Oil Cloth *n* Any fabric treated with linseed-oil varnish to make it waterproof. It comes in plain colors and printed designs and is most commonly used for table covers or shelf covering. It has now been widely replaced by plastic coated fabrics.

Oil Color *n* (1539) Oil paint containing a high concentration of colored pigment, commonly used for tinting paint. Syn: ▶ Oil Paste.

Oil Content See ▶ Oil Length.

Oil, Debloomed *n* Mineral oils which show a decided blue or purple cast when examined in a test tube or when spread in a thin film over a dark colored surface, are said to possess a bloom. When an oil has been so treated as to remove this color effect, it is said to be debloomed. Debloomed rubbing oil gives a clearer, sharper, rubbed surface than that which shows a bloom. See ▶ Bloom.

Oil-Drying *n* Any unsaturated oil which will react with oxygen from the air and dry to a relatively hard, tough, elastic substance when spread out in a thin film.

Oil, Fixed See ▶ Nondrying Oil.

Oil Green See ▶ Chromium Oxide Green.

Oil, Heat-Bodied n Drying or semidrying oil which has been heated to an elevated temperature and held at this temperature for the time required to produce a specific increase in viscosity, as distinct from oil polymerized by a process involving oxidation.

Oil Length n Ratio of oil to resin in a medium. For an oleoresinous varnish, the oil length may be expressed in terms of parts by weight of oil to 1 part by weight of resin or, in American practice, in terms of USA gallons of oil per 100 lb of resin. Thus, a 25 gal varnish would mean, in American usage, a varnish composed of 25 USA gallons of oil to 100 lb of resin. For alkyd resins, the oil length is expressed as the percentage of oil weight in the resin. (Gooch JW (2002) Emulsification and polymerization of alkyd resins. Kluwer Academic/Plenum, New York) See ▶ Long Oil, ▶ Medium Oil Varnish, and ▶ Short Oil Varnish.

Oil, Long See ▶ Long Oil.

Oil Melt n Solution of resin in oil, obtained by heating.

Oil-Modified Alkyd See ▶ Alkyds.

Oil, Nondrying n (1905) See ▶ Nondrying Oil.

Oil of Lemon n Volatile oil expressed from fresh peel of *Citrus limonum*. It contains limonene, terpinene, phellandrene, and pinene.

Oil of Mirbane See ▶ Nitrobenzene.

Oil of Pine Tar n Certain heavier fractions of the volatile oil recovered by distilling pine-tar oil to convert it into pine tar.

Oil of Turpentine n Pharmaceutical name for spirits of turpentine which conforms to the requirements of the National Formulary.

Oil Paint n (1790) (1) Paint that contains drying oil as the sole film forming ingredient. (2) A paint that contains drying oil, oil varnish, or oil-modified resin as the basic vehicle ingredient. (3) Commonly (but technically incorrect), any paint soluble in organic solvents (deprecated).

Oil Pastes n Very concentrated mixtures of pigment and oil, which are of paste-like consistency. They may be used for tinting purposes, or for preparation of oil paints by the simple addition of more vehicle and/or thinners, with the necessary driers. *These are also known as Colors in Oil.*

Oil Polish n Finished produced on wood by successive thin oats of linseed oil, accompanied by rubbing motion at the time of application. Also refers to any polishing material containing oil as one of the essential ingredients.

Oilproof n A term describing fabrics that are impervious to oil.

Oil-Reactive Phenolic Resins n Phenol-formaldehyde resins which react with drying oils on heating to yield products with special properties.

Oil-Repellent n A term applied to fabrics that have been treated with finishes to make them resistant to oil stains.

Oil Resistance n The ability of a material to withstand contact with an oil without deterioration of physical properties, or geometric change.

Oil Rubbing Process n of rubbing the dried film of a finished material with oil and pumice or some other suitable abrasive. Linseed oil thinned with naphtha or turpentine is sometimes used, as are also light mineral oils, such as neutral oil, straw oil, and paraffin oil.

Oil, Semidrying See ▶ Semidrying Oil.

Oil Shale n (1873) A rock from which oil can be recovered by distillation.

Oil Soluble n Materials capable of being dissolved in vegetable or mineral oils.

Oil-Soluble Resin n A resin that is capable of dissolving in or reacting with drying oils at moderate temperatures. Such resins are used for producing homogeneous coatings with modified characteristics.

Oil Stain n Solution of dyes and/or pigment dispersions blended with oil or varnish and reduced with hydrocarbon solvents. Those that are absorbed into the surface of the substrate are called penetrating stains, while nonpenetrating stains contain higher levels of nonvolatile vehicle and/or pigments. See ▶ Stain.

Oil, Stand See ▶ Stand Oil.

Oil Varnish See ▶ Oleoresinous Vehicle.

Oil, Vegetable See ▶ Vegetable Oil.

Oiticica Oil \ˌói-tə-ˈsē-kə-\ n [Portuguese, fr. Tupi] (1901) Obtained from the nuts of *Licania rigida*, a tree native to Brazil. It is unique by reason of the fact that its main constituent acid is licanic acid shown below. This acid is highly unsaturated because of its three conjugated double bonds and it also contains a ketonic group. Oiticica oil is similar to tung oil in that it has a high specific gravity, high refractive index, polymerizes rapidly on heating, dries quickly and possesses good water resistance. Like tung oil, it webs, frosts and wrinkles unless adequately processed. It is used as a partial substitute for tung oil and special care is necessary when making varnished from it. Sp gr. 0.982/20°C; iodine value, 145, saponification value, 190.

Oleamide n *cis*-$CH_3(CH_2)_7CH=CH(CH_2)_7CONH_2$. An ivory-colored powder used in low percentages as a slip agent for film-grade polyethylenes, as is its *trans*- isomer, Elaidamide.

Olefin \ˈō-lə-fən\ *n* [ISV, fr. F (*gaz*) *oléfiant* ethylene, fr. L *oleum*] (1860) (1) Any of the class of monounsaturated, aliphatic hydrocarbons of the general formula C_nH_{2n}, and named after the corresponding paraffins by changing their –*ane* endings to –*ene* or –*ylene*. Examples are ethylene (ethene), propylene, and butanes. The class of polymers of olefins is called *polyolefins* or *olefin plastics*. (2) The term is sometimes taken to include aliphatics containing more than one double bond in the molecule such as a diolefin or diene. *Butadiene* is a typical member and an important comonomer for plastics. (3) A general name for fibers from at least 85% other olefins, excluding rubber. (Strong AB (2000) Plastics materials and processing. Prentice Hall, Columbus, OH)

Olefin Fiber *n* A manufactured fiber in which the fiber-forming substance is any long chain synthetic polymer composed of at least 85% by weight of ethylene, propylene, or other olefin units. Olefin fibers combine lightweight with high strength and abrasion resistance, and are currently being used in rope, indoor–outdoor carpets, and lawn furniture upholstery. Also see ▶ Polyethylene Fiber and ▶ Polypropylene Fiber.

Olefinic *n* A reaction between two molecules, both containing carbon–carbon double bonds.

Olefin Plastic See ▶ Polyolefin.

Oleic Acid \ˈō-lē-ik-\ *n* (1819) $CH_3(CH_2)_7CH{=}CH(CH_2)_7COOH$. (*cis*-9-Octadecenoic acid) Fatty acid found in most vegetable oils, whether drying or nondrying. Used to some extent in nondrying and nonyellowing alkyd resins. Mp, 140°C; acid value, 199; iodine value, 90.1 A commercial grade is known as red oil. AS Registry Number: 112-80-1, CAS Name: (*Z*)-9-Octadecenoic acid. Molecular Formula: $C_{18}H_{34}O_2$. Molecular Weight: 282.46. Percent Composition: C 76.54%, H 12.13%, O 11.33%. Obtained by the hydrolysis of various animal and vegetable fats and oils. Prepn from olive oil: (1952) Biochem Prepn 2:100. Separation from olive oil by double fractionation via urea adducts: Rubin, Paisley (1962) Biochem Prepn 9:113. Stereochemistry: Thieme (1905) Ann 343:354. Synthesis: Robinson, Robinson (1925) J Chem Soc 127:175. ^{13}C-NMR studies: Stoffel W et al. (1972) Z Physiol Chem 353:1962; Batchelor JG et al. (1974) J Org Chem 39:1698. Toxicity data: Orö L, Wretlind A (1961) Acta Pharmacol Toxicol 18:141. Exptl use of ^{131}I-labelled oleic acid in myocardial imaging: Bonte FJ et al. (1973) Radiology 108:195. Review of diagnostic use of ^{3}H-oleic acid in pancreatic function: Pedersen NT (1987) Digestion 37(Suppl. 1):25–34. Properties: Pure oleic acid is a colorless or nearly colorless liq (above 5–7°). d_{25}^{25} ~0.895. Solidifies to cryst mass, mp 4°. bp_{100} 286°. At atm pressure it dec when heated at 80–100°. n_D^{18} 1.463; n_D^{26} 1.4585. Iodine no. 89.9; acid value 198.6. On exposure to air, especially when impure, it oxidizes and acquires a yellow to brown color and rancid odor. Practically insol in water. Sol in alcohol, benzene, chloroform, ether, fixed and volatile oils, and keep well closed, protected from light. From Merck Index: LD_{50} i.v. in mice: 230 ± 18 mg/kg. Several grades of the acid are available in commerce, varying in color from pale yellow to red-brown and, depending on the amount of saturated acid present, becoming turbid at 8–16°. The acid of commerce usually contains 7–12% saturated acids, e.g., stearic, palmitic; also some linoleic, etc., unsaturated acids. Melting point: mp 4°. Boiling point: bp_{100} 286°. Index of refraction: n_D^{18} 1.463; n_D^{26} 1.4585. Density: d_{25}^{25} ~0.895. Toxicity data: LD_{50} i.v. in mice: 230 ± 18 mg/kg. (Merck Index, 13th edn. Merck, Whitehouse Station, NJ, 2001)

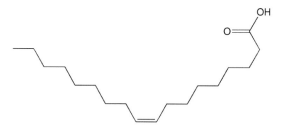

Oleo Chemical *n* A fatty acid, ester, amide, or a mixture of those compounded into thermoplastic resins, usually in fractional percentages, to enhance their release from injection molds and, in films, to reduce surface friction and stickiness.

Oleophobic *n* A term describing a substance that has a strong affinity for oil.

Oleoresin *n* \ˌō-lē-ō-ˈre-zᵊn\ *n* [ISV] (ca. 1846) Pine gum, the nonaqueous secretion of resin acids dissolved in a terpene hydrocarbon oil, which is: (1) produced or exuded from the intercellular resin ducts of a living tree; (2) accumulated, together with oxidation products, in the dead wood of weathered limbs and stumps. NOTE – The term "oleoresinous," however, is not restricted to products of pine gum. See ▶ Oleoresinous Vehicle.

Oleoresinous *adj* Indicating a material which has been made by the combination of an oil and a resin.

Oleoresinous Vehicle *n* A vehicle prepared by the addition of a resin to a drying oil. These two components may or may not be further processed to obtain specified properties. Alkyd resins are sometimes, but not generally, included in this category. See ▶ Varnish.

Olibanum See ▶ Gum Thus.

Oligomer \ə-ˈli-gə-mər\ *n* (1952) A substance composed of only a few monomeric units repetitively linked to each other, such as a dimmer, trimer, tetramer, etc., or their mixtures (ASTM D 883). Other definitions in the literature place the upper limit of repeating units in an oligomer at about ten. The term *telomere* is sometimes used synonymously with oligomer.

Ombré \ˈäm-ˌbrā\ *adj* [F, pp of *ombrer* to shade, fr. It *ombrare*, fr. *ombra* shade, fr. L *umbra*] (ca. 1896) A color effect in which the shade is changeable from light to dark, generally produced by using warp yarns of different tones. Ombré effects may also be produced by printing.

Omega Vibration Direction *n* (ω) Any vibration direction in the plane of the a axes for uniaxial crystals.

OMR Inks See ▶ Optical Mark Recognition Inks.

ONB *n* Abbreviation for ▶ *o*-Nitrobiphenyl.

Ondule *n* A general term for plain-weave fabrics of silk, cotton, or manufactured fiber having a wavy effect produced by weaving the warp or filling, but usually the filling, in a wavy line. An ondule reed is generally used to produce this effect, often in a leno weave to emphasize the wave. Ondule is used for dress fabrics.

One-Compartment Coating *n* Crosslinking systems which can be stored in a single compartment, as opposed to a two-compartment coating. *Also called One-Pot Coating.* See also ▶ Two-Compartment Coating.

One-Pot Coating See ▶ One-Compartment Coating.

One-Shot Molding *n* A process for molding polyurethane foam in which the reactants, usually an isocyanate, a polyol, and a catalyst, are fed in separate streams to a mixing head from which the mixed reactants are discharged into a mold. The polyol and catalyst are sometimes combined along with other additives prior to mixing, but the isocyanate is always fed separately to the mixing head.

Onguka Oil See ▶ Isano Oil.

On-Off Control *n* A simple type of process control in which an instrument, **On-Stream** *n* The state of having been brought into production. The term is usually used for chemical and metallurgical plants or processes.

O.P Abbreviation for ▶ Overproof.

Opacity \ō-ˈpa-sə-tē\ *n* [F *opacité* shadiness, fr. L *opacitat-*, *opacitas*, fr. *opacus* shaded, dark] (ca. 1611) A general term to describe the degree to which a material obscures a substrate, as opposed to transparency which is the degree to which a material does not obscure a substrate. (Saleh BEA, Teich MC (1991) Fundamentals of photonics. Wiley, New York) Opacity is sometimes described in terms of the contrast ratio. Opacity is the quality or state of a body that makes it impervious to rays of light. The resistance to transmission of radiant energy, usually visible light, in terms of percent opacity where 100% opacity is non-light transmitting and 0% opacity is completely light transmitting. Section 08 of ASTM Standards contains some nine tests dealing with optical properties of plastics (ASTM, www.astm.org) is a method for determining the opacity of papers, some of which are used as backings for plastic sheets. See ▶ Contrast Ratio and ▶ Hiding Power.

Opalescence \ˌō-pə-ˈle-sᵊnt\ *adj* (ca. 1813) (1) The limited clarity of vision through a sheet of transparent plastic at any angle, because of diffusion within or on the surface of the plastic. (2) Of a plastic material, the quality of having inner, tiny colored lights resembling those of opals. (Solomon DH, Hawthorne DG (1991) Chemistry of pigments and fillers. Krieger, New York)

Opaque \ō-ˈpāk\ *adj* [L *opacus*] (1641) (1) Impervious to light or not translucent. (2) Adjective to describe complete opacity. (3) Opaque to transmitted light, that is, 100% absorbing throughout the visible light range.

Opaque Ink *n* An ink that does not allow the light to pass through it and has good hiding power. It does not permit the paper or previous printing to show through.

Opaqueness See ▶ Opaque and ▶ Opacity.

Opaque Stains *n* All stains that are not classified as semitransparent stains.

Open Assembly Time See ▶ Time, Assembly.

Open-Cell Foamed Plastic *n* ▶ A Cellular Plastic in which most of the cells are interconnected in a manner such that gases can flow freely from one cell to another.

Open-Cell Foams *n* Forms are a very useful class of materials used as thermal and acoustic insulators, furniture, flotation devices and others. Foams could be of the closed or open cell type. Both types are characterized by a significant fraction of the material being voids (gas). When these voids are connected to one another it is called an open-cell structure. If the voids are individually surrounded by the plastic matrix such polyurethane, then is it is a closed-call structure.

Open Coat *n* In coated abrasives, when the individual abrasive grains are spaced at predetermined distance from one another the open coat covers about 50–70% of the coated surface with abrasive. Open coat has greater flexibility and resistance to filling or clogging than closed coat.

Open Grain *n* Common classification for woods with large pores such as oak, chestnut, and walnut. *Also known as Coarse Textured.*

Opening *n* (1) A preliminary operation in the processing of staple fiber. Opening separates the compressed masses of staple into loose tufts and removes the heavier impurities. (2) An operation in the processing of tow that substantially increases the bulk of the tow by separating the filaments and deregistering the crimp.

Open-Mold Process *n* Any technique for fabricating reinforced plastics in which a one-sided male or female mold is used, with no or low pressure being required. See ▶ Hand Layup, ▶ Sprayup, and ▶ Bag Molding.

Oppanol B *n* Poly(isobutylene), manufactured by BASF, Germany.

Oppanol C *n* Poly(vinylisobutyl ether), manufactured by BASF, Germany.

Oppanol O *n* Copolymer from 90% isobutene and 10% styrene, manufactured by BASF, Germany.

Optical Brightener *n* (1) A colorless compound that, when applied to fabric, absorbs the ultraviolet radiation in light but emits radiation in the visible spectrum. (2) Fluorescent materials added to polymer in manufactured fiber production that emit light in the visible spectrum, usually with a blue cast. See ▶ Brightener and ▶ Brightening Agents.

Optical Character Recognition Inks *n* **(Or OCR Inks)** Inks composed of lowest reflectance pigments, such as carbon black, which can be read by optical scanners (OCR readers). Non-readable inks, though visible to the human eye, cannot be read by OCR readers, because they present no reflectance contrast to the machine. (Printing ink handbook. National Association of Printing Ink Manufacturers, 1976)

Optical Composite *n* A composite that is transparent or nearly so. Making one requires that the refractive indices of the resin and reinforcing fiber, themselves transparent, be closely equal. This might be feasible with polymethyl methacrylate and a carefully chosen glass, since the refractive index of PMMA (n_D^{20}) is 1.49, that of fused quartz is 1.458, and most common glasses are above 1.51. Some fairly clear composites have been made with glass-reinforced polyurethanes.

Optical Density *n* The degree of ▶ Opacity of any translucent medium, sometimes expressed as the logarithm of the opacity.

Optical Dispersion *n* Of a transparent material, the difference between the refractive indices of the material for two different wavelengths of light: in particular, the wavelengths of red and violet lights (ca. 650 and 410 nm, respectively). Dispersion is an important property in the design of compound lenses.

Optical Distortion *n* Any apparent alteration of the geometric shape of an object as seen either through a transparent material or as a reflection from mirror surface.

Optical Fibers *n* (1962) Optical fibers are made from both glass and plastic and are being used increasingly to replace copper wire for carrying computer date and electronic signals because of their high speed and expansion capabilities over conventional wiring.

Optical Filter See ▶ Filter, Optical.

Optical Isomer *n* An isomer which will rotate the plane of polarized light.

Optically-Active Polymers *n* Examples of these materials are organic polymers, metallorganic optical materials and biomolecular optical materials. These materials may be used for various nonlinear optical and electroptical applications, principally in communications, information processing, memory/storage, and optical limiting.

Optical Mark Recognition Inks *n* **(Or OMR Inks)** Similar to OCR inks. Optical scanners (OMR readers) detect the presence of bar marks rather than data characters. Thus, OMR is generally less demanding in print quality than OCR, but more demanding in positional accuracy and space. (Printing ink handbook. National Association of Printing Ink Manufacturers, 1976)

Optical Properties *n* A general term used to refer to the relations of yarn or fibers with light. It includes such parameters as birefringence, refractive index, reflectance, optical density, etc.

Optic Axial Angle *n* The actual angle between two optic axes of a biaxial crystal in the plane of α and γ. It is constant for and characteristic of any particular substance.

Optic Axial Plane *n* The plane containing the optic axes as well as α and γ.

Optic Axis *n* (1664) In crystallography, the direction through an anisotropic crystal along which unpolarized light travels without suffering double refraction. Uniaxial crystals have one optic axis; biaxial crystals have two optic axes.

Optic Sign *n* For uniaxial materials, by definition, the sign is positive if the value ε − ω is positive and negative if ε − ω is negative. For biaxial materials, the sign is positive if γ is the acute bisectrix, and negative if α is the acute bisectrix or positive if γ − β > β − α and negative is γ − β < β − α.

Optimum Twist *n* In spun yarns, a term to describe the amount of twist that gives the maximum breaking strength or the maximum bulk at strength levels acceptable for weaving or knitting.

Oral Lethal Dose *n* (LD_{50}). The amount of a substance taken by mouth that would kill within 14 days half (50%) of those test animals exposed. The dose is measured in milligrams per kilogram of body weight of the test animal.

Orange Chrome See ▶ Light and ▶ Deep.

Orange Lead *n* Lead oxide pigment similar to red lead but containing less litharge and more lead peroxide. It is used as a base for lakes and to some extent as a pigment. See ▶ Red Lead and ▶ Red Lead (Nonsetting).

Orange Mineral Lead oxide pigment which is finer and of a brighter color than red lead. Made by roasting white lead carbonate or sublimed litharge. *Also known as Mineral Orange.*

Orange Peel *n* (ca. 1909) (1) Surface condition characterized by an irregular waviness of the enamel, resembling an orange skin texture; sometimes considered a defect. The condition is often caused by uneven wear of the mold surface due to overpolishing, overheating, or overcarburizing of the mold cavity. It may also be caused by overspraying with mold releases. See ▶ Pockmarking and ▶ Spray Mottle. (2) A type of mottle that occurs in flexographic, gravure and tin printing.

Orange Shellac *n* A refined lac which is soluble in alcohol; contains some wax and resin; used as a coating on floors and other wood surfaces. See ▶ Shellac.

Orbital \ˈȯr-bə-tᵊl\ *n* [*orbital*, adj] (1932) A discrete electronic energy level; also, the spatial distribution of the electron probability density, ψ^2, for such a level.

Orbital Pad Sander *n* A portable sanding machine consisting of a backup pad that moves in small circles, at high speed. Used for sanding metal, the end grain of wood, plastics, and the undercoats for paint and lacquer finishes. See ▶ Straight-Line Pad Sander.

Organdy \ˈȯr-gən-dē\ *n* [F *organdi*] (1835) A very thin, transparent, stiff, wiry, muslin fabric used for dresses, neckwear, trimmings, and curtains. Swiss organdy is chemically treated and keeps its crisp, transparent finish through many launderings. Organdy without chemical treatment loses its crispness in laundering and has to be restarched. Organdy crushes or musses but is easily pressed. Shadow organdy has a faint printed design in self-color.

Organic \ȯr-ˈga-nik\ *adj* (1517) (1) Designation of any chemical compound containing carbon (some of the simple compounds of carbon, such as carbon dioxide, are frequently classified as inorganic compounds). To date, nearly one million organic compounds have been synthesized or isolated. Many occur in nature; others are produced by chemical synthesis. *Refer to Carbonaceous.* (2) In medicine, producing or involving alteration in the structure of an organ; opposed to functional.

Organic Chemistry *n* The chemistry of carbon compounds, usually containing hydrogen, and many containing oxygen, nitrogen, halogens, sulfur, phosphorus, and, occasionally, other elements. The term arose because such compounds were first obtained from living organisms. (Morrison RT, Boyd RN (1992) Organic chemistry, 6th edn. Prentice Hall, Englewood Cliffs, NJ)

Organic Peroxides *n* Peroxides used in the plastics industry are thermally decomposable compounds analogous to hydrogen peroxide in which one or both of the hydrogen atoms are replaced by an organic radical. As they decompose, they form free radicals which can initiate polymerization reactions and affect crosslinking. See ▶ Peroxide.

Organic Pigments *n* Any pigment derived from naturally occurring or synthetic organic substances, characterized by good brightness and brilliance and (usually) transparency. They are generally more resistant to chemicals than inorganic pigments, but are less resistant to that, light, and solvents. Example: Lithol Rubine.

Organic Yellow See ▶ Hansa Yellow.

Organize Yarn Two or more threads twisted in the singles and then plied in the reverse direction. The number of turns per inch in the singles and in the ply is usually in the range of 10–20 turns. Organize yarn is generally used in the warp.

Organometallic Compound \-mə-ˈta-lik-\ *n* Any compound containing carbon, hydrogen and a metal, excluding ordinary metallic carbonates such as sodium bicarbonate and also excluding metallic salts of common organic acids. Many organometallic compounds are used as polymerization catalysts and stabilizes, most notably the ▶ Organotin Stabilizers.

Organopolysiloxane See ▶ Silicone.

Organosol *n* A suspension of a finely divided resin in a plasticizer together with a volatile organic liquid (ASTM D 883). A somewhat tighter definition requires that the volatile liquid comprise at least 5% of the total weight of the suspension. The resin used is most frequently PVC, but the term applies to such suspensions of any resin. An organosols can be prepared from a *plastisol* merely by adding a volatile diluent or solvent that serves to lower viscosity and evaporates when the compound is heated.

Organotin Stabilizer *n* Any of an important class of stabilizers for PVC, notable for their high efficiency, compatibility, and imparting of clarity. The family includes sulfides and oxides of tin-alkyls or –aryls, organotins salts of carboxylic acids, organotins mercaptides, and trialkyl or triaryl tin alcoholates. Certain dioctyltin mercaptides and maleate compounds have been approved for food contact use. See also ▶ Di-*n*-Octyltin Maleate Polymer and ▶ Di-*n*-Octyltin-s,s'bis(isooctyl mercaptoacetate).

Organza \ór-ˈgan-zə\ *n* [prob. alter. of *Lorganza*, a trademark] (1820) A stiff, thin, plain weave fabric made of silk, nylon, acrylic, or polyester, organza is used primarily in evening and wedding attire for women.

Oriental Blue *n* Synthetic ultramarine blue. See ▶ Ultramarine Blue.

Orientation \ˌōr-ē-ən-ˈtā-shən\ *n* (1839) The process of stretching a hot plastic article to align the molecular chains in the direction(s) of stretching, thus improving modulus and strength in those direction(s). When the stretching is applied to one direction or two perpendicular directions, the process is called *uniaxial* or *biaxial orientation*, respectively. Upon reheating, an oriented film will shrink in the direction(s) of orientation. This property is useful in applications such as shrink packaging, and for improving the strength of formed or extruded articles such as lids of dairy-product tubs, pipe, filaments, film, and sheet. (Kadolph SJJ, Langford AL (2001) Textiles. Pearson Education, New York; Chung DD (1994) Carbon fiber composites. Elsevier Science and Technology Books, New York)

Orientation Equipment *n* Equipment which draws and strains a film or fiber in one direction as it is extruded from a die.

Orientation-Release Stress *n* The internal stress remaining in a plastic sheet after orientation, which can be relieved by reheating the sheet to a temperature above that which it was oriented. This stress is measured by heating the sheet and determining the force per unit cross-sectional area exerted by the sheet as it attempts to revert to its preorientation dimensions.

Oriented Film Extruders *n* Screw type extruders with film dies and drawing equipment which strain or orient the film as it taken-up on a reel.

Orifice \ˈór-ə-fəs, ˈär-\ *n* [ME, fr. MF, fr. LL *orificium*, fr. L *or-*, *os* mouth + *facere* to make, do] (15c) (1) A small, usually cylindrical passage in a die, as in a strand die or an orifice-type rheometer. (2) A beveled, sharp-edged, usually circular hold in a thin metal plate that is inserted between flanges in a pipeline. By measuring the fluid pressure up- and downstream near the plate, one can calculate the rate of flow of the fluid. The method has long been used with simple gasses and liquids and is useful for dilute polymer solutions, but not for melts.

Orlon \ˈór-ˌlän\ Poly(acrylonitrile). Manufactured by DuPont, US.

Ormolu Varnish \ˈór-mə-ˌlü-\ *n* A varnish having the appearance of gold or gilded bronze.

Orpiment \ˈór-pə-mənt\ *n* [ME, fr. MF, fr. L *auripigmentum*, fr. *aurum* + *pigmentum* pigment] (14c) As_2S_3. Used to a small extent as a pigment, but is also used in shellac to mask its natural color. *Refer to Realgar. Known also as King's Yellow.*

Orr's White *n* Syn: ▶ Lithopone.

Ortho \ˈór-(ˌ)thō\ (1) Short for ▶ Orthogonal Weave. (2) (*ortho-* The relation of two adjacent carbon atoms in the benzene ring, abbreviated *o-* and ignored in alphabetizing lists of compounds.

Ortho-chlor-para Nitraniline *n* Pigment Red 4 (12085). This pigment has general properties similar to those of para red, although it is slightly superior in tint lightfastness to para red. See ▶ Chlorinated Para Reds.

Orthogonal Weave *n* (1) A type of reinforcing cloth in which fibers are equally distributed in three principal directions (0°, 60°, and 120°) to make a sheet whose properties are nearly equal in all planar directions. (2) A type of three-dimensional reinforcement weave in which fibers are distributed equally in all three principal directions to generate a perform that is then impregnated with resin and cured.

Orthophthalate Plasticizer *n* (phthalate-ester plasticizer) Any of a family of plasticizers derived by reacting phthalic anhydride with an alcohol. They include the widely used DOP, DIDP, and DTDP.

Orthorhombic \ˌór-thə-ˈräm-bik\ *adj* [ISV] (ca. 1859) One of the six crystal systems, in which the three principal axes are mutually perpendicular but the atomic spacings are different on all three.

Orthoscopic Observation *n* The normal way of viewing an object microscopically (cf., conoscopic observation). With Köhler illumination the field diaphragm and the ocular front focal plane as well as the specimen will be in focus.

Orthotropic *n* Having three mutually perpendicular planes of elastic symmetry, as in composites having fibers running in two perpendicular directions, or biaxially oriented sheet. If the fibers or orientation are unidirectional, the material is still orthotropic but also isotropic in the two directions perpendicular to the fibers or oriented polymer chains. (Complete textile glossary. Celanese Corporation, Three Park Avenue, NY; Brandrup J, Immergut EH (eds) (1989) Polymer handbook, 3rd edn. Wiley-Interscience, New York)

Oscillating Die *n* (1) A blown-film die that slowly rotates about its axis in one direction about 90°, then reverses to rotate as far in the opposite direction. The effect of the rotation is to distribute evenly over the width of the rolled-up film any slight differences in film thickness at different points around the die. (2) Flat or cylindrical strand dies in which the one die lip moves to and fro so as to cross the flow channels and produce nonwoven, knotless netting.

OSHA *n* See (1) Abbreviation for Occupational Safety and Health Administration, the Federal Agency established by the Department of Labor, Bureau of Labor Standards, to enforce occupational safety and health standards. The standards are known as Part 1910 of amended Chapter XVII of Title 29 of the Code of Federal Regulations established on April 13, 1971 (36 F R 7006) and as amended thereafter. (2) Abbreviation for ▶ Occupational Safety and Health Act.

OSHRC *n* Abbreviation for Occupational Safety and Health Review Commission.

Osmometer \äz-ˈmä-mə-tər, äs-\ *n* [*osmo*sis + -*meter*] (1854) An instrument for measuring osmotic pressure (OP). The essential elements are a membrane which is permeable to solvents but impermeable to polymer molecules of a specific size range, reservoirs on each side of the membrane containing respectively the polymer solution and pure solvent, means for holding the temperature of the reservoirs constant, and means for measuring the differential osmotic pressure between the solution and the solvent. Osmotic pressure is proportional to the *number* of molecules per unit volume of the solution. From the mass per unit volume and the molar concentration estimated from the OP, one finds the value of the ▶ Number-Average Molecular Weight. The useful range of the method is for M_n from 20,000 to 1,000,000. Membrane materials include cellophane, cellulose acetate, polyvinyl alcohol, and polychlorotrifluoro-ethylene.

Osmometry \ˌäz-mə-ˈme-tri\ *adj* [*osmo*sis + -*meter*] (1854) Measurement of molecular weight of polymers based on osmotic pressure. A plot of reduced osmotic pressure, Π/c vs. c and extrapolation to c = 0, then $1/M_n$ condition exists. (Collins EA, Bares J, Billmeyer FW Jr (1973) Experiments in polymer science. Wiley-Interscience, New York) See ▶ Osmotic Pressure.

Osmometry, Membrane *n* A semipermeable membrane used in an osmo-meter to allow solvent to penetrate, but not polymer.

Osmosis \äz-ˈmō-səs, äs-\ *n* [NL, short for *endosmosis*] (1867) The passage of solvent molecules through a semipermeable membrane from a solution of higher solvent concentration (lower solute concentration) to one of lower solvent concentration (higher solute concentration).

Osmotic Coefficient *n* Quantity characterizing the deviation of the solvent from ideal behavior referenced to Raoult's law. The osmotic coefficient on an amount fraction basis is given by

$$\varphi = (\mu_A^* - \mu_A)/(RT \ln X_A)$$

where

μ_A^* and μ_A are the chemical potentials of the solvent as a pure substance and in solution, respectively.
x_A is the amount fraction.
R is the gas constant and T is the temperature.

Osmotic Pressure *n* (1888) General and Biological: The hydrostatic pressure produced by a solution in a space divided by a semipermeable membrane due to a differential in the concentrations of solute. Osmoregulation is the homeostasis mechanism of an organism to reach balance in osmotic pressure.

Osmotic potential is the opposite of water potential with the former meaning the degree to which a solvent (usually water) would want to stay in a liquid.

- Hypertonicity is a solution that causes cells to shrink. It may or may not have a higher osmotic pressure than the cell interior since the rate of water entry will depend upon the permeability of the cell membrane.
- Hypotonicity is a solution that causes cells to swell. It may or may not have a lower osmotic pressure than the cell interior since the rate of water entry will depend upon the permeability of the cell membrane.
- Isotonic is a solution that produces no change in cell volume.

When a biological cell is in a hypotonic environment, the cell interior accumulates water, water flows across the cell membrane into the cell, causing it to expand. In plant cells, the cell wall restricts the expansion, resulting in pressure on the cell wall from within called turgor pressure. The osmotic pressure π of a dilute solution can be calculated using the formula,

$$\Pi = im\,RT$$

where

i is the van't Hoff factor
M is the molarity
R is the gas constant, where $R = 0.08206$ L atm $mol^{-1}\,K^{-1}$
T is the thermodynamic temperature (formerly called absolute temperature)

Note the similarity of the above formula to the ideal gas law and also that osmotic pressure is not dependent on particle charge. This equation was derived by van't Hoff. Osmotic pressure is the basis of reverse osmosis, a process commonly used to purify water. The water to be purified is placed in a chamber and put under an amount of pressure greater than the osmotic pressure exerted by the water and the solutes dissolved in it. Part of the chamber opens to a differentially permeable membrane that lets water molecules through, but not the solute particles. The osmotic pressure of ocean water is about 27 atm. Reverse osmosis desalinators use pressures around 50 atm to produce fresh water from ocean salt water.

Osmotic pressure is necessary for many plant functions. It is the resulting turgor pressure on the cell wall that allows herbaceous plants to stand upright, and how plants regulate the aperture of their stomata. In animal cells which lack a cell wall however, excessive osmotic pressure can result in cytolysis.

- Cell wall
- Cytolysis
- Gibbs-Donnan effect
- Osmosis
- Pfeffer cell
- Plasmolysis
- Turgor pressure

For the calculation of molecular weight by using colligative properties, osmotic pressure is the most preferred property.

Potential osmotic pressure is the maximum osmotic pressure that could develop in a solution if it were separated from distilled water by a selectively permeable membrane. It is the number of solute particles in a unit volume of the solution that directly determines its potential osmotic pressure. If one waits for equilibrium, osmotic pressure reaches potential osmotic pressure.

Polymer solutions: The osmotic pressure developed between any dilute solution and its solvent will shown to be a colligative property; it is therefore dependent only on the concentration of the solution and on the properties of the solvent; a semipermeable membrane separates the solvent from solution, and allows the solvent to flow to the solution; Van't Hoff's equation is:

$$\Pi = RTc/M_n$$

where,

Π = osmotic pressure, mmHg
M_n = number average molecular weight
c = concentration of small molecules (g/100 mL)
R = gas constant, and
T = absolute temperature K

and where the solute molecular weight (or size) is many hundreds of times larger than the solvent, the equation becomes,

$$\Pi/c = (RT/M_n) + B_c,$$

$B = RTA_2$ and A_2 is the second virial coefficient

$$(\Pi/RTc)_c = 0 = 1/M_n$$

(Collins EA, Bares J, Billmeyer FW Jr (1973) Experiments in polymer science. Wiley-Interscience, New York.)

Osnaburg *n* A coarse cotton or polyester/cotton fabric, often partly of waste fiber, in a plain weave, medium to heavy in weight, that looks like crash. Unbleached osnaburg is used for grain and cement sacks, and higher grades are used as apparel and household fabrics.

Ostwald Color System *n* System of classifying and designating colors in terms of their full color content, their white content, and their black content. The system was devised by Wilhelm Ostwald, based on additive color mixture, by means of spinning a Maxwell Disc and was used as the basic principle in making the Color Harmony Manual.

Ostwald-de Waele Model See ▶ Power Law.

Ostwald-Fenske Viscometer A capillary viscometer for measuring time v. flow time a liquid or solution; viscosity molecular weight of a polymer can **Ostwald Ripening** *n* The process in an oil-in-water emulsion by which oil diffuses from the smaller droplets, across the aqueous phase, and into the larger droplets. The driving force for diffusion is the minimization of interfacial free energy by accumulating oil in larger droplets. Ostwald ripening can be mitigated by the addition of a costabilizer. The concentration of the costabilizer caused by monomer diffusion counters the reduction in interfacial area, causing an equilibrium condition. See ▶ Dilute-Solution Viscosity.

Ostwald U-Tube *n* Viscometer in the form of a "U," in which the liquid flows from a bulb at a higher level on one side of the "U," through a capillary, to a receiving bulb on the other side. The time is measured for a given volume of liquid to pass from one side to the other.

Otsu-ye *n* (Japanese) A rough broadsheet painting, of small size on paper: the precursor of the print. *Also spelled "Otsue."*

Ottoman \ä-tə-mən\ *n* (1605) Heavy, large, filling rib yarns, often of cotton, wool, or waste yarn, covered in their entirety by silk or manufactured fiber warp yarns, characterized this fabric used for women's wear and coats.

Ouricuri Wax *n* Hard wax obtained from the leaves of the Bahia palm, and used as a substitute for carnauba wax. *Also spelled "Ouricury."*

Ouropardo Wax *n* Unusual hard wax of exceptionally high sp gr (1.23), and possessing the property of resinifying on heating, without melting.

Outflow Quench *n* Air for cooling extruded polymer that is directed radically outward from a central dispersion device around which the filaments descend.

Outgassing *n* A term used in the vacuum-metallizing industry for the evaporation under vacuum of a volatile substance such as moisture, solvent, or plasticizer, from plastic articles to be coated with metal. Outgassing can cause pressure increase (loss of vacuum), also darkening and poor adhesion of the metal coating.

Outsert Molding *n* A term coined to distinguish the process of molding small plastic parts in a large metal plate from the "insert molding", in which small metallic parts are incorporated into a larger plastic molding. The place is indexed in the injection mold by a pre-and-hole system, with the plastic parts injected through blanks prepunched in the metal plate. The process makes it feasible to use plastics-metal combinations where economics formerly directed all-metal components.

Outside Finish *n* The surface treatment or decorative trim on the exterior of a building.

Ovaloid *n* A surface of revolution symmetrical about the polar axis that forms the end closure for a filament-wound cylinder.

Oven-Dry Weight *n* The constant weight of a specimen obtained by drying in an oven under prescribed conditions of temperature and humidity.

Overall Conductance *n* (overall heat-transfer coefficient) In heat-transfer engineering, the reciprocal of the total ▶ Thermal Resistance, for heat flow through plane walls or tube walls. It is defined by the equation: $U = q/A \, \Delta T$, where q = the rate of heat flow through (and normal to) the surface of area A, and ΔT is the fall in temperature through the layer in the direction of q. This is a modification of Fourier's law, invented to deal conveniently with heat flow through stagnant fluid films adjacent to walls, films whose thicknesses are difficult or impossible to measure. For curved surfaces such as pipes, U must be referred to either the inside or outside area, usually the later (U_o). The SI unit for overall conductance is $W/(m^2 \, K)$.

Overbaking *n* An exposure of the coating to a temperature moderately higher or for a longer period of time, or both, than recommended by the manufacturer of the coating for normal curing. This condition is in contrast to "heat resistance" which is a parameter relating to the service life of a coating.

Overcoating *n* In extrusion coating, the practice of extruding a web beyond the edges of the substrate web.

Overcure *n* Caused by an aftercure or by being subjected to too high a temperature or too long a period at a proper temperature and resulting in a product less resistance to aging.

Overcut *n* A staple fiber that is longer than nominal length. Usually, the length is a multiple of 2, 3, or more times the nominal length. An overcut is caused by the failure of filaments to be cut to the desired length during staple manufacture.

Overflow Groove *n* A small groove used in molds to allow material to escape freely to prevent flash and low part density, and to dispose of excess material.

Overlap See ▶ Lap.

Overlapping Seam *n* Method of hanging wallpaper in which only one selvage is trimmed.

Overlay Sheet *n* (top sheet, surfacing mat) A nonwoven fibrous mat of glass or synthetic fiber used as the surfacing sheet in decorative laminates. Its function is to provide a smoother finish, hide the fibrous pattern of the laminate, and/or to provide a decorative motif when printed on the underside.

Overlength See ▶ Overcut.

Overprint *n* The printing of one impression over another.

Overprint Varnish *n* A clear varnish applied over a printed job to improve its gloss and/or mar resistance, etc.

Overproof *n* Containing a greater proportion of alcohol than proof spirit, especially, containing more than 50% alcohol by volume. Abbreviation is O.P.

Overspray *n* The roughness of a film of paint or lacquer due to dry particles deposited on, but not melded into a previously sprayed, semidried film. Overspray is encountered particularly with surfaces in more than one plane, such as auto bodies, television cabinets, etc. The remedy for overspray is a well-balanced solvent system containing enough high boiler to prevent the drying out of spray droplets before they land on an overshot surface.

Over-the-Counter *adj* (1921) A term that usually refers to direct sales to a retail customer in a store, as opposed to wholesale marketing.

Overtone See ▶ Mass Color.

Oxalic Acid \(ˌ)äk-ˈsa-lik-\ *n* [F (*acide*) *oxalique*, fr. L *oxalis*] (1791) $(COOH)_2 \cdot 2H_2O$. Dicarboxylic acid, used as a bleaching agent. Mp, 99°C; sp gr, 1.65; mol wt. 126.07.

Oxamide *n* $H_2NCOCONH_2$. A white powder, used as a stabilizer for cellulose nitrate.

CAS Registry Number: 471-46-5, CAS Name: Ethanediamide. Additional Names: oxalamide; oxalic acid diamide; ethanedioic acid diamide. Molecular Formula: $C_2H_4N_2O_2$. Molecular Weight: 88.07. Percent Composition: C 27.28%, H 4.58%, N 31.81%, O 36.33%. Line Formula: $H_2NCOCONH_2$. Literature References: Prepd from formamide by glow-discharge electrolysis: Brown et al. (1962) J Org Chem 27:3698. Crystal structure: DeWith G, Harkema S (1977) Acta Crystallogr 33B:2367. Metabolized in body to form oxalic acid. Properties: Triclinic needles, dec 350°. d_4^{20} 1.667. Sparingly sol in hot water, alcohol. Density: d_4^{20} 1.667.

Oxanilide *n* $C_6H_5NHCOCONHC_6H_5$. A plasticizer with bp, 320°C; mol wt, 240.11; mp. 245°C. *Known also as Camphol.*

Oxford Cloth *n* A soft but stout shirting fabric in a modified basket weave with a large filling yarn having no twist woven under and over two single, twisted warp yarns. The fabric is usually made from cotton or polyester/cotton blends and is frequently given a silk like luster finish.

Oxidation \ˌäk-sə-ˈdā-shən\ *n* [F, fr. *oxider, oxyder* to oxidize, fr. *oxide*] (1791) (1) The formation of an oxide or, more generally, any increase in valence of an element. (2) In coatings, the introduction of oxygen into a molecule thereby producing a cured film. Rapid oxidation accompanied by flame is called *burning*. Oxidation is always exothermic.

Oxidation Inhibitor See ▶ Anti-Oxidant.

Oxidation Number (Oxidation State) *n* (1926) A somewhat arbitrary but useful way of indicating the approximate electrical characteristic of an atom.

Oxidation Potential *n* A measure of the tendency of an oxidation half-reaction to occur; expressed as the voltage produced by a cell employing the half-reaction at its anode and using the standard hydrogen electrode as its cathode.

Oxidation-Reduction *n* (1909) A chemical reaction in which one or more electrons are transferred from one atom or molecule to another.

Oxidation-Reduction Reaction *n* An electron-transfer reaction.

Oxidative Aging, Elastomers *n* Breaking down of an elastomer through the action of oxygen on the polymer itself or an other ingredients of the compound. The process may be signaled by change of color, visible deterioration of the part surface, or lowered performance in service.

Oxidative Coupling *n* A process defined as a reaction of oxygen with active hydrogen atoms from different molecules, producing water and a dimerized molecule. If the hydrogen-yielding substance has two active hydrogen atoms polymerization results. This process is used in the polymerization of phenols, particular polyphenylene oxide.

Oxidative Degradation *n* Breaking down of a polymer or plastic product through the action of oxygen on the polymer itself or on other ingredients of the compound. The process may be signaled by change of color, visible deterioration of the part surface, or lowered performance in service.

Oxidative Dehydrogenation *n* A chemical process used in making monomers such as styrene, butadiene, and vinyl chloride. Such "oxydehydro" processes involve either removal of hydrogen, from a hydrocarbon by oxygen, forming water; or removal of hydrogen from a hydrocarbon by a halogen to form the hydrogen halide, then regeneration of the halogen with oxygen.

Oxide of Chromium See ▶ Chromium Oxide Green.

Oxidized Asphalt See ▶ Blown Bitumen.

Oxidized Bitumen See ▶ Blown Bitumen.

Oxidized Rosin *n* Resin of much higher melting point than ordinary rosin and characterized by good solubility in industrial alcohol.

Oxidized Rubber *n* Rubber which has been subjected to oxidation by air, in the presence of catalysts. Oxidized rubber differs considerably from the parent rubber in that it is a dark reddish-brown color, and is much more compatible with solvents, drying oils, etc. It has good heat resisting properties and yields useful films on baking.

Oxidized Turpentine *n* Turpentine which has been thickened by blowing or oxidation with air. *Also called Fat Turpentine.*

Oxidizing Agent *n* A substance or species which gains electrons in a reaction. An electron acceptor.

Oxime \ˈäk-ˌsēm\ *n* [ISV *ox-* + *-ime* (from *imide*)] (ca. 1890) Compound of the general type characterized by the grouping C = NOH. Butyraldoxime and methyl ethyl ketoxime are used as antiskinning agents and as anti-oxidants for dip-tank stability, etc.

Oxirane *n* C_2H_4O. (1) Epoxide. Describing the oxygen atom of the epoxides ring. (2) Ethylene oxide. Syn: ▶ Ethylene Oxide.

Oxirane Group Syn: Epoxy Group.

Oxirane Value *n* (oxirane oxygen. The percent of oxygen absorbed by an unsaturated raw material during epoxidation; a measure of the amount of epoxidized double bond in a material. The oxirane value and the ▶ Iodine Value are used in evaluating epoxy plasticizers. A high oxirane value and low iodine value are considered to be essential for good performance, but these are not the only criteria.

Oxonium Ion *n* Hydronium ion; H_3O^+.

Oxo Process *n* A chemical process utilizing a reaction known as *oxonation* or *hydroformylation*, in which hydrogen and carbon monoxide are added across an olefinic bond to produce an aldehyde containing one more carbon atom than the olefin. The aldehydes produced by this process can be reduced to alcohols which are used for making many ester-type plasticizers.

Oxy- A prefix denoting the –O– radical or (primarily in Europe) the –OH radical.

***p*-Oxybenzoyl Copolyester** *n* Any of a family of readily moldable polyester copolymers consisting of mixtures of *p*-oxybenzoÿl with units from aromatic carboxylic acids and aromatic diphenols.

***p*-Oxybenzoyl Polymer** *n* A polymer based on *p*-hydroxybenzoic acid (derived from phenol and carbon dioxide). Technically a thermoplastic, the polymer retains good stiffness at temperatures up to 315°C and, at temperatures around 425°C, undergoes a second-order transition and becomes malleable so that it can be forged like ductile metals. Some other properties are high dielectric strength, elastic modulus, thermal conductivity, resistance to wear and solvents, self-lubricity, and good machinability. It has also been blended with metals to form alloys. Copolymers (see preceding entry) sacrifice some of the high-temperature performance to gain moldability.

4,4′-Oxybis(Benzenesulfonylhydrazide) *n* (OBSH) The most important of the Sulfonyl hydrazide family of blowing agents, a white crystalline solid melting at 164°C and yielding nitrogen upon decomposition. It is used widely in rubber-resin blends due to its ability to simultaneously foam the blends and act as a crosslinking agent, but it is also used in polyethylene, PVC, phenolics, and epoxy resins. Favorable properties are low odor, nontoxicity, and freedom from discoloration.

Oxygen *n* Element, gas at ambient temperature, atomic number 8 and molecular weight 15.999 g/mole in column VIA of the periodic table. Oxygen Radicals: A radical (sometimes called a "free radical") is a molecule that has an unpaired electron (represented by a dot next to the chemical structure, e.g. A·). Some radicals are stable and long-lived (the most common example is O_2, the oxygen in the air we breath, which is a "biradical" as it has two unpaired electrons), but most radicals are highly reactive and thus unstable. As a rule, a radical needs to pair its unpaired electron with another, and will react with another molecule in order to obtain this missing electron. If a radical achieves this by "stealing" an electron from another molecule, that other molecule itself becomes a radical (Reaction 1), and a self-propagating chain reaction is begun (Reaction 2).

$$A \cdot + B : \rightarrow A : + B \cdot \quad (1)$$

$$B \cdot + C : \rightarrow B : + C \cdot \quad (2)$$

If a radical pairs its unpaired electron by reacting with a second radical, then the chain reaction is terminated, and both radicals "neutralize" each other (Reaction 3).

$$A \cdot + B \cdot \rightarrow A : B \quad (3)$$

Radicals are produced by normal aerobic (oxygen-requiring) metabolism and are necessary to life. However, excess amounts of radicals are harmful because of their reactivity. Radicals can be formed easily by compounds that readily give up a single electron (for example, polyunsaturated fatty acids). Radicals are also produced by processes other than normal metabolism-by ionizing radiation, smoking and other pollutants, herbicides and pesticides, and are even found in certain types of food (e.g., deep-fat fried foods). Radicals can damage lipids (fats, which make up cell membranes), proteins, and DNA. Radical damage can be significant because it can proceed as a chain reaction. An example of radical damage is that caused by the hydroxyl radical (described in more detail below). Singlet Oxygen: Oxygen in the air we breathe is in its "ground" (not energetically excited) state and is symbolized by the abbreviation $^3O_2$. It is a free radical-in fact it is a diradical, as it has two unpaired electrons. Electrons "spin" (that is, rotate about an axis passing through the electron). Molecules whose outermost pair of electrons have parallel spins (symbolized by ↑↑) are in the "triplet" state; molecules whose outermost pair of electrons have antiparallel spins (symbolized by ↑↓) are in the "singlet" state. Ground-state oxygen is in the triplet state (indicated by the superscripted "3" in $^3O_2$) - its two unpaired electrons have parallel spins, a characteristic that, according to rules of physical chemistry, does not allow them to react with most molecules. Thus, ground-state or triplet oxygen is not very reactive. However, triplet oxygen can be activated by the addition of energy, and transformed into reactive oxygen species.

If triplet oxygen absorbs sufficient energy to reverse the spin of one of its unpaired electrons, it forms the singlet state. Singlet oxygen, abbreviated $^1O_2^*$, has a pair of electrons with opposite spins; though not a free radical it is highly reactive. (The * symbol is used to indicate that this is an excited state with excess energy. It should be emphasized that neither triplet- nor singlet-state molecules are necessarily in the excited state; the designation of "triplet" or "singlet" refers only to the spin state. For example, carotenoids exist in an unexcited, ground singlet state S_0 and can be excited by light absorption to a higher-energy singlet state S_1, with no change in spin state (Christophersen et al. 1991).

·O—O·   triplet oxygen (↑↑) (ground state)
↓ energy
O—O: singlet oxygen (↑↓) (highly reactive)

This reaction can also be written in this form:

$$^3O_2 + energy \rightarrow {}^1O_2^*$$

Singlet oxygen is produced as a result of natural biological reactions and by photosensitization by the absorption of light energy. According to rules of physical chemistry, the "relaxation" (excess energy loss) of singlet oxygen back to the triplet state is "spin forbidden" and thus singlet oxygen has a long lifetime for an energetically excited molecule, and must transfer its excess energy to another molecule in order to relax to the triplet state. Superoxide: Triplet oxygen can also be transformed into a reactive state if it is accepts a single electron. This process of accepting an electron is called reduction, and in this case, is "monovalent" reduction because only one electron is involved. The molecule that gave up the electron is oxidized. The result of monovalent reduction of triplet oxygen is called superoxide, abbreviated $O_2^{·-}$. Superoxide is a radical. It is usually shown with a negative sign, indicating that it carries a negative charge of −1 (due to the extra electron, e^-, it gained).

·O—O·   triplet oxygen (ground state)
↓ monovalent reduction
·O—O: superoxide

This reaction can also be written in this form:

$$^3O_2 + e^- \rightarrow O_2^-$$

Superoxide can act both as an oxidant (by accepting electrons) or as a reductant (by donating electrons). However, superoxide is not particularly reactive in biological systems and does not by itself cause much oxidative damage. It is a precursor to other oxidizing agents, including singlet oxygen, *peroxynitrite*, and other highly reactive molecules. However, superoxide is not all bad-in fact it is necessary for health. For example, certain cells in the human body produce superoxide (and the reactive molecules derived from it) as a antibiotic "weapon" used to kill invading microorganisms. Superoxide also acts as a signaling molecule needed to regulate cellular processes. Under biological conditions, the main reaction of superoxide is to react with itself to produce hydrogen peroxide and oxygen, a reaction known as "dismutation". Superoxide dismutation can be spontaneous or can be catalyzed by the enzyme superoxide dismutase ("SOD").

$$2O_2^- + 2H^+ \rightarrow H_2O_2 + {}^3O_2$$

Superoxide is also important in the production of the highly reactive hydroxyl radical (HO·) (discussed in more detail below). In this process, superoxide actually acts as a reducing agent, not as an oxidizing agent. This is because superoxide donates one electron to reduce the metal ions (ferric iron or Fe^{3+} in the example below) that act as the catalyst to convert hydrogen peroxide (H_2O_2) into the hydroxyl radical (HO·).

$$O_2^- + Fe^{3+} \rightarrow {}^3O_2 + Fe^{2+}$$

The reduced metal (ferrous iron or Fe^{2+} in this example) then catalyzes the breaking of the oxygen-oxygen bond of hydrogen peroxide to produce a hydroxyl radical (HO·) and a hydroxide ion (HO⁻):

$$Fe^{2+} + H_2O_2 \rightarrow Fe^{3+} + HO· + HO^-$$

Superoxide can react with the hydroxyl radical (HO·) to form singlet oxygen ($^1O_2^*$) (not a radical but reactive nonetheless):

$$O_2^- + HO· \rightarrow {}^1O_2^* + HO^-$$

Superoxide can also react with nitric oxide (NO·) (also a radical) to produce peroxynitrate (OONO⁻), another highly reactive oxidizing molecule.

$$O_2^- + NO· \rightarrow OONO^-$$

Hydrogen Peroxide: Superoxide (O_2^-) can undergo monovalent reduction to produce peroxide (O_2^{-2}), an activated form of oxygen that carries a negative charge of –2. Usually *peroxide* is termed "hydrogen peroxide" (H_2O_2) since in biological systems the negative charge of –2 is neutralized by two protons (two hydrogen atoms, each with a positive charge).

·O—O: superoxide
↓ monovalent reduction
H:O—O:H   hydrogen peroxide

Hydrogen peroxide is important in biological systems because it can pass readily through cell membranes and cannot be excluded from cells. Hydrogen peroxide is actually necessary for the function of many enzymes, and thus is required (like oxygen itself) for health. Hydrogen peroxide is not as reactive as a product it can form, the hydroxyl radical. Hydroxyl Radical: Hydrogen peroxide, in the presence of metal ions, is converted to a hydroxyl radical (HO·) and a hydroxide ion (HO⁻). The metal ion is required for the breaking of the oxygen–oxygen bond of peroxide. This reaction is called the Fenton reaction, and was discovered over a 100 years ago. It is important in biological systems because most cells have some level of iron, copper, or other metals which can catalyze this reaction.

H:O—O:H   hydrogen peroxide
↓ Metal ion
HO·   hydroxyl radical
and
HO: hydroxide ion

This reaction can also be written (with iron as the metal):

$$Fe^{2+} + H_2O_2 \rightarrow Fe^{3+} + HO· + HO^-$$

A hydroxyl radical can also react with superoxide to produce singlet oxygen and a hydroxide ion:

$$O_2^- + HO· \rightarrow {}^1O_2^* + HO^-$$

Like hydrogen peroxide, the hydroxyl radical passes easily through membranes and cannot be kept out of cells. Hydroxyl radical damage is "diffusion rate-limited". Simply put, the rate at which hydroxyl radicals can damage other molecules is limited chiefly by how far and fast it can diffuse (travel) in the cell. This highly reactive radical can add to an organic (carbon-containing) substrate (represented by R below)-this could be for example a fatty acid-forming a hydroxylated adduct that is itself a radical.

$$HO\cdot + R \rightarrow HOR\cdot$$

This adduct can be further oxidized (e.g., by metal ions or oxygen) by a one-electron transfer (monovalent reduction), resulting in a oxidized but stable product. In the first case, the extra electron is transferred to the metal ion, and in the second case, to oxygen (forming superoxide).

$$HOR\cdot + Fe^{3+} \rightarrow HOR + Fe^{2+} + H^+$$

$$HOR\cdot + {}^3O_2 \rightarrow HOR + O_2^{\cdot -} + H^+$$

Two adduct radicals can also react with each other, forming a stable, crosslinked-but oxidized-product, with water as a byproduct.

$$HOR\cdot + HOR\cdot \rightarrow R-R + 2H_2O$$

The *hydroxyl radical* can also oxidize the organic substrate by "stealing" or abstracting an electron from it.

$$HO\cdot + R \rightarrow R\cdot + H_2O$$

The resulting oxidized substrate is again itself a radical, and can react with other molecules in a chain reaction. For example, it could react with ground-state oxygen to produce a peroxyl radical (ROO·).

$$R\cdot + {}^3O_2 \rightarrow ROO\cdot$$

The *peroxyl radical* again is highly reactive, and can react with another organic substrate in a chain reaction (Christophersen AG, Jun H, Jørgensen K, Skibsted LH (1991) Photobleaching of astaxanthin and canthaxanthin: quantum-yields dependence of solvent, temperature, and wavelength of irradiation in relation to packageing and storage of carotenoid pigmented salmonoids. Z Lebensm Unters Forsch 192:433–439.).

$$ROO\cdot + RH \rightarrow ROOH + R\cdot$$

Oxygen-Index Flammability Test See Flammability, ASTM D 2863.

Oxygen-Transmission Rate See ▶ Permeability.

Oxymethylene *n* (1) Syn: ▶ Formaldehyde. (2) (oxane) The group (–OCH$_2$–) the chain unit of ▶ Acetal Resins.

$$\cdot H_2C \longrightarrow O\cdot$$

Oxyns *n* Name given to the oxidation products of drying oils, which are obtained by the addition of oxygen to the double bonds of the unsaturated fatty acids.

Ozokerite \ō-zō-ˈkir-ˌīt\ *n* [Gr *Ozokerit*, fr. Gk *ozein* to small + *kēros* was] (ca. 1837) Natural mineral wax of variable properties, found in Germany, America, South Africa, and Austria. Its mp varies from 59° to 90°C and its sp gr from 0.900 to 0.960.

Ozonation \ˌō-(ˌ)zō-ˈnā-shən\ *n* (1854) Chemical reaction with ▶ Ozone.

Ozone \ˈō-ˌzōn\ *n* [Gr *Ozon*, fr. Gk *ozōn*, pp of *ozein* to smell] (ca. 1840) An allotropic form of oxygen, O$_3$, a faintly blue, irritating gas and a powerful oxidant.

Ozone Fading *n* The fading of a dyed textile material, especially those in blue shades, caused by atmospheric ozone (O$_3$).

Ozone Resistance *n* The ability of a plastic or elastomer to withstand, without diminution of useful properties, the chemical action (strong oxidation) of ozone.

Ozonolysis *n* Oxidation of an organic material by means of ozone.

P

p *n* \pē\ (1) Abbreviation for SI prefix ▶ Pico-. (2) *p*-Abbreviation for ▶ Para-.

P *n* (1) Chemical symbol for the element phosphorus. (2) Abbreviation for SI prefix ▶ Peta-. (3) Symbol for pressure or permeability.

Pa *n* SI abbreviation for ▶ Pascal.

PA *n* (1) Abbreviation for ▶ Polyamide. See ▶ Nylon. (2) Abbreviation for ▶ Phthalic Anhydride.

PAA *n* Abbreviation for Polyarylic Acid.

PABM See ▶ Polyaminobismaleimide Resin.

PAC *n* Poly(acrylonitrile) fiber.

Pack \pak\ *n* {*often attributive*} [ME, pf LGr or D origin; akin to MLGr & MD *pak* pack] (13c) (1) The complete assembly of filters and spinneret through which polymer flows during extrusion. (2) A unit of weight for wool, 240 lb. (Shenoy AV (1996) Thermoplastics melt rheology and processing. Marcel Dekker, New York)

Package Build *n* A general term that applies to the shape, angles, tension, etc., of a yarn package during winding. Package build affects performance during subsequent processing.

Package Dyeing See Dyeing, Yarn Dyeing.

Package Stability *n* The ability of a liquid such as paint or varnish, to retain its original quality after prolonged storage.

Packing Braid *n* A braid of reinforcing fiber having a fully filled, square cross section.

Pack Life *n* The time during which a pack assembly can remain in use and produce good quality yarn.

Padding *n* The application of a liquor or paste to textiles either by passing the material through a bath and subsequently through squeeze rollers, or by passing it between squeeze rollers, the bottom one of which carries the liquor or paste.

Paddle Agitator *n* One of the simple types of mixing equipment for plastics in the form of dispersions, pastes, and dough's. The most common form comprises a set of rotating blades driven by a vertical shaft and intermeshing with a set of fixed blades.

Paddle Dyeing Machine *n* A machine used for dyeing garments, hosiery, and other small pieces that are packaged loosely in mesh bags. The unit consists of an open tank and revolving paddles that circulate the bags in the dyebath.

PAI Abbreviation of ▶ Polyamide-imide Resin.

Paint \pānt\ (1) *v* To apply a thin layer of a coating to a substrate by brush, spray, roller, immersion, or any other suitable means. (2, *n*) Any pigment liquid, liquefiable, or mastic composition designed for application to a substrate in a thin layer which is converted to an opaque solid film after application. Used for protection, decoration or identification, or to serve some functional purpose such as the filing or concealing, of surface irregularities, the modification of light and heat radiation characteristics, etc. (3) The dispersion of pigment in a liquid vehicle that may be applied to surfaces to form a thin adherent protective or decorative coating. The liquid vehicle usually consists of a film-forming resin dissolved in a solvent, or resin latex. Resins must frequently used in paints are phenolics, polyesters, urea's, melamine's, cellulosics, acrylics, vinyls, alkyds, and epoxies. (ASTM Special Technical Publication No, 500. American Society for Testing and Materials, Philadelphia, PA, 1972; Tatton WH, Drew EW (1964) Industrial paint application. Hart, New York; Tatton WH, Drew EW (1964) Industrial paint application. Van Nostrand, Princeton, NJ; Taubes, Frederic (1971) The painter's dictionary of materials and methods. Watson-Guptill, New York; Sward GG (ed). Paint testing manual: Physial and chemical examination of paints, varnishes and lacques, and colors, 13th edn.)

Paint and Varnish Remover *n* Liquid, principally solvents, sometimes with wax or thickeners, which is applied to a coated surface in order to soften the old coating and bring it to such a condition that it can be easily removed. *Also called Stripper.*

Paint Base *n* The vehicle into which pigment is mixed to form a paint; commonly alkyd, latex, and acrylic. (Talen HW (1962) Some consideration on the formation of films. J Oil Colour Chem Assoc 45(6):387–415)

Painter's Mitt *n* A glove-like device that slips over the hand. Used for coating areas inaccessible with a brush or roller such as fences and walls behind radiators.

Painter's Naphtha See ▶ VM&P Naphtha.

Painter's Putty See ▶ Putty.

Paint Harling *n* Process of throwing paint-coated granite chips on to a sticky paint film previously applied to a surface, to give a roughcast effect. This gives a thick durable finish which had been used, e.g., on steel clad houses.

Painting of Plastics *n* Plastic articles are painted not only to enhance their appearance, but also to provide desired surface properties lacking in the unpainted articles. For example, electrical properties and resistance to water, solvents, chemicals, and abrasion resistance may be improved by painting. Adhesion of paints to plastics is achieved by intermolecular attraction, solvent etching, or a combination of both. In the case of plastics of low surface energy, e.g., polyethylene, an oxidative pretreatment is mandatory for good coating adhesion. The methods used to apply paints to plastics are spraying (with or without masks). ▶ Dip Coating, Flow Coating, Roller Coating, Screen Printing; and Spray-And-Wipe Painting, all of which see.

Paint Mitt See ▶ Painter's Mitt.

Paint Remover See ▶ Paint and Varnish Remover.

Paint Research Institute n (PRI) Fundamental research-granting agency of the coatings industry; supported mainly by the Federation of Societies for Coatings Technology (FSCT), and by donations from individuals, corporations, and many associations within the coatings industry.

Paint Roller n A cylindrical tube which is coated on the outside with Nonwoven fibers such as nylon, mohair, and lamb's wool, and mounted on a roller with a handle; used for application of paint or varnish.

Paint System See ▶ Coat.

Paint Thinner See ▶ Thinner.

Paisley \pāz-lē\ *adj* {*often capitalized*} [Paisley, Scotland] (1824) A drop-shaped pattern that is extremely popular for men's ties and women's wear.

PAK n Abbreviation for Polyester Alkyd Resin. See ▶ Alkyd Resin.

Pale Boiled Oil n Lightly blown linseed oil containing a small amount of driers, such as lead and manganese.

Pale Crepe Light, unsmoked natural rubber.

Palette \pa-lət\ n [F, fr. MF, dim. of *pale* spade, fr. L *pala*; prob. akin to L *pangere* to fix] (1622) (1) Surface in which an artist sets out, and mixes his pigments and range of colors which he uses. (2) The typical color range of a school or group. (3) The gamut of colors possible when mixing a prescribed group of colorants. (Gair A (1996) Artist's manual. Chronicle Books LLC, San Francisco, CA)

Palette Knife n (1759) A spatula, usually smaller and slightly more flexible than the kind used for domestic and laboratory purposes, which serves to mix the artist's paint.

Palmitic Acid \( )pal-mi-tik-\ n [ISV, fr. *palmitin*] (1857) $CH_3(CH_2)_{14}COOH$. Bp, 278 °C; mp, 62 °C; sp gr, 0.846/8 °C; acid value, 218.8.

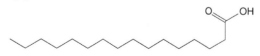

Palm Oil Pitch See ▶ Fatty Acid Pitches.

Pan \pan\ n [ME *panne*, fr. OE (akin to OHGr *phanna* pan), fr. L *patina*, fr. Gk *patanē*] (before 12c) In coil coating, the open container at the coater which holds the paint, and also the place where the pickup roller revolves in the paint.

PAN n Abbreviation for Poly(Acrylonitrile).

Panel Decoration n A type of wallpaper which flourished in the second half of the eighteenth century, related to the wood-paneled walls then in use. Wallpaper panels of the present day are thought of more as "spot" decorations.

Paneling n Distortion of a filled or partly full plastic container occurring during aging or storage, due to outward diffusion of solvent that causes reduced pressure inside the container.

Panels (Hosiery) n Knitted panels used for testing purposes.

Panné Satin n A satin fabric with an unusually high luster because of the application of very heavy roll pressure in finishing. Panné satin is made of silk or one of the manufactured fibers.

Pantone Printing n Planographic printing proves using plates having mercury- or amalgam-coated nonprinting areas.

PAPA Abbreviation for ▶ Polyazelaic Polyanhydride.

Paper, Building See ▶ Building Paper.

Paper Chromatography \-krō-mə-tä-grə-fē\ n (1948) The original ▶ Chromatography.

Papermarker's Felt n Formerly, a heavy, wide, coarse, worsted, or woolen fabric that was threaded between the rolls of the papermaking machine to form an endless conveyer belt for pulp or wet paper in its passage through the machine. These products are now also made of various constructions, woven, and nonwoven, of manufactured fibers and monofilaments.

PAPI (PMPPI) Abbreviation for Polymethylenepolyphenylene Isocyanate. See ▶ Diisocyanate.

Papier-mâché \pā-pər-mə-shā\ n [F, literally, chewed paper] (1753) Paper which is wet, molded, and hardened into forms used for decorative purposes.

Para- \pär-ə\ n [ME, fr. MF, fr. L, fr. Gk fr. *para*; akin to Gk *pro* before] (1) A chemical prefix from the Greek word meaning beside or beyond, denoting a relation "alongside" another compound such as a higher hydrated form of an acid or a polymeric form, as in paraldehyde. (2) (italicized, *p*-) Denoting the relation of opposite carbon atoms in the benzene ring, the 4-position in a singly substituted benzene. In this use, the prefix is ignored in alphabetizing compounds.

Para Brown *n* Bright brown pigment of good staining power, made by treating para red with copper salts.

Parachlor Nitraniline Red See ▶ Parachlor Red.

Para-chlor-ortho-nitraniline *n* Pigment Red 6 (12090). The parachlor variety of chlorinated para red is redder (less yellow) in shade and more transparent compared to the ortho-chlor but much superior in lightfastness. It is superior in tint lightfastness to toluidine red. Special precautions are required in drier addition to parachlor red paints since it discolors (dark and dull) with cobalt and/or iron driers. Syn: ▶ Parachlor Red.

Parachlor Red See ▶ Para-Chlor-Ortho-Nitraniline.

Paracoumarone-Indene Resins *n* Resin obtained from coumarone by polymerization with sulfuric acid. See ▶ Coumarone-Indene Resins.

Paraffin \ˈpar-ə-fən\ *n* [Gr, fr. L *parum* too little (akin to Gk *pauros* little, *paid-*, *pais* child) + *affinis* bordering on] (1838) (1) A Syn: ▶ Alkane. (2) A colorless, translucent wax obtained from petroleum-refining residues, a mixture of most saturated, straight-chain hydrocarbons melting between 49°C and 63°C. (3) In Britain and its former possessions, kerosene.

Paraffin Oil *n* An oil either pressed or dry-distilled from paraffin distillate.

Paraffin Wax *n* Inert hydrocarbon wax derivative of crude petroleum. Paraffin waxes are distinguished by their melting points. Their main uses include the conferring of water resistance, slip, or solvent retention in special types of compositions.

Paraformaldehyde *n* \ˌpar-ə-fór-ˈmal-də-ˌhīd\ *n* (1894) A low-molecular-weight, linear polymer of formaldehyde $HOCH_2(OCH_2)_nOCH_2OH$, a white solid that is easily depolymerized by mild heating to yield anhydrous formaldehyde gas. It is therefore a convenient form in which to handle and ship formaldehyde for industrial processes such as the manufacture of ▶ Acetal Resins, it high-molecular-weight, stable homologs. See also ▶ 1,3,5-Trioxane.

Paralac Polyester resin, manufactured by ICI, Great Britain.

Paralleling *n* The process of aligning fibers to produce a more uniform, smoother, stronger yarn.

Parallel Laminated \ˈpar-ə-ˌləl ˈla-mə-ˌnāt-\ *n* Pertaining to a laminate in which all layers of reinforcement are oriented approximately parallel with respect to the length or the direction of applied tensile stress. See ▶ Laminated, Parallel.

Parallel-Plate Viscometer *n* (1) An instrument consisting of two circular parallel plates, the lower one stationary, the upper one rotatable, the disk-shaped specimen being confined between the plates. An example of a this type of viscometer is the ▶ Mooney Viscometer.

Parallels *n* (1) Spacers placed between the steam plate and press platen to prevent the middle section of a compression mold from bending under pressure. (2) Pressure pads or spacers between the steam plates of a mold when the land area is too small. The pads control the height when the mold is closed and thus prevent crushing parts of the mold.

Paramagnetic Materials \ˌpar-ə-mag-ˈne-tik-\ *n* Those wit\hin which an applied magnetic field is slightly increased by the alignment of electron orbits. The slight diamagnetic effect in materials having magnetic dipole moments is overshadowed by this paramagnetic alignment. As the temperature increases this Paramagnetism disappears leaving only diamagnetism. The permeability of paramagnetic materials is slightly greater than that of empty space. (Handbook of chemistry and physics, 52nd edn. Weast RC (ed). The Chemical Rubber, Boca Raton, FL)

Paramagnetism *n* A weak attraction into a magnetic field, a result of the presence of unpaired electrons in a substance.

Parameter \pə-ˈram-ə-tər\ *n* [NL, fr. *para-* + Gk *metron* measure] (1656) (1) Loosely, a system factor or variable that may take on a range of values as decided by the observer or operator of the system. Example: hydraulic-line pressure and cylinder temperature are parameters in injection molding. (2) A defining constant of a statistical distribution, such as the mean or standard deviation of a normal distribution, and distinct from estimates of same calculated from sample measurements. (3) An independent variable through whose functions relations between other factors may conveniently be expressed.

Parameter of Specular Gloss *n* A measurement of specular gloss dependent on glossmeter geometry.

Para Nitraniline Red See ▶ Para Reds.

Paraphthalate Plasticizer *n* Any of a family of plasticizers derived by reacting terephthalic acid with an alcohol. They are similar in plasticizing capacity to the ▶ Orthophthalate Plasticizers while offering improved performance in areas such as volatility, low temperature flexibility, electrical, and lacquer-marring. With the exception of dioctyl terephthalate (DOTP), a liquid plasticizer suitable for plastisols, most paraphthalates are solids when prepared from alcohols having an average chain length over six carbon atoms.

Para Red *n* Pigment Red 1 (12070). Series of red pigments made by coupling reactions involving diazotized *p*-nitroaniline and alkaline β-naphthol. Adjustments in shade are made by substituting part of the β-naphthol.

Para Toner See ▶ Para Reds.

Paraxylylene See ▶ Parylene.

Parfocal Objectives \pär-fō-kəl-əb-jek-tiv\ *n* Objectives which are mounted so that only small adjustment of the bodytube and stage is necessary to focus after changing from one objective to any of the others. They are mounted in such a way that the second conjugate plane is in the same position on the optical axis of the microscope for each objective. Objectives used on a rotating nosepiece are usually parfocal. Eyepieces are also parfocal within any given manufacturer's series.

Parison *n* The hollow tube or other preformed shape of molten thermoplastic that is inflated inside the mold in the process of blow molding. Most commonly, the parison is extruded immediately before blowing, but parisons are also injection molded and may also be chilled and stored, to be reheated before blowing. In the earliest application of blow molding, a pair of calendered sheets joined along the edges was used as the parison.

Parkerized *adj* Descriptive of iron or steel which has received a rust-proofing treatment by being dipped in a boiling solution of manganese dihydrogen phosphate; this protective coating also improves the bonding of paints and lacquers.

Parkesine *n* The name given to the historic first (commercially unsuccessful) thermoplastic, made by plasticizing ▶ Cellulose Nitrate. The polymer was dissolved in a solvent, castor oil was mixed in, and the solvent was evaporated. The product was developed by Alexander Parkes and was the forerunner of Celluloid, which was perfected in 1870 by John Wesley Hyatt, who used camphor as the plasticizer.

Parlon *n* Chlorinated rubber, manufactured by Hercules Powder, US.

Parquet \pär-kā\ *n* [F, fr. MF, small enclosure, fr. *parc* park] (1818) A wood floor inlaid with geometric patterns. A type of decoration dating from seventeenth century France.

Parqueting *vt* (1678) Method of reinforcing the backs of wood panels with battens to prevent them from warping.

Parquetry \pär-kə-trē\ *n* (ca. 1842) Patterned wood inlay, especially for floors.

Partial Aromatic Solvents See ▶ Aromatic Solvents.

Partially Oriented Staple *n* Staple fibers cut from tow that has been drawn less than normal so that only partial longitudinal orientation of the polymer molecules exists.

Partially Oriented Yarns *n* (**Poly**) Filament yarns in which the draw ratio is less than normal resulting in only partial longitudinal orientation of the polymer molecules.

Partial Molar Quantities *n* The partial molar quantity of the substance A in a mixture of n components is the change in a thermodynamic property of such mixture per mole of component A. The partial molar quantities show the contribution of component A to the total thermodynamic properties of the mixture.

Partial Pressure *n* The pressure which a gas in a mixture would exert on the walls of a container if no other gases were present. See ▶ Raoult's Law.

Particle Board *n* A panel material composed of small discrete pieces of wood or other lignocellulosic materials that are bonded together in the presence of heat and pressure by a synthetic resin adhesive. (Handbook of adhesives. Skeist I (ed). Van Nostrand Reinhold, New York, 1990) Particle boards are further defined by the method of pressing. When the pressure is applied in the direction perpendicular to the faces, as in a conventional multiplaten hot press, they are defined as flatplaten pressed, and when the applied pressure is parallel to the faces, they are defined as extruded. *Also called Flake Board.*

Particle Size *n* (**Rayleigh, Mie, Frankhoffer**) (1) Solid particles of matter and their dimensions (usually average diameter) which industrial materials are composed; carbon black, clays, etc., the fundamental theories for studying the scattering of light by the Tyndel effect produced by increasing sizes of particles from "molecular to to colloid to visible" are Rayleigh, Mie and Frankhoffer, respectively. (2) In paints, the diameter of a pigment or latex particle; usually expressed in mils or micrometres. (Particle size distribution II: Assessment and characterization. Provder T (ed). Oxford University Press, New York, 1991)

Particle size conversion				
Sieve designation		Nominal sieve opening		
Standard	Mesh	Inches	mm	Microns
25.4 mm	1 in.	1.00	25.4	25,400
22.6 mm	7/8 in.	0.875	22.6	22,600
19.0 mm	3/4 in.	0.750	19.0	19,000
16.0 mm	5/8 in.	0.625	16.0	16,000
13.5 mm	0.530 in.	0.530	13.5	13,500
12.7 mm	1/2 in.	0.500	12.7	12,700
11.2 mm	7/16 in.	0.438	11.2	11,200
9.51 mm	3/8 in.	0.375	9.51	9,510
8.00 mm	5/16 in.	0.312	8.00	8,000
6.73 mm	0.265 in.	0.265	6.73	6,730

Particle size conversion				
Sieve designation		Nominal sieve opening		
Standard	Mesh	Inches	mm	Microns
6.35 mm	1/4 in.	0.250	6.35	6,350
5.66 mm	No. 3 1/2	0.223	5.66	5,660
4.76 mm	No. 4	0.187	4.76	4,760
4.00 mm	No. 5	0.157	4.00	4,000
3.36 mm	No. 6	0.132	3.36	3,360
2.83 mm	No. 7	0.111	2.83	2,830
2.38 mm	No. 8	0.0937	2.38	2,380
2.00 mm	No. 10	0.0787	2.00	2,000
1.68 mm	No. 12	0.0661	1.68	1,680
1.41 mm	No. 14	0.0555	1.41	1,410
1.19 mm	No. 16	0.0469	1.19	1,190
1.00 mm	No. 18	0.0394	1.00	1,000
841 µm	No. 20	0.0331	0.841	841
707 µm	No. 25	0.0278	0.707	707
595 µm	No. 30	0.0234	0.595	595
500 µm	No. 35	0.0197	0.500	500
420 µm	No. 40	0.0165	0.420	420
354 µm	No. 45	0.0139	0.354	354
297 µm	No. 50	0.0117	0.297	297
250 µm	No. 60	0.0098	0.250	250
210 µm	No. 70	0.0083	0.210	210
177 µm	No. 80	0.0070	0.177	177
149 µm	No. 100	0.0059	0.149	149
125 µm	No. 120	0.0049	0.125	125
105 µm	No. 140	0.0041	0.105	105
88 µm	No. 170	0.0035	0.088	88
74 µm	No. 200	0.0029	0.074	74
63 µm	No. 230	0.0025	0.063	63
53 µm	No. 270	0.0021	0.053	53
44 µm	No. 325	0.0017	0.044	44
37 µm	No. 400	0.0015	0.037	37

Larger sieve openings (1–1/4 in.) have been designated by a sieve "mesh" size that corresponds to the size of the opening in inches. Smaller sieve "mesh" sizes of 3 1/2–400 are designated by the number of openings per linear inch in the sieve.

The following convention is used to characterize particle size by mesh designation:

- A "+" before the sieve mesh indicates the particles are retained by the sieve.
- A "−" before the sieve mesh indicates the particles pass through the sieve.
- Typically 90% or more of the particles will lie within the indicated range. For example, if the particle size of a material is described as −4 +40 mesh, then 90% or more of the material will pass through a 4-mesh sieve (particles smaller than 4.76 mm) **and** be retained by a 40-mesh sieve (particles larger than 0.420 mm). If a material is described as −40 mesh, then 90% or more of the material will pass through a 40-mesh sieve (particles smaller than 0.420 mm).

This information is also provided on page T848 of the Aldrich 2003–2004 Catalog/Handbook of Fine Chemicals.

Particle Size Distribution n The relative percentage by weight or number of each of the different size fractions of particulate matter.

Particulate n \pär-ˈti-kyə-lət also -ˌlāt\ (1871) (1, adj) Existing as minute, separate particles. (2, n, plural) A particulate substance, a powder.

Particulate Composite n A plastic filled with solid particles of one or more substances that do not melt during processing. See ▶ Filler.

Particulates n Finely divided solid or liquid particles in the air or in an emission. Particulates include dust, smoke, fumes, mist, spray, and fog.

Parting Agent n (release agent, mold lubricant, mold release) A lubricant, often wax, silicone oil, or fluorocarbon fluid or solid, used to coat a mold cavity to prevent the molded piece from sticking to it, and thus to facilitate its removal from the mold. Parting agents are often packaged in aerosol cans for convenience in application.

Parting Line n (flash line) (1) A line marked on a three-dimensional model from which a mold is to be prepared, to indicate where the mold is to be split into two halves or several components. (2) The mark on a molded or cast article caused by slight flow of material into the crevices between mold parts. If the amount of material is sufficient that it must be removed, it is called Flash.

Partition \par-ˈti-shən\ n (15c) (1) Any building component that divides space such as a wall, door, window, roof, or floor-ceiling assembly; or a combination of components such as a wall containing a door and a window. (2) A solvent system in which the solute is divided between two components of the liquid phase, the ratio of their concentration (partition coefficient) being constant for a given substance.

Parylene n (poly-p-xylylene) Generic name for a group of film-forming thermoplastics introduced in 1965 by Union Carbide. The basic member, trade named Parylene N, is a polymer of p-xylylene, which has the structure shown below. Parylenes C and D contain one and two chlorine atoms on the benzene ring, and other types have recently become available. The polymer is formed on a receiving surface by pyrolyzing the dimmer of p-xylylene, a white powder, in vacuum (0.13 Pa abs) to its monomer vapor, which then flows to the room-temperature deposition chamber and forms a polymeric film with the structure $(-C_6H_4-CH_2CH_2-)_n$. Objects to be coated are usually mounted on rotatable racks to improve the uniformity of coating thickness. The films are tough, have excellent chemical resistance, low permeability, high thermal stability, and dielectric strength, and have been used to protect electronic assemblies and other critical parts from atmospheric oxygen and moisture. Very thin films – down to 25 nm (1 μm) – can be formed without pinholes, and thickness uniformity is good even on irregular surfaces. The rather high cost of the material and processing still limits use.

Parylen C *n* Poly(monochyloro-*p*-xylylene), manufactured by Union Carbide, US.

$$*-\left[-H_2C-\bigcirc-CH_2-\right]_x-*$$

Parylen N *n* Poly(*p*-xylylene), manufactured by Union Carbide, US.

Pascal \pas-kal\ *n* [Blaise *Pascal*] (1956) (Pa) the SI unit of pressure and stress, equal to 1 newton per square meter (N/m^2). The pascal and its multiples are intended to supersede all other units of force per unit area such as pounds per square inch, atmospheres, torrs, etc. 1 sp = 6.894,757 kPa, 1 atm = 101.3250 kPa, and 1 torr = 133.322 Pa.

Pascal Second *n* (pascal-second, Pa s) The SI unit of dynamic (absolute) viscosity, equal to 1 N s/m^2. Some conversions of older viscosity units (of which there is a bewildering plethora) to Pa s are given in the appendix. When shear stress τ assumes its alternate identify, ▶ Momentum Flux, the pascal-second is interpreted as 1 kg/m s).

Pascal's Law *np* Externally applied pressure on a confined fluid is transmitted equally in all directions.

Pass *n* Motion of spray gun in one direction only.

Passivation \pa-si-vā-shən\ *n* (1913) Act of making inert or unreactive.

Paste \pāst\ *n* [ME, fr. MF, fr. LL *pasta* dough, paste] (14c) An adhesive composition having a characteristic plastic-type consistency, that is, a high order or yield value, such as that of a paste prepared by heating a mixture of starch and water and subsequently cooling the hydrolyzed product. See also ▶ Adhesive, ▶ Glue, ▶ Mucilage and ▶ Sizing.

Paste Blue See ▶ Iron Blue.

Paste Driers *n* Driers made by grinding or mixing suitable inorganic drying compounds of lead or manganese into drying oil, together with some cheap filler to provide bulk. The resultant product is a stiff paste. Rarely made today. *Known also as Patent Driers.*

Paste Filler See ▶ Filler.

Pastel \pa-stel\ *adj* (1884) (1) Light tint; masstone to which white has been added. (2) Soft delicate hue. (3) In art, a picture made with a crayon of pigment with some binding medium such as gum.

Pastel Painting ▶ See Pastel.

Paste, Pigment *n* Pigment dispersion concentrate to permit substantial reduction (let down) with solvent, water, or vehicle.

Paste, PVC A term sometimes used for ▶ Plastisol.

Paste Resin A term sometimes used for PVC resins used in making vinyl dispersions such as plastisols. See ▶ Dispersion Resin.

Pasteurizing Varnishes \pas-chə-rīz-eŋ vär-nish\ *n* Varnishes for food containers which are capable of withstanding immersion in water at pasteurizing temperature for half an hour or more. Pasteurizing temperature is about 70°C, but food canners often require varnishes to withstand immersion in boiling water, or water boiling under pressure, without softening.

Paste Water Emulsion Wax *n* Similar to liquid water emulsion wax except furnished in paste form. Must be polished for luster. Occasionally pigmented. May contain some solvent.

Pastiche \pas-tēsh\ *n* [F, fr. It *pasticcio*] (1878) Work of art composed in the style of another person.

PAT Abbreviation for ▶ Polyaminotriazole.

Patent Driers See ▶ Paste Driers.

Patent Vermillion See ▶ Mercuric Sulfide.

Pathogenicity \pa-jə-ni-sə-tē\ *n* [ISV] (1852) The state or condition of producing disease or capable of doing so.

Pathogen, Pathogene \pa-thə-jən\ *n* [ISV] (1880) A parasite or virus capable of causing disease.

Patina \pə-tē-nə\ *n* [It, fr. L, shallow dish] (1748) (1) The film (usually greenish) which forms on copper or copper alloys through chemical action. (2) The gloss on wooden and other surfaces. (3) The softening of color that develops with age.

Pat-Out See ▶ Tap-Out.

Pattern \pa-tərn\ *n* [ME *patron*, fr. MF, fr. ML *patronus*] (14c) (1) An arrangement of form; a design or decoration such as the design of woven or printed fabrics. (2) A model, guide, or plan used in making things, such as a garment pattern.

Pattern Staining *n* Patterns on surfaces caused by deposits of dust, the amount of which varies according to the relative thermal conductivity of different pats of the structural background. The more highly conductive (colder) areas collect more dust that those of poor conductivity, thus the pattern of the underlying structure shows. The dust is often carried towards the surface by convection currents.

Pattern Wheel *n* In a circular-knitting machine, a slotted device for controlling individual needles so that patterns can be knit in the fabric.

Pb *n* Chemical symbol for the element lead (Latin: *plumbum*).

PB *n* Abbreviation for ▶ Poly-1-Butene. See ▶ Polybutylene Resin.

PBAN *n* Abbreviation of Polybutadiene-Acrylonitrile copolymer. See ▶ NBR and ▶ Acrylonitrile-Butadiene Copolymer.

PBI *n* Abbreviation for ▶ Polybenzimidazole, manufactured by Celanese, US.

PBMA *n* Abbreviation for Poly-*n*-Butyl Methacrylate.

PBS *n* Abbreviation for Polybutadiene-Styrene copolymer. See ▶ Styrene-Butadiene Thermoplastic.

PBTP *n* Abbreviation for ▶ Polybutylene Terephthalate.

PC *n* Abbreviation for Polycarbonate Resin.

PCB *n* Poly(acrylonitrile) fibers (EEC abbreviation). See ▶ Chlorinated Diphenyls.

PCF *n* Poly(trifluorochloroethylene) fiber.

PCL *n* Abbreviation for Polycarproylactone.

PCT See Polycyclohexylene Dimethylene Terephthal-Ate.

PCTFE *n* Poly(trifluorochloroethylene) fiber. See ▶ Polychlorotri-Fluoroethylene.

PCTFE Fluoroplastics *n* [CF_2CFCl]. Common name for Ethane, chlorotrifluoro-, homopolymer. It is also commonly known as Halocarbon oil. PCTFE is prepared by free radical polymerization in aqueous solution. It is used in seals, gaskets and in certain electrical applications.

PDAP *n* Abbreviation for Poly(Diallyl Phthalate). The abbreviation DAP is widely used for both the monomeric and polymeric foams of ▶ Diallyl Phthalate.

PDCA *n* Abbreviation for Painting and Decorating Contrac-Tors of America.

PDMS *n* Abbreviation for ▶ Polydimethyl Siloxane.

PE *n* (1) Polyester fiber (EEC abbreviation). (2) Abbreviation for ▶ Poly-Ethylene. (3) Abbreviation for ▶ Pentaery-Thritol.

Peacock Blue *n* (1881) Lake of aid glaucine blue dye on alumina hydrate. Structurally, the dye is alpha, alpha-bis [*N*-ethyl-*N*-(4-sulfobenzyl) amino phenyl] – alpha-hydroxy-ortho toluenesulfonic acid sodium salt.

Peanut-Hull Flour See ▶ Nutshell Flour.

Peanut Oil *n* (1882) A fixed oil that is yellow to greenish yellow. It is a typical nondrying oil of olive oil type. It is soluble in ether, petroleum ether, carbon disulfide and chloroform; insoluble in alkalies, but saponified by alkali hydroxides with formation of soaps; insoluble in water; slightly soluble in alcohol. It is principally glycerides of oleic and linoleic acids, with lesser amounts of the glycerides of palmitic, stearic, archidic, behenic, and lignoceric acids. Peanut oil is made by pressing ground peanut meats or by extraction with hot or cold solvents. Constants: Sp gr, 0.912–0.920 (25°C); solidifying point, −5°C to +3°C; saponification value, 186–194; iodine value, 99–98; refractive index, 1.4625–1.4645 (40°C). (Bailey's industrial oil and fat products. Shahidi F, Bailey AE (eds). Wiley, New York, 2005; Paint: Pigment, drying oils, polymers, resins, naval stores, cellulosics esters, and ink vehicles, vol 3. American Society for Testing and Material, 2001) Also known as *Archis Oil and Groundnut Oil*.

Pearl \\ˈpər(-ə)l\\ *n* See ▶ Purl (2).

Pearlescent \\ˌpər-ˈle-s°nt\\ *adj* (1936) An appearance resembling that of natural pearls or mother-of-pearl; it results from the specular reflectance of alternating thin layers of differing refractive index; similar to nacreous, interference color; the pigment particles are transparent, thin platelets of high refractive index which partially reflect and partially transmit incident light; simultaneous reflection from many layers of oriented platelets creates a sense of depth that is characteristic of nacreous luster. Syn: ▶ Nacreous. See ▶ Interference Color. (Bailey's industrial oil and fat products. Shahidi F, Bailey AE (eds). Wiley, New York, 2005; Paint/coatings dictionary. Compiled by Definitions Committee of the Federation of Societies for Coatings Technology, 1978)

Pearlescent Pigment *n* (pearl-essence pigment, nacreous pigment) A pigment with crystalline, transparent particles in the form of parallel platelets that impart an appearance of mother-of-pearl to plastics. The thin platelets have a high refractive index. Each crystal reflects only a portion of incident light reaching it, transmitting the remaining light to the crystal below. The simultaneous reflection of light from many parallel layers produces the characteristic pearly luster, the brilliance of which depends on the uniformity and parallelism of the crystals. Natural pearlescent pigments are composed primarily of guanine crystals derived from fish scales. They are expensive but nontoxic. The synthetic pearlescents are based on crystallized lead or bismuth compounds or platelets of mica coated with a dye or pigment. (Bailey's industrial oil and fat products. Shahidi F, Bailey AE (eds). Wiley, New York, 2005) See ▶ Nacreous Pigment, ▶ Interference Pigments.

Pearl Essence *n* (ca. 1909) Extract from fish scales used for obtaining the mother-of-pearl effect.

Pearlite \\ˈpər(-ə)-ˌlīt\\ *n* [F *perlite*, fr. *perle* pearl] (1888) The lamellar mixture of ferrite and cementite in slowly cooled iron-carbon alloys occurring normally as a principal constituent of both steel and cast iron. Same as perlite.

Pearl Lacquer See ▶ Pearlescent.

Pearl Moss See ▶ Carrageen.

Pearl Polymerization See Granular Polymerization and Suspension and Bead Polymerization.

Pearlstone See ▶ Perlite, Expanded.

Pear Oil See ▶ Amyl Acetate.

Peat Wax *n* Wax, resembling molten wax, obtained by the solvent extraction of pet. Mp, 70–100°C.

Peau de Soie *n* A heavyweight, soft satin of silk or manufactured fiber with a fine cross rib and a dull luster. The term is French for "skin of silk."

Pebble Mill *n* A rotating porcelain, buhrstone (or other nonmetallic lined) cylinder containing pebbles or porcelain balls or rods as the grinding media. In the manufacture of pigments or paints it is used to grind and/or disperse. See ▶ Ball Mill.

Pebble-Weave Fabric *n* A fabric with an irregular or rough surface texture formed by either a special weave or by the use of highly twisted yarns that shrink when they are wet.

PE CE *n* Post-chlorinated vinyl chloride polymer. The post-chlorination process increases chlorine content from 57% to 64%. The resulting polymer is soluble in acetone and can be wet spun. Manufactured by BASF, Germany.

Pectin \ˈpek-tən\ *n* [F *pectine*, fr. *pectique*] (1838) A water-soluble plant polysaccharide, mainly d-galacturonic acid, but also containing other sugar units. CAS Registry Number: 685-73-4. Molecular Formula: $C_6H_{10}O_7$. Molecular Weight: 194.14. Percent Composition: C 37.12%, H 5.19%, O 57.69%. Literature References: Obtained by hydrolysis of pectin where it is present as polygalacturonic acid: Ehrlich (1917) Chem Ztg 41:197; Ehrlich, Guttmann (1933) Biochem Z 259:100; Ber 66:220 (1933); Niemann, Link (1934) J Biol Chem 95:203; 104, 743 (1934); Morell, Link (1933) ibid. 100:385; Anderson, King (1961) J Chem Soc 5333. Isoln from mustard seeds: Goering, US 2987448 (1961 to Oil Seed Prod.).

Derivative Type: α-Form. Properties: Monohydrate, needles, mp 159°. Soluble in water; slightly sol in hot alcohol. Practically insol in ether.

Melting point: mp 159°. Optical Rotation: $[\alpha]_D^{20}$ +98.0° → +50.9° (water). (Merck index, 13th edn. Merck, Whitehouse Station, NJ, 2001)

Pectin is polygalacuronic acid: Composed of repeat units of D-Galacturonic acid,(2S, 3R, 4S, 5R)-2,3,4,5-Tetrahydroxy-6-oxo-hexanoic acid

Pedion *n* In crystallography, a single face having no equivalent.

PEEK *n* Abbreviation for Polyetheretherketone.

Peel Adhesion *n* The force required to delaminate a structure or to separate the surface layer from a substrate. Peel adhesion is the usual measure of the strength of the bond between fiber reinforcements and rubber in tires and other mechanical rubber goods.

Peeler *n* (1) A machine for slitting large rolls or blocks of foamed plastics into thin sheets, by rotating the blocks into a horizontally mounted band saw blade. Sheets as thin as 1.5 mm may be produced by this method. (2) In beaming, a defect caused by a portion of an end sticking or remaining on the beam, causing the filament to strip back or peel until it is broken. Although they are often associated with ringers, peelers are not necessarily defects that will circle the beams.

Peeling *n* Spontaneous removal in ribbons or sheets of a paint, varnish, or lacquer film from a surface due to loss of adhesion.

Peel Ply *n* The outside layer of a laminate that is removed or sacrificed to achieve improved bonding of additional layers.

Peel Test *n* See Scotch-Tape Test.

PEG *n* Abbreviation for Polyethylene Glycol.

PEI *n* Abbreviation for Polyetherimide.

PEK *n* Abbreviation for Polyetherketone.

Pelerine \ˌpe-lə-ˈrēn\ *n* [obs. F, neckerchief, fr. F *pèlerine*, feminine of *pèlerin* pilgrim, fr. LL *pelegrinus*] (1744) A device for transferring stitches from the cylinder to the dial or vice versa on a circular-knitting machine.

Pelletier's Green See ▶ Hydrated Chromium Oxide.

Pelletization \ˌpe-lə-tə-ˈzā-shən\ *n* (1942) Processing of pigments or other chemical products into very small, free-flowing beads, which eliminates the dust nuisance.

Pelletizers *n* Equipment which forms pellets for plastic molding or other.

Pellets \ˈpe-lət\ *n* [ME *pelote*, fr. MF, fr. (assumed) VL *pilota*, dimin. of L *pila* ball] (14c) (molding powder) Granules or tablets of uniform size, consisting of resins or mixtures of resins with compounding additives, that have been prepared for molding and extrusion by shaping in a pelletizing machine or by extrusion into strands that are cut while hot or after solidifying in a water bath.

Peltier Effect \ˈpel-tē-ər, ˌpel-tē-ˈā\ *n* The physical phenomenon occurring at the junctions, in electric circuits, of dissimilar metals that is the inverse of the

Hall effect (see ▶ Thermocouple). When a current flows through the junction of two unlike metals it causes an absorption or liberation of heat, depending on the direction of flow and the metals. (Handbook of chemistry and physics, 52nd edn. Weast RC (ed). The Chemical Rubber, Boca Raton, FL)

Pencil Hardness *n* A measure of coating hardness based on the scratching of the film with pencil leads of known hardness. The result is reported a the hardest lead which will not scratch or cut through the film to the substrate. (Paint and coating testing manual (Gardner-Sward handbook) MNL 17, 14th edn. ASTM, Conshohocken, PA, 1995)

Pencilling *n* Painting mortar joints of brickwork with white paint to bring out the contrast between the joints and the brickwork. (Paint/coatings dictionary. Federation of Societies for Coatings Technology, Philadelphia, Blue Bell, PA, 1978)

Pendulum \ˈpen-jə-ləm\ *n* [NL, fr. L, neuter of *pendulus*] (1660) For a simple pendulum of length l, for a small amplitude, the complete period,

$$T = 2\pi\sqrt{\frac{l}{g}} > \text{ or } g = 4\pi^2\frac{l}{T^2}$$

T will be given in seconds if l is in centimeter and g in centimeter per square second. For a sphere suspended by a wire of negligible mass where d is the distance from the knife edge to the center of the sphere whose radius is r, the length of the equivalent simple pendulum,

$$l = d + \frac{2r^2}{5d}$$

If the period is P for an arc θ, the time of vibration in an infinitely small arc is approximately

$$T = \frac{P}{1 + \frac{1}{4}\sin^2\frac{\theta}{4}}$$

For a compound pendulum, if a body of mass m be suspended from a point about which its moment of inertia is I with its center of gravity a distance h below the point of suspension, the period

$$T = 2\pi\sqrt{\frac{I}{mgh}}.$$

(Handbook of chemistry and physics, 52nd edn. Weast RC (ed). The Chemical Rubber, Boca Raton, FL)

Penetrating Finish *n* A low-viscosity oil or varnish which penetrates wood, leaving a very thin film at the surface.

Penetration \ˌpe-nə-ˈtrā-shən\ *n* (1605) The entering of an adhesive into an adherend. This property of a system is measured by the depth of penetration of the adhesive into the adherend. It is also the ability of a liquid (ink, varnish, or solvent) to be absorbed into the paper or other printing substrate. (Handbook of adhesives. Skeist I (ed). Van Nostrand Reinhold, New York, 1990)

Penetration Number *n* Figure used, chiefly for bituminous type products, as a measure of their softness. It is obtained by allowing a weighted needle of specified dimensions to penetrate into the material under test, at a definite temperature. The penetration figure is usually recorded as the number of units of depth which the needle penetrates in a given time. (Asphalt science and technology. Usmani AM (ed). Marcel Dekker, New York, 1997)

Penetrometer \ˌpe-nə-ˈträ-mə-tər\ *n* [L *penetrare* + ISV -*meter*] (1905) Apparatus for measuring penetration number of a solid. (ASTM, www.astm.org)

Penné Velvet *n* Velvet of silk or a manufactured fiber, with a finish in which the pile is flattened and laid in one direction. Panné velvet is a lustrous, lightweight fabric. (Humphries M (2000) Fabric glossary. Prentice-Hall, Upper-Saddle River, NJ)

Pensky-Martens Closed Flash Tester *n* A device used in determining the flash point of liquids which have a viscosity of 45 SUS or more at 37.8°C, (100°F), or contain suspended solids and require stirring to obtain uniform distribution of heat, or have a tendency to form a surface film under test conditions.

Pentachloroethane \ˌpen-tə-ˌklōr-ˈe-ˌthän\ *n* $CHCl_2CCl$. Nonflammable solvent. Bp, 161°C; vp, 5 mmHg/20°C; sp gr, 1.685. CAS Registry Number: 76-01-7. Additional Names: Pentalin. Molecular Formula: C_2HCl_5. Molecular Weight: 202.30. Percent Composition: C 11.87%, H 0.50%, Cl 87.62%. Line Formula: CCl_3CHCl_2. Literature References: Toxicity data: Barsoum GS, Saad K (1934) Quart J Pharm Pharmacol 7:205. Properties: Liquid; chloroform-like odor. d_4^{25} 1.6712; bp 161–162°. mp −29°. n_D^{15} 1.5054. Insol in water. Miscible with alcohol, ether. MLD (mg/kg) in dogs: 1,750 orally; 100 i.v.; in rabbits: 700 s.c. (Barsoum, Saad). Melting point: mp −29°. Boiling point: bp 161–162°.

Index of refraction: n_D^{15} 1.5054. Density: d_4^{25} 1.6712. Toxicity data: MLD (mg/kg) in dogs: 1,750 orally; 100 i.v.; in rabbits: 700 s.c. (Barsoum, Saad). CAUTION: Potential symptoms of overexposure in exptl animals are irritation of eyes, skin; weakness, restlessness, irregular respiration, muscle incoordination; liver, kidney, lung changes. (Merck index, 13th edn. Merck, Whitehouse Station, NJ, 2001)

Pentachlorophenol \ ˌpen-tə-ˈklōr-ə-ˈfē-ˌnōl\ *n* (1879) C$_6$HCl$_5$O. A toxic, oil-soluble chemical; widely used as a wood preservative for protection against decay and insects. CAS Registry Number: 87-86-5, Additional Names: Penta; PCP; penchlorol. Trademarks: Santophen 20 (Monsanto). Molecular Formula: C$_6$HCl$_5$O. Molecular Weight: 266.34. Percent Composition: C 27.06%, H 0.38%, Cl 66.56%, O 6.01%. Literature References: Prepd by the chlorination of phenol in the presence of a catalyst. Toxicity study: Gaines TB (1969) Toxicol Appl Pharmacol 14:515. Properties: Needle-like crystals, mp 190–191°; bp ∼309–310° (dec). d$_4^{22}$ 1.978. Very pungent odor only when hot. Sublimes in needles. Almost insol in water (8 mg in 100 mL). Freely sol in alc, ether; sol in benzene; slightly sol in cold petr ether. LD$_{50}$ in male, female rats (mg/kg): 146, 175 orally (Gaines). Melting point: mp 190–191°. Boiling point: bp ∼309–310° (dec). Density: d$_4^{22}$ 1.978. Toxicity data: LD$_{50}$ in male, female rats (mg/kg): 146, 175 orally (Gaines). Derivative Type: Sodium salt. Additional Names: Sodium Pentachlorophenate; sodium pentachlorophenoxide. Trademarks: Santobrite (Monsanto); Dowicide G (Dow). Properties: Sol in water. CAUTION: Potential symptoms of overexposure are irritation of eyes, nose, throat; sneezing, cough; weakness, anorexia, weight loss; sweating; headache, dizziness; nausea, vomiting; dyspnea, chest pain; high fever. Direct contact may cause dermatitis. See NIOSH pocket guide to chemical hazards (DHHS/NIOSH 97–140, 1997) p 242; Patty's industrial hygiene and toxicology, vol 2B, Clayton GD, Clayton FE (eds). Wiley-Interscience, New York, 1994, pp 1603–1613. Use: Insecticide for termite control; pre-harvest defoliant; general herbicide. Antimicrobial preservative and fungicide for wood, wood products, starches, textiles, paints, adhesives, leather, pulp, paper, industrial waste systems, building materials. Surface disinfectant. (Merck index, 13th edn. Merck, Whitehouse Station, NJ, 2001) (See ▶ Pentachlorophenol)

Pentacite An alkyd resin formed by using pentaerythritol as the polyhydric alcohol.

Pentaerythritol (tetramethylol methane) 2,2-bis-hydroxymethyl)-1,3-propanediol. A white crystalline powder derived by reacting acetaldehyde with an excess of formaldehyde in an alkaline medium. It is used in the production of alkyd resins and chlorinated polyethers. It is used in the manufacture of alkyds, copal ester, rosin esters, synthetic resins, and lubricants. Bp, 276°C; mp, 262°C; sp gr, 1.399. Abbreviation: PE.

1,5-Pentanediol (pentamethylene glycol) HOCH$_2$(CH$_2$)$_3$CH$_2$OH. A colorless liquid used in the production of polyester and urethane resins. CAS Registry Number: 111-29-5, Additional Names: Pentamethylene glycol; 1,5-dihydroxypentane. Molecular Formula: C$_5$H$_{12}$O$_2$. Molecular Weight: 104.15. Percent Composition: C 57.66%, H 11.61%, O 30.72%. Line Formula: HOCH$_2$(CH$_2$)$_3$CH$_2$OH. Literature References: Prepd by hydrogenolysis of tetrahydrofurfuryl alcohol in the presence of copper chromite: Connor, Adkins (1932) J Am Chem Soc 54:4678; Kaufman D, Reeve W (1955) Org Syn coll vol III:693. Toxicity study: Smyth HF et al. (1962) Am Ind Hyg Assoc J 23:95. Properties: Viscous, oily liquid. Bitter taste. d^{20} 0.9941. mp −18°. bp$_{760}$ 239°, bp$_{3.0}$ 120°. n_D^{20} 1.4499. Flash pt 125°C (275°F). Miscible with water, methanol, alc, acetone, ethyl acetate. Soly in ether (25°): 11% w/w. Limited soly in benzene, trichloroethylene, methylene chloride, petr ether, heptane. LD$_{50}$ orally in rats: 5.89 g/kg (Smyth). Melting point: mp −18° Boiling point: bp$_{760}$ 239°; bp$_{3.0}$ 120°. Flash point: Flash pt 125°C (275°F). Index of refraction: n_D^{20} 1.4499. Density: d^{20} 0.9941. Toxicity data: LD$_{50}$ orally in rats: 5.89 g/kg. Use: As plasticizer in cellulose products and adhesives, in brake fluid compositions. Forms esters and polyesters which can be used as plasticizers, emulsifying agents and resin intermediates. (Plasticizer's data base. Wypych G (ed). Noyes, New York, 2003; Merck index, 13th edn. Merck, Whitehouse Station, NJ, 2001)

HO~~~~~~OH

Penta Resin *n* Ester gum made from rosin and pentaerythritol.

Penton® *n* Poly(2,2-dichloromethyl trimethylene oxide. Trade name for ▶ Chlorinated Polyether. Manufactured by Hercules Powder, US.

Penultimate Unit *n* \pi-ˈnəl-tə-mət-\ The unit adjacent to the radical endgroup (next to last unit) on a growing copolymer chain; prepenulti-mate is second to last unit, antepenulti-mate unit. (Odian GC (2004) Principles of polymerization. Wiley, New York)

PEO *n* Abbreviation for ▶ Polyethylene Oxide.

Peppery See ▶ Bitty.

Peptization *n* Process of bringing a solid into a colloidal solution. (Becher P (1989) Dictionary of colloid and surface science. Marcel Dekker, New York)

Peptizing Agents *n* (1) Materials which, when added in relatively small amounts, are capable of causing substantial reduction in viscosity, as the result of depolymerization, or dispersion. Peptization can occur in aqueous media, such as when agents are added to prevent flocculation or aggregation of particles. (2) Substances that act as a chemical plasticizer for natural and synthetic rubbers. They act as catalysts for oxidation breakdown of rubber during the milling or mastication period. Additional plasticization can be accomplished without further milling by heating the rubber containing the peptizing agent, thus reducing power consumption in breakdown. (Becher P (1989) Dictionary of colloid and surface science. Marcel Dekker, New York)

Peracetylated Rubber *n* Product prepared from raw rubber by treatment with acetic acid and hydrogen peroxide. Improved compatibility is obtained.

Perbunan C *n* Poly(chloroprene), manufactured by Bayer, Germany.

Perbunan N *n* Butadiene/acrylonitrile copolymer, manufactured by Bayer, Germany.

Percale \(ˌ)pər-ˈkā(ə)l\ *n* [Persian *pargālah*] (1840) A closely woven, plain-weave, spun fabric used for dress goods and sheeting, generally 80 × 80 threads per inch or better.

Percentage Elongation *n* 100 × ▶ Elongation.

Percent by Mass (Mass Percent) *n* A concentration unit: 100 times the mass of one component divided by the total mass of the solution.

Perceptible \pər-ˈsep-tə-bəl\ *adj* (1603) Capable of being perceived; discernible.

Perception \pər-ˈsep-shən\ *n* [L *perception-*, *perceptio* act of perceiving, fr. *percipere*] (14c) The process of mental functions that includes the combination of different sensations and the utilization of past observations in recognizing the objects and "facts" from which the stimulation arises. This is the phenomenon of "experience" that serves the engineer, scientist, researcher for performing competent work. In other words, there is no substitute for experience.

Perceptual *n* Adjective pertaining to or involving perception.

Perching Inspection of cloth for defects while it is run over a roller.

Perchloroethylene \(ˌ)pər-ˌklōr-ō-ˈe-thə-ˌlēn\ *n* (1873) Used for vapor degreasing and dry cleaning. Bp, 121°C; vp, 15 mmHg/20°C.

Perchloromethane See ▶ Carbon Tetrachloride.

Perchloropentacyclodecane *n* $C_{10}Cl_{12}$. A bridged bicyclic, saturated compound, a solid filler used as a flame retardant in epoxy resins, often in conjunction with antimony trioxide.

Perduren *n* Thioplasts, manufactured by Hoechst, Germany.

Perfect Diffuser *n* Theoretical ideal white substance which reflects 100% of the incident light in a perfectly diffuse way; official CIE reference white to which colors being measured are compared; a Lambert surface, following the Lambert cosine law perfectly. See ▶ Lambert's Law of Reflection.

Perfecting Press Any printing press that prints both sides of paper in one operation. (Saleh BEA, Teich MC (1991) Fundamentals of photonics. Wiley, New York)

Perfluoroalkoxy Resin *n* (PFA) A class of melt-processable fluoroplastics in which perfluoroalkyl side chains are connected to the fluorocarbon backbone of the polymer through flexible oxygen linkages. PFA resins have the desirable properties associated with fluoroplastics plus superior creep resistance, and are more easily processed by extrusion and injection molding.

Perfluoroelastomer *n* (tetrafluoro-perfluoromethyl vinyl ether copolymer) Introduced by DuPont in 1977 as Kalrez®, this elastomer combines the properties of a conventional fluoroelastomer, such as vinylidene fluoride-hexafluoropropylene copolymer, with those of a fluorocarbon resin such as polytetrafluoroethylene. It has found application for O-rings and seals that must withstand strong chemicals and solvents at high temperatures.

Perfluoroethylene *n* Syn: ▶ Tetrafluoroethylene. (See ▶ Perfluoroethylene)

Perforating *n* Any process by which plastic film, sheet, or tubing is provided with holes ranging from relatively large diameters for decorative effects (by means of punching or clicking) to very small, even invisible sizes. The latter are achieved by passing the material between rollers or plates, of which one of the pair is equipped with loosely spaced, fine needles; or by spark erosion.

Perilla Oil \pə-ˈri-lə\ *n* (1917) A drying oil obtained from the seed of the perilla plants, *Perilla ocymoides* and *Perilla nankinensis*, natives of the Orient. Its main constituent acids are linoleic and linolenic, and it has the highest iodine value of all known vegetable oils except chia. It is superior to linseed oil, both from the point of view of drying rate, especially in the form of stand oil, and also of polymerization rate. Sp gr, 0.933–0.937/15°C; iodine value, 194, saponification value 192. (Paint: Pigment, drying oils, polymers, resins, naval stores, cellulosics esters, and ink vehicles, vol 3. American Society for Testing and Material, 2001)

Period *n* (1) A horizontal series (row) of elements in the periodic table. (2) In uniform circular motion is the time of one complete revolution. In any oscillatory motion it is the time of a complete oscillation. Dimension – [*t*].

Periodic Law *n* (1872) Elements when arranged in the order of their atomic weights or atomic numbers show regular variations in most of their physical and chemical properties. (Whitten KW, Davis RE, Davis E, Peck LM, Stanley GG (2003) General chemistry. Brookes/Cole, New York)

Perishing *n* Degradation of a film with age, resulting in loss of film strength, flexibility, adhesion, etc., and which may be shown by the development of chalking, checking, cracking, and flaking, etc. (Degradation and stabilization of polymers. Zaiko GE (ed). Nova Science, New York, 1995)

Periston *n* Poly(vinyl pyrrolidone), manufactured by Bayer, Germany.

Perlenka *n* Poly(caprolactam), manufactured by AKU, The Netherlands.

Perlite \ˈpər-ˌlīt\ *n* [F, fr. *perle* pearl] (1833) A siliceous lava which, when heated to 720–1,090°C, expands to 10–20 times its original volume, forming tiny, hollow, spherical bubbles. Perlite is much used as an ingredient of lightweight concrete and as a density-lowering filler for plastics. See also ▶ Microspheres.

Perlite, Expanded *n* $1Na_2O \cdot 1\ K_2O \cdot 2.5\ Al_2O_3 \cdot 19.5\ SiO_2$. A unique form of siliceous lava that is characterized by many spherical and convoluted cracks. The interior or internal structure causes it to leak up into small spheres or "pebbles." Perlite expands when heated at a range of 720°C (1,500°F) to 1,090°C (2,000°F) to 10–20 times its initial volume. Used as insulation and filler, also extender. Density, 2.2 g/cm^3 (1.87 lb/gal); particle size, 44 μm. Syn: ▶ Pearlstone, ▶ Ground Perlite, ▶ Pearlite.

Perlon *n* Generic name for polyamides from caprolactam (nylon 6).

Perlon U *n* Polyurethane, manufactured by Bayer, Germany.

Perm *n* (1) A unit of measurement of water vapor transmission or permeance; a metric perm, 1 gr/24 h/m^2/mmHg or US unit, 1 grain/h ft^2 in.Hg; used to express the resistance of a material to the penetration of moisture. (2) A process involving ammoniacal chemicals and, usually, heat, by which straight strands of keratinous fiber are rendered into circular, spiral, and wave-like forms.

Permachor *n* A concept of M. Salame (1961), who discovered that the permeation rates of many (though not all) organic liquids through polyethylene (PE) films could be estimated with the equation:

$$\log Pf = 16.55 - 3700/T - 0.22\,\pi$$

in which *Pf* = the "permeability factor," T is absolute temperature, (K), and π is the *permachor* calculated for the permeating compound from a table of empirically determined, additive, atomic and structural contributions. Pf was measured by sealing the test liquid in a PE bag and measuring the loss of weight over a period of days, and was computed as (rate of weight loss, g/day) (film thickness, mil)/(film area, 100 in.^{2}), so has the units g mil/(day 100 in.^{2}), not the same as ▶ Permeability defined below. The concept worked well for some other film materials, too. Later (1967), Salame extended it to polymer permeation by oxygen, nitrogen, and carbon dioxide, this time computing the permachor for the polymer rather than the permeant. (Paint and coating testing manual (Gardner-Sward handbook) MNL 17, 14th edn. ASTM, Conshohocken, PA, 1995; Paint/coatings dictionary. Federation of Societies for Coatings Technology, Philadelphia, Blue Bell, PA, 1978)

Permanence $n \setminus$ˈpər-mə-nən(t)s\ n (15c) The resistance of any material to change with regard to its properties and performance with age or exposure to deleterious conditions such natural weathering. (Paint and coating testing manual (Gardner-Sward handbook) MNL 17, 14th edn. ASTM, Conshohocken, PA, 1995)

Permanent Deformation n The change in length of a sample after removal of an applied tensile stress and after the removal of any internal strain (e.g., by boiling off the sample and allowing it to dry without tension). The permanent deformation is expressed as a percentage of the original sample length.

Permanent Fast Violet See ▶ Mineral Violet.

Permanent Finish n A term for various finishing treatments, chemical and/or mechanical, applied to fabric so that it will retain certain properties, such as glaze of chintz, crispness of organdy, smoothness of cotton table damask, and crease, crush, and shrinkage resistance of many apparel fabrics during the normal period of wear and laundering.

Permanent Inks n Inks that do not readily fade or change color when exposed to light and weather. (Printing ink manual, 5th edn. Leach RH, Pierce RJ, Hickman EP, Mackenzie MJ, Smith HG (eds). Blueprint, New York, 1993)

Permanent Set n The deformation remaining after a specimen has been stressed in tension, compression, or shear for a specified time period and released for a specified time period. (Shah V (1998) Handbook of plastics testing technology. Wiley, New York)

Permanent Violets n (1) A light resistant, tungstated or molybdated methyl violet pigment used in printing inks. (2) Carbazole violet.

Permanent White See ▶ Barium Sulfate, Natural.

Permanent Yellow FGL n Pigment Yellow 97 (US Pat. 2,644,814). A monoazo pigment that has gained significant use in trade sales finishes. It is very similar in color to dalamar yellow and somewhat more bleed-resistant.

Permeability \ˈpər-mē-ə-ˈbi-lə-tē\ n (1759) (Polymeric mateials) The ease with which a gas or vapor passes through a membrane, e.g., a plastic sheet or film. It has been proved that permeability is equal to the product of *diffusivity* times the *solubility* of the gas or vapor in the plastic. *Coefficient of permeability* (*permeability coefficient*) = the rate of permeation of a gas or vapor per unit cross-sectional area of the film, divided by the concentration gradient through the film. Because of the wide choice that has existed for units of permeation rate, film area, and concentration, many different "convenient" sets of units have been used. ASTM (www.astm.org) defines permeability as the product of ▶ Permeance times the thickness of a film. Permeabilities of polymeric films to atmospheric gases and carbon dioxide vary from about 0.009–2.5 pmol/(m s Pa).

Permeance n The ratio of the ▶ Gas-transmission Rate to the difference in partial pressure of the gas on two sides of a sheet or film. The SI unit is mol/(m^2 s Pa), but much smaller submultiples are convenient. The test conditions must be stated. See ▶ ASTM D 1434 (Section 15.09), also Permeability (1).

Permeation \ˈpər-mē-ˈā-shən\ n (ca. 1623) The passage of gas, vapor, or liquid molecules through a film or membrane, usually without physically or chemically changing it, except that permeation involves solubility of the vapor in the film. See ▶ Permeability. (See ▶ ASTM, www.astm.org)

Permittivity \ˈpər-ˈmi-ˈti-və-tē\ n [^{1}permit + -ivity (as in *selectivity*)] (1887) See ▶ Dielectric Constant.

Permselective Membrane n A thin film that will preferentially permit gases of different kinds to pass through the film at different rates. For common gases such as hydrogen, oxygen nitrogen, and carbon dioxide, silicon rubber is the most permeable polymer.

Peroxide \pə-ˈräk-ˌsīd\ n [ISV] (1804) A polymerization *initiator* containing at least one pair of oxygen atoms bonded by a single covalent bond. Organic peroxides, analogous to H_2O_2 in which either or both of the H atoms have been replaced by organic radicals, are thermally unstable and are widely used as initiators in polymerizations. As they decompose, they form free two free radicals that can initiate polymerization reactions and effect crosslinking. A hydroperoxide provides only one free radical available for initiation. The rate of decomposition can be controlled by means of promoters or accelerators or by inhibitors when it is desired to slow the rate. Peroxides are used to cure thermosetting resins. (Odian GC (2004) Principles of polymerization. Wiley, New York; Whittington's dictionary of plastics. Carley James F (ed). Technomic, 1993)

$$HO\text{——}OH$$
peroxide

Peroxyester *n* (perester, *t*-alkyl peroxyester) Any of a family of liquid initiators used for crosslinking of polyethylene, polymerization of vinyls, high-temperature crosslinking of diallyl phthalate-modified polyesters, curing of styrene-modified polyesters, and styrenation of alkyd paints. They are aliphatic, not prone to yellowing and bleaching, and have good solubility and compatibility characteristics. One example of the numerous compounds in the family is *t*-butyl peroxypentanoate.

Persorption *n* The adsorption of a substance in pores only slightly wider than the diameter of adsorbed molecules of the substance.

Perspex Poly(methyl methacrylate). Manufactured by ICI, Great Britain.

Persulfonium Ion \-ˌsəl-ˈfō-nē-əm-\ *n* The ionic mechanism (*initiator*) proposed for sulfur vulcanization involves the reaction of a highly polarized sulfur bond R-S$^+$-S$^-$R, present in either elemental sulfur or an organic polysulfide with a double bond to form an intermediate persulfonium ion; an initiator for sulfur vulcanization. (Odian GC (2004) Principles of polymerization. Wiley, New York)

Perylene Pigments *n* A group of vat pigments, nearly all of which are N,N'-substituted peryzene-3,4,9,10-tetracarboxylic diimides. Scarlet and vermilion varieties, resistant to bleeding, light, heat, and chemicals are used in plastics.

PES *n* Polyester fiber *n* Abbreviation for ▶ Polyethersulfone.

PET *n* (PETP) Abbreviation for ▶ Polyethylene Terephthalate.

Peta- (P) The SI prefix meaning × 10^{15}.

PETP *n* (PET) Abbreviation for ▶ Polyethylene Terephthalate.

Petrifying Liquid *n* Usually a dilute emulsion of drying oil and/or resin in water, used as a sealing coat before applying an oil-bound water paint to a porous surface. It may also be used in place of water to thin the first coat of water paint before application to a porous surface.

Petrochemical \ˌpe-trō-ˈke-mi-kəl\ *n* (1942) Any chemical derived directly or indirectly from petroleum or natural gas.

Petrolatum \ˌpe-trə-ˈlā-təm, ˈlä-\ *n* [NL, fr. ML *petroleum*] (1887) Purified mixture of semisolid hydrocarbons with unctuous nature derived from petroleum. Syn: ▶ Petroleum Jelly.

Petroleum Bitumens \pə-ˈtrō-lē-əm bə-ˈtyü-mən\ *n* Distillation residues derived from crude petroleum.

Petroleum Ethers *n* Low boiling aliphatic fractions derived from crude petroleum by fractional distillation. The boiling ranges are fairly restricted, and products known as 40–60°C and 60–80°C petroleum ethers are common. The low boiling ranges and exceptionally high evaporation rates restrict the use of the solvents in the paint trade considerably. The petroleum ethers bear no resemblance chemically to the true ethers. The only resemblance is the high evaporation rate.

Petroleum Hydrocarbons *n* Hydrocarbons of aliphatic type obtained from crude petroleum and including VM&P naphtha, mineral spirits, and paraffins.

Petroleum Jelly *n* (1897) See ▶ Petrolatum.

Petroleum Naphtha \-ˈnaf-thə\ *n* A generic term applied to refined, partly refined, or unrefined petroleum products and liquid products of natural gas, not less than 10% of which distills below 175°C (347°F) and not less than 95% of which distills below 240°C (464°F), when subjected to distillation in accordance with AST Method D 86, Test for Distillation of Petroleum Products.

Petroleum Resin See ▶ Hydrocarbon Plastics.

Petroleum Spirits See ▶ Mineral Spirits.

PF *n* Abbreviation for Phenol-Formaldehyde Resin.

PFA See ▶ Perfluoroalkoxy Resin.

PFA Fluoroplastic Resins See ▶ Perfluoroalkoxy Resin.

PFEP *n* Copolymer from tetrafluoroethylene and hexafluoropropylene.

PF Resins See ▶ Phenolic Resins.

Pfund Hardness *n* A method of measuring ▶ Indentation Hardness of paints and related coatings in which a small hemispherical indenter made of quartz or sapphire is pressed into the coating (see ▶ ASTM, www.astm.org).

Pfund Hardness Number *n* The indentation hardness determined with a Pfund indenter and calculated as follows:

$$\text{PHN} = \frac{L}{A} = \frac{4L}{\pi d^2} = 1.27 \frac{L}{d^2}$$

where L = load in kilograms applied to the indenter, A = area of projected indentation in square millimetres, and d = diameter of projected indentation in millimeters. Abbreviation: PHN. (Handbook of physical polymer testing, vol 50. Brown R (ed). Marcel Dekker, New York, 1999)

Pfund Indenter *n* A hemispherical quartz or sapphire indenter of prescribed dimensions used for testing indentation hardness of organic coatings. (Handbook of physical polymer testing, vol 50. Brown R (ed). Marcel Dekker, New York, 1999)

pH *n* A measure of the acidity or alkalinity of an aqueous solution or mixture, defined as the logarithm to base 10 of the reciprocal of the effective hydrogen-ion concentration (H^+) in gram-equivalents per liter, moles per liter (moles/L). The equation for calculating pH is:

$$\text{pH} = -\log[H^+] = \log\frac{1}{H^+}$$

Pure water has a pH of 7 and is neutral. A one-unit drop in pH corresponds to a tenfold rise in H^+.

Phase \ˈfāz\ *n* [NL *phasis*, fr. Gk, appearance of a star, phase of the moon, fr. *phainein* to show (middle voice, to appear)] (1812) A physical distinct region with a uniform set of properties throughout.

Phase Angle *n* In many cyclic processes, particularly ones at high frequency, there is usually a time lag between the impulses driving the processes and the responses to those impulses. The lag time divided by the cycle period and multiplied by 2π is the phase angle (radians). This occurs in oscillatory rheometry, where the shear applied to the rotating element results in a lagging torque, which has two components, an elastic part, in phase with the displacement, and the viscous, out-of-phase component. The same situation exists in fatigue testing of plastics.

Phase Equilibrium *n* Liquids tend to evaporate and a liquid–vapor equilibrium can be established in which the rate of evaporation equals the rate of condensation. Such liquid–vapor equilibria are but one type of equilibrium which can exist between two different phases. Solids also tend to transform directly into the gaseous state. This process, analogous to evaporation of a liquid, is called *sublimation*. Sublimation occurs because at any temperature a certain fraction of the molecules at the surface of a crystal are moving energetically enough to fly off. The rate of sublimation of most common solids is low under ordinary temperature and pressure conditions. Snow, for example, can be observed to disappear gradually over a period of several days during which the temperature never rises to "freezing." Mothballs are composed of crystals of naphthalene or *para*-dichlorobenzene, which very slowly sublime to release toxic vapors. Just as the rate of evaporation of a liquid can be increased by lowering the external pressure on it, so also can the rate of sublimation of a solid similarly be increased. Fruits, vegetables, meats, and beverages such as coffee can be *freeze-dried* by first lowering the temperature far below 0°C and then subjecting them to a vacuum. The water then rapidly sublimes, leaving a dehydrated product which can be reconstituted later by restoring the water.

When a subliming solid is enclosed in a container, redeposition of gas molecules on the surface is possible and a state of equilibrium is reached: Solid $\rightleftharpoons$ gas The pressure of the gas in equilibrium with its solid is called the *equilibrium vapor pressure*, or *sublimation pressure*, of the solid. Just as with a liquid, the vapor pressure of a solid increases with increasing temperature. At the lower left of Vapor Pressure Curves figure is shown the variation of the vapor pressure of ice with temperature. The curve rises to the right until it stops at 0.01°C, at which temperature the vapor pressure is still only 4.58 mmHg. It stops at this point, because ice cannot exist above this temperature, called the *triple-point temperature* (for reasons which will become apparent shortly).

The *Clausius-Clapeyron equation* applies to solid-gas equilibria as well as to the liquid–gas equilibria. Thus if we plot the logarithm of the vapor pressure of a solid against the reciprocal of the temperature, we get a straight line of slope $\frac{-\Delta H_{sub}}{2.303R}$ where ΔH_{sub} is the heat of sublimation of the solid.

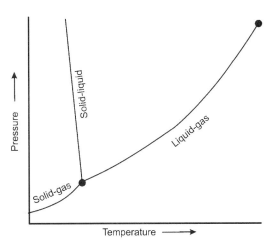

Conditions for phase equilibrium in water

Conditions for Phase Equilibrium shows a plot of the vapor-pressure curve, it can be viewed from a slightly different viewpoint: the points on each curve represent all the temperature–pressure combinations at which two phases (solid + gas, or liquid + gas, depending on the curve) can be at equilibrium. Liquid and gaseous water, for example, can be at equilibrium *only* at a temperature and pressure represented by a point which is on the liquid–gas equilibrium line.

Phase of Oscillatory Motion *n* The fraction of a whole period which has elapsed since the moving particle last passed through its middle position in a positive direction.

Phases Beam *n* A beam on which each of the ends is wound from the same depth of each of the bobbins on the creel. Phased beams are prepared when yarn properties vary from the inside to the outside of the bobbins in order to prevent warp streakiness in the finished fabric. (Kadolph SJJ, Langford AL (2001) Textiles. Pearson Education, New York)

Phase Separation *n* Separation of two or more phases due to temperature changes.

Phenol \ˈfē-ˌnōl\ *n* [ISV *phen-* +³-*ol*] (ca. 1852) (Carbolic Acid) C_6H_5OH. A corrosive poisonous crystalline acid compound present in coal tar and wood tar that in dilute solution is used as a disinfectant. The simplest phenol. Used for the preparation of phenol-formaldehyde resins. Bp, 182°C; mp, 42°C. Phenols are a class of aromatic compounds containing $-OH$ groups attached directly to the benzene ring and which are used in the manufacture of epoxy resins, phenol-formaldehyde resins, plasticizes, plastics and wood preservatives. The specific name for C_6H_5OH (carbolic acid, phenylic acid, hydroxybenzene). Phenol was first derived from coal tar but today is most commonly synthesized from benzene or toluene. (Morrison RT, Boyd RN (1992) Organic chemistry, 6th edn. Prentice Hall, Englewood Cliffs, NJ)

Phenol-Aldehyde Polymers *n* Common name for phenol, polymer with formaldehyde. It is also commonly known as Phenol-aldehyde resin. See ▶ Phenolic Resins.

Phenol-Aldehyde Resins See ▶ Phenolic Resins.

Phenol-Aralkyl Resin *n* Any of several thermosetting resins produced by the condensation of aralkyl ethers and phenols.

Phenol Coefficient *n* Joseph Lister introduced phenol (carbolic acid) as a disinfectant in 1967. It has been the standard disinfectant to which other disinfectants are compared under the same conditions. The result of this comparison is the phenol coefficient. Salmonella typhi, a pathogen of the digestive system, and Staphlococcus aureus, a common wound pathogen, are typically used to determine phenol coefficients. A disinfectant with a phenol coefficient of 1.0 has the same effectiveness as phenol (Dorland's Illustrated Medical Dictionary). A coefficient less than 1.0 means that the disinfectant is less effective than phenol and greater than 1.0 is more effective than phenol. Phenol coefficients are reported separately for the different test organisms. The procedure for determining the phenol coefficient is pervasive throughout the published literature including a summariy description by Black (2005), and examples of phenol coefficients are listed in the following table. The phenol coefficient provides an acceptable means of evaluating the effectiveness of chemical agents derived from phenol, but it is was acceptable for other agents. Another problem is that the materials on or in which organisms may affect the usefulness of a chemical agent by complexing it or inactivating it. These effects are not reflected in the phenol coefficient number. References: Black JG (2002) Microbilogy, 5th edn. Wiley, New York, pp 315–316; Dorland's Illustrated Medical Dictionary (2004).

Phenol coefficients of common agents (Black 2002)

Chemical agent	Staphlococcus aureus	Salmonella typhi
Phenol	1.0	1.0
Chlorlamine	133.0	100.0
Cresols	2.3	2.3
Ethyl alcohol	6.3	6.3
Formalin	0.3	0.7
Hydrogen peroxide	–	0.01
Lysol™	5.0	3.2
Mercury chloride	100.0	143.0
Tincture of iodine	6.3	5.8

Phenol-Formaldehyde *n* A condensation product of reaction of phenol and formaldehyde. If the reaction is carried out under acidic conditions, the product formed is Novolac. Under alkaline conditions, with an excess formaldehyde, a C-Stage resin is formed. (Odian GC (2004) Principles of polymerization. Wiley, New York; Polymeric materials encyclopedia. Salamone JC (ed). CRC, Boca Raton, FL, 1996)

Phenol-Formaldehyde Resin *n* (PF resin, phenolic resin) The most important of the ▶ Phenolic Resins. Made by condensing phenol with formaldehyde, these were the first synthetic thermosetting resins to be developed (LH Baekeland 1907) and were marketed under the trade name Bakelite. (Polymeric materials encyclopedia. Salamone JC (ed). CRC, Boca Raton, FL, 1996)

Phenol-Furfural Resin A type of ▶ Furan Resin.

Phenolic \fi-ˈnō-lik\ (1) *n* Any of several types of thermoset plastics obtained by the condensation of phenol or substituted phenols with aldehydes such as formaldehyde and furfural. (2) *adj* Containing or pertaining to phenol.

Phenolic Novolac *n* (novolac, novolak) Thermoplastic, water-soluble resins obtained by reacting a phenol with an aldehyde, usually form-aldehyde, in the proportion of less than one mole of the phenol with one mole of aldehyde, in the presence of an acid catalyst. When a source of methylene groups is added, linkage between the methylenes and the phenolic rings occurs and the resins can react with diamines or diacids (e.g., hexa-methylenetetramine) to form thermosetting, insoluble resins. Absent a source of methylene groups, the resin remains permanently thermoplastic. (Polymeric materials encyclopedia. Salamone JC (ed). CRC, Boca Raton, FL, 1996) See also ▶ Resinoid and Epoxy-Novolac Resin.

Phenolic Plastics *n* Plastics based on resins made by the condensation of phenols, such as phenol and cresol, with aldehydes.

Phenolic Resin *n* Any of a wide range of thermosetting resins made by reacting a phenol with an aldehyde, followed by curing and cross-linking. Known also as *Phenol-Formaldehyde or PR Resins*.

Phenol Red *n* (1916) $C_{19}H_{14}O_5S$. A red crystalline compound used as an acid-base indicator.

Phenoplast *n* Another name for ▶ Phenol-Formaldehyde Resin.

Phenoxide \fi-ˈnäk-ˌsīd\ *n* (1888) A salt of a phenol in its capacity as a weak acid.

Phenoxy *n* Copolymer from bisphenol A + epichlorohydrin, manufactured by Union Carbide, US.

Phenoxy Resin *n* (polyhydroxyether) Any linear thermoplastic resin made by reacting an exact equivalent of epichlorohydrin with bisphenol A and sodium hydroxide in dimethyl sulfoxide.

Phenyl Benzoate *n* $C_6H_5COOC_6H_5$. A plasticizer. Bp, 314°C. (Plasticizer's data base. Wypych G (ed). Noyes, New York, 2003)

Phenylethylene Syn: ▶ Styrene.

Phenylformic Acid *n* Syn: ▶ Benzoic Acid.

Phenyl Group *n* (phenyl radical) The group C_6H_5-, existing only in combination.

Phenylsilane Resin *n* Any thermosetting copolymer or silicone and phenolic resins, available in solution form.

Phosphate \ˈfäs-ˌfāt\ *n* [F, fr. *acide phosphorique* phosphoric acid] (1795) (1) A salt or ester of a phosphoric acid. (2) The trivalent anion PO_4^{3-} derived from phosphoric acid H_3PO_4. (3) An organic compound of phosphoric acid in which the acid group is bound to nitrogen or a carboxyl group in a way that permits useful energy to be released.

Phosphate Plasticizer *n* Any of a group of plasticizers derived from phosphoric acid and aliphatic alcohols and phenols, and used in conjunction with others to impart flame resistance. (Troitzsch J (2004) Plastics flammability handbook: Principle, regulations, testing and approval. Hanser-Gardner, New York; Plasticizer's data base. Wypych G (ed). Noyes, New York, 2003)

Phosphating *n* Pretreatment of steel or certain other metal surfaces by chemical solutions containing metal phosphates and phosphoric acid as the main ingredients, to form a thin, inert, adherent, corrosion-inhibiting phosphate layer which serves as a good base for subsequent paint coats.

Phosphatize (Phosphate) Coat See ▶ Phosphating.

Phosphazene Polymer *n* Any of a family of experimental resins built on long chains of alternating phosphorus and nitrogen atoms. The general structure is $(-PX_2=N-)_n$, where X may be a halogen or organic radical. They have been used in fuel hoses that remain flexible in subzero, arctic climates, prosthetics for reconstruction surgery, and fabric waterproofing. In solution these polymers can be reacted with various nucleophilic agents to form a range of thermoplastics and elastomers. An important potential use may be as flame retardants for textiles. They can also be foamed.

Phosphomolybdic Pigments *n* A series of colored pigments derived from the interaction of basic dyes and complex acids such as phosphomolybdic, phosphotungstic, and Tungstomolybdic. They are characterized by good resistance to alkali, heat, light, and water.

Phosphorescence \-ˈre-sᵊn(t)s\ *n* (1796) The ability of certain substances to continue to emit light long after the source of excitation energy has been removed.

Phosphorescent Paint See ▶ Luminous Paint.

Phosphorescent Pigment *n* One of a family of pigments, generally an inorganic sulfide crystal of fairly large and controlled size, which absorbs the energy of incident light then slowly re-emits it as radiation of a color specific to each pigment. The phosphorescence gradually dims in darkness, to be renewed by the next light restimulation.

Phosphoric Acid *n* (1791) An inorganic acid having the formula (H_3PO_4).

Photochemical Catalysis *n* Chemical reaction stimulated by action of light. (Fouassier J-P (1995) Photoinitiation, photopolymerization and photocuring. Hanser-Gardner, New York)

Photochemically-Reactive Organic Material *n* Any organic material which will react with oxygen, excited oxygen, ozone and/or other free radicals generated by the action of sunlight on components in the atmosphere, giving rise to secondary contaminants and reaction intermediates in the atmosphere which can have detrimental effects. (Hare CH (2001) Paint film degradation - mechanisms and control. Steel Structures Paint Council; Degradation and stabilization of polymers. Zaiko GE (ed). Nova Science, New York, 1995)

Photochemically-Reactive Solvent *n* Any solvent with an aggregate of more than 20% of its total volume composed of the chemical compounds classified below or which exceeds any of the following individual percentage composition limitations, referred to the total volume of solvent: (1) A combination of hydrocarbons, alcohols, aldehydes, esters, ethers or ketones having an olefinic or cycloolefinic type of unsaturation: 5%. (2) A combination of aromatic compounds with eight or more carbon atoms to the molecule except ethylbenzene: 8%. (3) A combination of ethylbenzene, ketones having branched hydrocarbon structures, trichloroethylene or toluene: 20%. (As defined by Rule 66). (Degradation and stabilization of polymers. Zaiko GE (ed). Nova Science, New York, 1995)

Photochemical Oxidants n Secondary pollutants formed by the action of sunlight on the oxides of nitrogen and hydrocarbons in the air. They are the primary contributors to photochemical smog.

Photochromatic n Materials exhibiting the phenomenon of photochromatism. (Fox AM (2001) Optical properties of solids. Oxford University Press, UK; Meeten GH (1986) Optical properties of polymers. Springer, New York)

Photochromatism n Phenomenon of reversible change exhibited by some materials when exposed to light.

Photochromic \ˌfō-tə-ˈkrō-mik\ adj [phot- + chrom- + ^{1}-ic] (1953) Capable of changing color on exposure to radiant energy (as light). (Saleh BEA, Teich MC (1991) Fundamentals of photonics. Wiley, New York)

Photoconductive \-kən-ˈkək-tiv\ adj (1929) Becoming electrically conductive when irradiated by light or ultraviolet light. (Optical and electrical properties of polymers: Materials research society symposium proceedings, vol 24. Emerson JA, Torkelson JM (eds). Materials Research Society, 1991) See ▶ Poly (N-Vinylcarbazole).

Photocurable Coating n A polymerizable mixture that can be applied as a thin film to a substrate and polymerized at a rapid rate by exposure to actinic light. (Fouassier J-P (1995) Photoinitiation, photopolymerization and photocuring. Hanser-Gardner, New York)

Photodegradation adj (1971) Breakdown of plastics due to the action of visible or ultraviolet light. Most polymers tend to absorb high-energy radiation in the ultraviolet portion of the spectrum, which elevates their electrons to higher reactivity and causes oxidation, chain cleavage, and other destructive reactions. Delay of photodegradatioin may be accomplished by incorporation of UV light absorbers. The two most common methods of promoting degradation by UV light are incorporation of additives that act as photoinitiators or photosensitizers, and copolymerization for deliberately incorporating weak links with the polymer chain. (Degradation and stabilization of polymers. Zaiko GE (ed). Nova Science, New York, 1995).

Photoelasticity n Changes in optical properties of isotropic, transparent materials when subjected to stress. Mainly, they become birefringent, a property that has been widely used to detect frozen-in stresses in sheets and other products. (Meeten GH (1986) Optical properties of polymers. Springer, New York)

Photoengraving n (1872) A process for producing an alpha-numeric script or graphical image on a sensitized metal plate by placing a transparent negative (i.e., light sensitive emulsion) between the plate and a source of light. The areas not rendered water-insoluble by light exposure are washed and etched with acid solutions. (Printing ink handbook. National Association of Printing Ink Manufacturers, 1976)

Photogenic Property \ˌfō-tə-ˈje-nik-\ n The tendency of a pigment to darken on exposure to sunlight and to be rebleached on being placed in the dark.

Photographic Density The density D of silver deposit on a photographic plate or film is defined by the relation

$$D = \log O$$

where O is the opacity. If I_o and I are the incident and transmitted intensities respectively the opacity is given by I_o/I. The transparency is the reciprocal of the opacity or I/I_o. (Printing ink handbook. National Association of Printing Ink Manufacturers, 1976)

Photographic Printing See ▶ Printing.

Photographing n Brush marks or other irregularities in the previous coat or substrate that show through the dried topcoat. Also called Telegraphing or Show-Through.

Photogravure \ˌfō-tə-grə-ˈvyúr\ n [F, fr. phot- + gravure] (1879) Process for making prints from photomechanically prepared intaglio plates. (Printing ink handbook. National Association of Printing Ink Manufacturers, 1976)

Photoinitiation n Initiation of a free radical polymerization by irradiation with ultraviolet light (or other frequency of light), causing photopolymerization. (Odian GC (2004) Principles of polymerization. Wiley, New York)

Photoinitiator n (1) A substance which, by absorbing light, becomes energized into forming free radicals which promote secondary radical reactions. (2) Initiators which create initiating species for polymerization reactions when exposed to light; cationic free radical, etc. (Fouassier J-P (1995) Photoinitiation, photopolymerization and photocuring. Hanser-Gardner, New York)

Photolithography \-li-ˈthä-grə-fē\ *n* [ISV] (1856) Process utilizing surfaces which, under the selective action of light, are transformed from a water-receptive to a water-repellant and ink-receptive nature. (Photoinitiated polymerization. Belfield KD, Crivello JV (eds). American Chemical Society Publications, 2003)

Photometer \fō-ˈtä-mə-tər\ *n* [NL *photometrum*, fr. *phot-* + *-metrum* –*meter*] (1778) An instrument for the measurement of the intensity of emitted, reflected, or transmitted light. For the measurement of luminous intensity, a visual receptor element (the eye), may be used in the measuring device or a physical receptor element may be used which can be related to the calculated response of a standard observer. (Harris DC (2002) Quantitative chemical analysis. W. H. Freeman, New York; Krause A, Lange A, Ezrin M (1988) Plastics analysis guide: Chemical and instrumental methods. Oxford University Press, UK)

Photometric \ˌfō-tə-ˈme-trik\ *adj* (ca. 1828) Pertaining to the measurement of the intensity of light. (Saleh BEA, Teich MC (1991) Fundamentals of photonics. Wiley, New York)

Photomicrograph \ˌfō-tə-ˈmī-krə-ˌgraf\ *n* (1858) A large photograph of a small object as through a microscope, the image of which is enlarged or magnified. (Loveland RP (1981) Photomicrograsphy. Krieger, New York)

Photon \ˈfō-ˌtän\ *n* [*phot-* + *²-on*] (1916) A photon (or γ-ray) is a quantum of electromagnetic radiation which has zero rest mass and an energy of h (Planck's constant) times the frequency of the radiation. Photons are generated in collisions between nuclei or electrons and in any other process in which an electrically charged particle changes its momentum. Conversely photons can be absorbed (i.e., annihilated) by any charged particle. A quantum of electromagnetic energy: $E_{photon} = h\nu$. (Handbook of chemistry and physics, 52nd edn. Weast RC (ed). The Chemical Rubber, Boca Raton, FL)

Photopic \fōt-ˈō-pik\ *adj* [NL *photopia*, fr. *phot-* + *-opia*] (1915) An adjective used to describe vision mediated by the cone receptors in the retina of the eye, which give rise to the sensation of color occurring at high and medium levels of luminance. (Stedman's medical dictionary, 27th edn. Lippincott Williams & Wilkins, New York, 2000)

Photopolymer \ˌfō-tō-ˈpä-lə-mər\ *n* (1932) A plastic compound containing an agent that undergoes a change, proportional to light intensity, upon exposure to light, so that images can be formed on its surface by a photographic process. Photopolymers play an important role in the manufacture of semiconductors.

Photopolymerization *n* An addition reaction brought about by exposure of the monomer or mixture of monomers to natural or artificial light, with or without a catalyst. Methyl methacrylate, styrene, and vinyl chloride are examples of monomers that can be photopolymerized. (Fouassier J-P (2005) Photoinitiation, photopolymerization and photocuring. Hanser-Gardner, New York)

Photopolymers These are photoinitiated and photo-crosslinked polymers; e.g., photoresists, printing inks, can coatings, etc. (Fouassier J-P (1995) Photoinitiation, photopolymerization and photocuring. Hanser-Gardner, New York; Printing ink manual, 5th edn. Leach RH, Pierce RJ, Hickman EP, Mackenzie MJ, Smith HG (eds). Blueprint, New York, 1993)

Photosensitizers \-ˈsen(t)-sə-ˌtīz-\ *n* (ca. 1923) A substance which, by absorbing light, passes its energy to another substance which then reacts. (Fouassier J-P (1995) Photoinitiation, photopolymerization and photocuring. Hanser-Gardner, New York)

Phr *n* Abbreviation for *parts* per *hundred* parts of *resin*, a fractional measure of composition, simpler than weight percentage that facilitates making up multicomponent compounds. For example, as used in plastics formulations, 5 phr means that 5 kg of an ingredient would be combined with 100 kg of resin.

Phthalate Ester \ˈtha-ˌlāt ˈes-tər\ *n* (*o*-phthalic ester) Any of a large class of plasticizers produced by the direct action of alcohol on phthalic anhydride. They are the most widely used of all plasticizers and are generally characterized by moderate cost, good stability, and good all-round properties.

Phthalic Acid \ˈtha-lik ˈa-səd\ *n* o-Benzenedicarboxylic acid. $C_8H_6O_4$; mol wt, 166.13, crystals; mp, about 230°C when heated, forming phthalic anhydride and water.

Phthalic Anhydride *n* (phthalic-acid anhydride) $C_6H_4(CO)_2O$. A white, odorless, crystalline compound derived by oxidation of naphthalene or o-xylene, shipped in flake or molten form. It major use is in the production of phthalate esters for plasticizing vinyl and cellulosic resins. It is also an important intermediate in the manufacture of alkyd and unsaturated polyester resins, and is a curing agent for epoxy resins. Bp, 284°C; mp, 130°C; sp gr, 1.53; acid value, 758.0.

Phthalic Anhydride Test *n* Phthalic anhydride reacts with primary alcohols when the mixture is refluxed in benzene. Secondary alcohols react less readily, usually requiring a reaction temperature of 100–200°C, whereas the tertiary alcohols do not react.

Phthalocyanine \ tha-lō- sī-ə- nēn\ *n* [ISV *phthalic acid* + -*o*- + *cyanine*] (1933) (Cyan) Blue and green ink pigments characterized by extreme light fastness and resistance to solvents, acid and alkali. The blue is now widely used in process inks.

Phthalocyanine Pigments *n* Series of organic pigments having as a structural unit four isoindole groups, $(C_6H_4)C_2N$, linked by four nitrogen atoms so as to form a conjugated chain. There are four commercially important modifications, including the basic compound: (1) Phthalocyanine (metal free), $(C_6H_4C_2N)_4 N_4$, blue-green; (2) Copper phthalocyanine, in which a copper atom is held by secondary valences of the isoindole nitrogen atoms; sp gr, 1.59; (3) Chlorinated copper phthalocyanine, green, in which 15–16 hydrogen atoms are replaced by chlorine; (4) Sulfonated copper phthalocyanine, water-soluble, green in which two hydrogen atoms are replaced by sulfonic acid, HSO_4, groups. (Solomon DH, Hawthorne DG (1991) Chemistry of pigments and fillers. Krieger, New York)

Phycocolloid *n* Any of several polysaccharide hydrocolloids from brown to red seaweeds. (Industrial gums: Polysaccharides and their derivatives. Whistler JN, BeMiller JN (eds). Elsevier Science and Technology Books, 1992) See ▶ Gum, Natural.

Physical Catalyst *n* Radiant energy capable of promoting or modifying a chemical reaction, as in ▶ Photopolymerization.

Physical Change *n* A change in which no new substances are formed.

Physical Property A property which can be described without referring to a chemical reaction.

PI *n* Abbreviation for the *trans*-1,4- type of ▶ Polyisoprene. See ▶ Polyimides.

PIA *n* Abbreviation for Plastics Institute of America, INC., headquartered at 333 Aiken Street, Lowell, MA 01854-3686. PIA sponsors seminars in plastics technology and supports graduate research in plastics engineering and undergraduate scholarships.

PIB *n* Abbreviation for Poly(Isobutylene).

pi (π) Bond *n* π A covalent bond in which the electron charge cloud of the shared pair of electrons is located in two regions on opposite sides of the bond axis.

Pick *n* A single filling thread carried by one trip of the weft-insertion device across the loom. The picks interlace with the warp ends to form a woven fabric. See Picking. Also see ▶ Filling.

Pick Count *n* The number of filling yarns per inch or per centimeter of fabric.

Pick Counter *n* (1) A mechanical device that counts the picks as they are inserted during weaving. (2) A mechanical device equipped with a magnifying glass used for counting picks (and/or ends) in finished fabrics.

Picker *n* (1) A machine that opens staple fiber and forms a lap for the carding process used in the production of spun yarns. (2) That part of the picking mechanism of the loom that actually strikes the shuttle.

Picker Lap *n* A continuous, considerably compressed sheet of staple that is delivered by the picker and wound into a cylindrical package. It is used to feed the card.

Picker Sticks *n* The two sticks that throw the shuttles from box to box at each end of the faceplate of the loom.

Picking *n* (1) The adherence of a sheet of paper to the plate due to the tack of the ink. (2) The removal of the surface of the paper during printing. It occurs when the pulling force (tack) of the ink is greater than the surface strength of the paper whether coated or uncoated.

Picking Up *n* (1) The blending of a coat of freshly applied paint with another over which it is applied. (2) The joining up of a wet edge. See ▶ Pulling Up.

Pickled Pine *n* A gray finish which duplicates the effect formerly produced by actually pickling the wood with nitric acid but now obtained by using a gray stain.

Pickling *n* (1) Treatment for the removal of rust and mill scale from steel by immersion in an acid solution containing an inhibitor. Pickling should be followed by thorough washing and drying before painting. (2) The process of removing paint and varnish with an alkaline preparation or strong solvents.

Pick Out Mark *n* A filling wise band or bar characterized by a chafed or fuzzy appearance due to pulled-out picks.

Pickup Groove See ▶ Hold-Down Groove.

Pick-up Roll *n* (1) In the coil coating industry, the roll which revolves within the pan and is partially immersed in the paint. This roll picks up paint from the pan and applies it to the transfer or applicator roll. (2) Spreading device where the roll for picking up the adhesive runs in a reservoir of adhesive.

Pico- (p) The SI prefix meaning $\times 10^{-12}$.

Picot *n* \ pē-( )kō, pē- \ *n* [F, literally, small point, fr. MF, fr. *pic* prick, fr. *piquer* to pick] (ca. 1882) (1) A small loop woven on the edge of ribbon, or a purl on lace. A picot edge may also be produced by a hemstitching machine. (2) A run-resistant loop usually found at the top of hosiery.

PIC Test *n* Abbreviation for Pseudoisochromatic Test for defective color vision.

Piece \ pēs\ *n* [ME, fr. OF, fr. (assumed) VL *pettia*, of Gaulish origin, akin to Welsh *peth* thing] (13c) A standard length of a fabric, such as 40, 60, 80, or 100 yards.

Piece Dyeing *vt* (1920) See ▶ Dyeing.

Piecing The joining of two or more ends of sliver, roving, yarn, etc.

Piezo-Electric Effect *n* The phenomenon exhibited by certain crystals of expansion along one axis and contraction along another when subjected to an electric field. Conversely, compression of certain crystals, generate an electrostatic voltage across the crystal. Piezoelectricity is only possible in crystal classes which do not possess a center of symmetry. (Handbook of chemistry and physics, 52nd edn. Weast RC (ed). The Chemical Rubber, Boca Raton, FL)

Piggyback *adj* (1823) A word used to designate a system of two extruders in which one discharges to the other. Such an arrangement has occasionally been used to separate the melt-generating (plasticating) function from the pressure-developing and shaping function of the standard extruder thus gaining more precise control over both functions and eliminating a main cause of surging at the die. For general extrusion this method of accomplishing that separation has been superseded by the cheaper and more compact combination of extruder with ▶ Gear Pump. Also see ▶ Extruder, Tandem for a special application of the piggyback concept.

Pigment \ pig-mənt\ *n* [L *pigmentum*, fr. *pingere* to paint] (14c) Finely ground, natural or synthetic, inorganic or organic, insoluble dispersed particles (powder) which, when dispersed in a liquid vehicle to make paint, may provide, in addition to color, many of the essential properties of a paint – opacity, hardness, durability, and corrosion resistance. The Color Index is a method of classifying pigments established under the joint partnership of the American Association of Textile Chemists and Colorist in the United States and the Society of Dyes and Colorist in the United Kingdom. (Leher LR, Salzsman M (1985) Color pigments. Applied Polymer Science, American Chemical Society, Washington, DC) An example is shown in the following figure The term "pigment" is used to include extenders, as well as white or color pigments. The distinction between powders which are pigments and those which are dyes is generally considered on the basis of solubility – pigments being insoluble and dispersed in the material, dyes being soluble or in solution as used. See also ▶ Colorant. (Vesce YC (1959) Exposure studies of organic pigments in paint systems. National Aniline Div., Allied Chemical Corp., New York; 1978 annual book of ASTM standards, Pt. 28: Pigments, resins, and polymers. American Society for Testing and Materials, Philadelphia, PA; Bentley KW (1960) The natural pigments. Interscience, New York; Pigments (in English and French), International Organization for Standardization (ISO), Geneva, Switzerland, 1963; Fuller WR, Love CH (1968) Inorganic color pigments, unit 8 of the federation series on coatings technology. Federation of Societies for Coatings Technology, Philadelphia, PA; Patton TC (1964) Paint flow and pigment dispersion. Interscience, New York; Synonyms for pigment names. Pamphlet of the British Standards Institution, London, 1957)

$$O^{--}$$
$$Ti^{+4} \quad O^{--}$$

Pigment/Binder Ratio *n* Ratio (generally, by weight in the US; by volume in Britain) of total pigment to binder solids in paint. *Also called Binder Ratio*.

Pig Wrack See ▶ Carrageen.

Pilaster \pi- las-tər\ *n* [MF *pilastre*, fr. It *pilastro*] (1575) Flat vertical projection from a wall, with the proportions, capital shaft, and base of a column.

Pile \ pī(ə)l\ *n* [ME, fr. L *pilus* hair] (15c) (1) A fabric effect formed by introducing tufts, loops, or other erect yarns on all or part of the fabric surface. Types are warp, filling, and knotted pile, or loops produced by weaving an extra set of yarns over wires that are then drawn out of the fabric. Plain wires leave uncut loops; wires with a razor-like blade produce a cut-pile surface. Pile fabric can also be made by producing a double-cloth structure woven face to face, with an extra set of yarn interlacing with each cloth alternately. The two fabrics are cut apart by a traversing knife, producing two fabrics with a cut-pile face. Pile should not be confused with nap. Corduroys are another type of pile fabric, where long filling floats on the surface are slit, causing the pile to stand erect. (2) In carpets, pile refers to the face yarn, as opposed to backing or support yarn. Pile carpets are produced by either tufting or weaving. (Complete textile glossary. Celanese Corporation, Three Park Avenue, New York, 2000) Also see ▶ Cut Pile and ▶ Loop Pile.

Pile Burning See ▶ Bin Cure.

Pile Crush *n* The bending of upholstery or carpet pile that results from heavy use or the pressure of furniture.

Pile Weave *n* A weave in which an additional set of yarns, either warp or filling, floats on the surface and is cut to form the pile. Turkish toweling is a pile weave fabric with uncut loops on one or both sides.

Pile Wire *n* A metal rod over which yarn is woven to generate a pile fabric.

Pill *n* A small accumulation of fibers on the surface of a fabric. Pills, which can develop during wear, are held to the fabric by an entanglement with surface fibers of the material, and are usually composed of the same fibers from which the fabric is made. A term sometimes used for Preform.

Pilling *n* (1) Behavior of a very quick-drying paint which, during application by brush, becomes so sticky that the resulting film is thick and uneven. (2) The buildup of ink on rollers, plate or blanket. Also see ▶ Caking.

Pilot \ ˈpī-lət\ *n* [ME *pilote*, fr. It *pilota*, alter. of *pedota*, fr. (assumed) MGk *pēdōtēs*, fr. Gk *pēda* steering oars, plural of *pēdon* oar, prob. akin to Gk *pod-*, *pous* foot] (1530) A woolen cloth generally made in navy blue and used for seamen's coats. It is usually a heavily milled 2/2 twill with a raised, brushed finish.

Pimelic Ketone *n* A Syn: ▶ Cyclohexanone.

Pimple \ ˈpim-pəl\ *n* [ME *pinple*] (14c) Small, conical protrusion on the surface. It also may refer to small surface blister.

Pin *n* Another name for ▶ Mandrel (2).

Pinacold *n* Two parallel faces of a crystal having no equivalents.

Pinacolone *n* $CH_3COC(CH_3)_3$. Methyl tert butylketone. Medium-boiling ketonic solvent. Bp, 106°C. *Also known as Pinacolin.*

Pinched Coating *n* In coated fabrics, a ridge or a wrinkle in the coating.

Pinch Effect *n* When an electric current, either direct or alternating, passes through a liquid conductor, that conductor tends to contract in cross-section, due to *electromagnetic* forces.

Pinch-Off *n* (1) In blow molding, a raised edge around the cavity in the mold that seals off the part and separates the excess material as the mold closes around the parison. (2) In making tubular film, the paired rollers (pinch rolls) at the top of the tower that flatten the tube and confine the air in the bubble.

Pinch-Off Blades *n* In blow molding, the mating parts of the mold at the bottom that come together first, to pinch off the parison at the bottom and serve the tail, then, to help form the bottom of the part as the parison is inflated.

Pinch-Off Land *n* The width of the ▶ Pinch-Off Blade that effects the sealing of the parison.

Pinch-Off Tail *n* In blow molding, the bottom tip of the parison that is pinched off and severed as the mold closes. See the three preceding entries.

Pin Drafting *n* Any system of drafting in which the orientation of the fibers relative to one another in the sliver is controlled by pins.

Pineholes *n* Film defect characterized by small pore-like flaws in a coating which extend entirely through the applied film and have the general appearance of pin pricks when viewed by reflected light. The term is rather generally applied to holes caused by solvent bubbling, moisture, other volatile products, or the presence of extraneous particles in the applied film. See ▶ Pores.

Pinene \ ˈpī-ˌnēn\ *n* [ISV, fr. L *pinus*] (1885) $C_{10}H_{16}$. A bicyclic terpene hydrocarbon, the principal constituent of all turpentines, and existing therein in two isomeric forms, alpha- and beta-piene. The latter is found in appreciable quantity only in gum spirits and sulfate wood turpentine. Pinene not otherwise described usually means alpha-pinene. Alpha-pinene: Bp, 155°C; sp gr, 0.858/20°C; refractive index, 1.466. Beta-piene: Bp, 163°C; sp gr, 0.869/20°C; refractive index, 1,476.

Pine Needle Oil *n* Essential oil of typical fragrance obtained by steam distillation of the leaves (needles) of certain species of pine or other coniferous trees.

Pinene Resin See ▶ Polyterpene Resin.

Pine Oil *n* Colorless to amber colored volatile oil with characteristic pinaceous odor, consisting principally of isomeric tertiary and secondary cyclic terpene alcohols, with variable quantities of terpene hydrocarbons, ethers, ketones, phenols and phenolic ethers, the amount and character of which depend on the source and method of manufacture. The four commercial kinds of pine oil are: (1) *Steam Distilled Pine Oil*. Obtained from pine wood by steam distillation or by solvent extraction followed by such distillation. (2) *Destructively Distilled Pine Oil*. Obtained from the lighter distillate from the destructive distillation (carbonization) of pine wood. (3) *Synthetic Pine Oil*. Obtained by chemical hydration of terpene hydrocarbons to form the terpene alcohols, or by dehydration of terpin hydrate. (4) *Sulfate Pine Oil*. A high boiling fraction obtained in the refining and fractional distillation of the condensed vapors released during the ingestion of wood by the sulfate processes.

Pine Pitch *n* Dark-colored to black solidified material, somewhat pliant and tenacious, obtained by distilling off practically all the volatile oil from a retort pine tar; the genuine contains no added free rosin.

Pine Tar *n*, Kiln Burned Heavy, oily liquid resulting from controlled carbonization (slow burning) of pine knots and stumpwood to charcoal in earth-covered piles or "kilns," with introduction of insufficient air to permit complete combustion; contains under composed resin acids along with the decomposition products. This product is sometimes called "country tar."

Pine Tar Oil *n* Oil obtained by condensing the vapors from the retorts in which resinous pine wood is destructively distilled (carbonized).

Pine Tar *n*, Retort Tar produced by removal of volatile oils from pine tar oil by stream distillation. Several grades are marketed, namely; thin, medium, heavy and extra heavy, so classified on the basis of viscosity, and depending upon the quantity of volatile oils removed.

Pine Tar *n*, Stockholm Kiln-burned pine tar produced in Scandinavian countries from wood of the Northern European pine, *Pinus sylvestris*.

Pinhead *n* (1593) A small pinhead-sized opening usually found about 10–12 in. from a selvage. Pinheads usually run in a fairly straight line along the warp and are formed by the shuttle pinching the filling, causing small kinks that show up as small holes in transmitted light.

Pinholing *n* (1) Failure of a printed ink to form a completely continuous film. Visible in the form of small holes or voids in the printed area. (2) The appearance of fine, pimply elevations or tiny holes on a coating.

Pinion Barré \\pin-yən bär\ *n* A fine, filling-wise fabric defect appearing as one or two pick bars in an even repeat. It is caused by a faulty loom pinion.

Pink Staining *n* A pink-colored stain that sometimes appears on vinyl-coated fabrics of white and pastel colors when they have lain on earth for a long time. It has been attributed to growth of fungi of the genus *Penicillium*, and to the bacterium. *Streptomyces rubrireticuli*. It can be prevented by treating the fabric with a fungicide, e.g., *N*-(trichloromethylthio) phthalimide or an arsenic compound.

Pin Mark See ▶ Clip Mark.

Pinning See ▶ Pin Drafting.

Pinolin See ▶ Rosin Spirit.

Pinpoint Gate *n* (pin gate) In injection molding, a very small orifice, generally 0.75 mm (30 mils) or less in diameter (or maximum lateral dimension), connecting the runner and mold cavity, and through which molten plastic flows into and fills the cavity. Such a gate leaves a small, easily removed mark on the part, but due to the tendency of the melt to freeze early in the pinpoint gate as flow slows, its use is limited to small parts and to resins with good fluidity. In multicavity molds, the dimensions of pinpoint gates must be held within very tight tolerances in order to fill all cavities at the same time and to avoid differences in dimensions among the parts extracted from the several cavities. See also ▶ Balanced Gating, Gate, and Restricted Gate.

Pinsonic® Thermal Joining Machine *n* A rapid, efficient quilting machine that uses ultrasonic energy rather than conventional stitching techniques to join layers of thermoplastic materials. The ultrasonic vibrations generate localized heat by causing one piece of material to vibrate against the other at extremely high speed, resulting in a series of welds that fuse the materials together.

Pipe Clay See ▶ Aluminum Silicate (Clay).

Pipe Die *n* An extrusion die whose lands form a circular annulus used in extrusion of plastic pipe or tubing. The outer shell of the die is usually called the *die*, the core is called the *mandrel*. Pipe dies may be side-fed or end-fed, and the mandrel may be supported by a trio of legs called a *spider*, or it may be supported from the rear of a side-fed die. It is easier to achieve circumferential uniformity of wall thickness with a spider die, but the splitting of the melt stream at the legs has sometimes caused weak welds because of insufficient knitting time before the pipe emerges and is chilled.

Piperidine \pi-per-ə-dēn\ *n* [ISV *piper*ine + *-idine*] (1854) A heterocyclic, secondary amine with a six-membered ring. $C_5H_{10}NH$, a slow-acting curing agent for thick-section epoxy castings or laminates, where faster curing would cause exotherm problems such as bubbling, distortion, or cracking.

Pipe Train *n* A term used in pipe extrusion that denotes the entire equipment assembly, i.e., extruder, die, external sizing means, cooling bath, haul-off, and coiler or cutter.

Piqué \pi-ˈkā\ *n* [F *piqué*, fr. pp of *piquer* to pick, quilt] (1852) (1) A medium weight to heavyweight fabric with raised cords in the warp direction. (2) A double-knit fabric construction knit on multifeed circular machines.

PIR Abbreviation for Polyisocyanurate.

Pirn \ˈpərn, *2 is also* ˈpirn\ *n* [ME] (15c) (1) A wood, paper, or plastic support, cylindrical or slightly tapered, with or without a conical base, on which yarn is wound. (2) The double-tapered take-up yarn package from drawtwisting of nylon, polyester, and other melt spun yarns.

Pirn Barré A fabric defect consisting of crosswise bars caused by unequal shrinkage of the filling yarn from different points on the original yarn package.

Piston \ˈpis-tən\ *n* [f, fr. It *pistone*, fr. *pistare* to pound, fr. OIt, fr. ML, fr. L *pistus*, pp of *pinsere* to crush] (1704) See ▶ Force Plug.

Pit *n* [ME, fr. OE *pytt* (akin to OHGr *pfuzzi* well), fr. L *puteus* well, pit] (before 12c) An imperfection, a small crater in the surface of the plastic, with its width of about the same size as the depth. See ▶ Pockmarking.

Pitch *n* (1) Of an extruder screw, the axial distance from a point on a screw flight to the corresponding point on the next flight. In a single-flight (single-start) screw, the pitch and lead are equal. In a screw having *n* parallel multiple flights, pitch = lead/*n*. In certain ▶ Solids-Draining Screws with two flights, the lead of one flight is slightly larger than that of the other. In such a screw, pitch varies continuously along the two-flighted section. (2) Any of various black or dark semisolid to solid materials obtained as residues from the distillation of tars, and sometimes including natural bitumen.

Pitch *n* Psychological response of the ear, primarily dependent upon the frequency of vibration of the air. The intensity of the sound also has a certain effect on the pitch. Pitch of a screw is the axial distance between adjacent turns of a single threat on the screw.

Pitch, Archangel *n* Originally a genuine pine pitch made from pine tar in the Archangel district of Russia; in this country a similar product is made from residues of pine origin, blended with various oils to make a pitch for caulking boats. Its acidity is due mainly to rosin acids.

Pitch, Brewer's *n* Type of pitch made by blending certain oils, waxes or other ingredients with rosin for the coating of beer barrels.

Pitch, Burgundy *n* Originally the solidified resin obtained by heating and straining the air-dried oleo-resin exuded by the Norway spruce (*Picea excelsa*) and European silver fir (*Abies petinata*); now denotes an artificial mixture made by heating rosin with certain fixed oils, the combination being used for adhesive plasters.

Pitch *n*, Navy Pitch obtained by melting rosin with pine tar, with or without rosin distillation residues.

Pitch Oil See Creosote.

Pitch Pocket *n* An opening between the growth rings containing resin in certain softwoods. Syn: ▶ Resin Pocket and ▶ Pitch Streak.

Pitch Streak See ▶ Pitch Pocket.

Pitting *n* Formation of holes or pits in the surface of a metal, by corrosion, or in the surface of a coating while it cures often because of the escape of gas or solvent.

pK *n* The negative logarithm of the dissociation constant K or –log K that serves as a convenient measure of an acid [K for acetic acid is 0.000018 or 1.8×10^{-5} from which pK is (5–0.25) or 4.75].

PKT See ▶ Potassium Titanate.

PL *n* Poly(ethylene) (EEC abbreviation).

Plain-Knit Fabric See ▶ Flat-Knit Fabric.

Plain Weave *n* One of the three fundamental weaves: plain, satin, and twill. Each filling yarn passes successively over and under each warp yarn, alternating each row.

Plaiting See ▶ Braiding.

Planar Helix Winding *n* A winding in which the filament path on each dome lies on a plane that intersects the dome, while a helical path over the cylindrical section is connected to the dome paths.

Planar Winding *n* A winding in which the filament lies on a plane that intersects the winding surface.

Planckian Radiator *n* A body which absorbs all radiation falling on it and reflects none. At each temperature to which it is heated, it radiates a precise magnitude per area of its surface at each wavelength throughout the spectrum, thus, a family of spectral curves is formed for the family of temperatures. The chromaticity coordinates of the integration of these individual curves trace a curve on a CIE chromaticity diagram which is called the Planckian Locus and is the chromaticity reference for designating color temperature and correlated color temperature. The temperatures are expressed in the

Kelvin (absolute) scale. See ▶ Correlated Color Temperature.

Planck's Constant \ˈpläŋk-\ *np* (*h*) (Max Plank, 1858–1947) A universal constant of nature which relates the energy of a quantum of radiation to the frequency of the oscillator which emitted it. It has the dimensions of action (energy × time). Classically expressed by

$$E = h\nu$$

where E is the energy of the quantum and ν (or sometimes f) is its frequency and h is Planck's constant. Its numerical value is $(6.6517 \pm 0.00023) \times 10^{-27}$ erg s or 10^{-34} J s) where ν is frequency, λ is wavelength (m), c is the velocity of light (3×10^8 m/s, it is calculated by

$$\nu = \frac{c}{\lambda}.$$

(Russell, Russell JB (1980) General chemistry. McGraw-Hill, New York)

Planetary-Screw Extruder See ▶ Extruder, Planetary-Screw.

Planishing \ˈpla-nish-shən\ *n* See ▶ Press Polishing.

Planography \plā-ˈnä-grə-fē\ *n* (1909) Printing processes which are dependent upon the fact that an oily ink will not adhere to water-moistened, nondesign areas of a level plate but will wet design areas treated with a greasy ink or varnish.

Plaskon *n* Urea-formaldehyde resin, manufactured by Allied Chemicals, US.

Plasma Etching *n* A process in which a plastic surface to be metal-plated is exposed to a gas plasma in a vacuum, producing chemical and physical changes that yield bondability and wettability equivalent to those produced in the past by stringent and hazardous chemical pretreatments. Although a variety of gases may be used, bottled oxygen has been found to be best. A radio-frequency source inside the high-vacuum chamber generates the plasma (an ionized gas consisting of an equal number of positive ions and electrons). The process has been effective on nylons, acrylonitrile-butadiene-styrene resins, and plastics based on phenylene oxide.

Plasma Flame See ▶ Flame Spray.

Plasma Polymerization *n* The process of polymerizing a gaseous monomer (e.g., tetrafluorethylene) in a radio-frequency (rf) field in a low vacuum whereby the monomer absorbs electromagnetic energy sufficiently to excite the π-bond thereby producing free radicals to initiate polymerization. The polymer adheres to the surfaces within the vacuum chamber (usally falls due to gravity on a surface). Poly(ethylene tetrafluoride) and other films can be deposited on metals, etc., using this process that would be difficult to deposit otherwise. (Plasma polymer films. Biederman H (ed). Imperial College Press, London; Plasma processing of polymers. D'Agostino, Fracassi F (eds). Kluwer Academic, New York, 1997)

Plasma-Spray Coating *n* A spray-coating process developed to apply sinterable plastics such as polytetrafluoroethylene to metals and ceramics. A special spray gun produces a rotating jet of hot, ionized gas particles (plasma) with laminar-flow characteristics. Plastic powder supplied to the gun is channeled within the gun so that it emerges as a layer on the periphery of the plasma jet where temperatures are lower than those in he center of the jet. The process is capable of producing coatings as thin as 2.5 μm on unprinted but clean substrates, without after-baking. Substrates must be capable of withstanding the sintering temperature of the polymer.

Plaster \ˈplas-tər\ *n* [ME, fr. OE, fr. L *emplastrum*, fr. Gk *emplastronomer*, fr. *emplassein* to plaster on, fr. *en-* + *plassein* to mold, plaster, perhaps akin to L *planus* level, flat] (before 12c) A paste-like material, usually a mixture of Portland cement, lime or gypsum with water and sand; fiber or hair may be added as a binder; applied to surfaces such as walls or ceilings in he plastic state; later it sets to form a hard surface.

Plasterboard See ▶ Gypsum Wallboard.

Plaster of Paris \-ˈpar-əs\ {*often capitalized 2d P*} [*Paris*, France] (15c) A group of cements which consist essentially of calcium sulfate and are produced by a partial dehydration of gypsum to the hemihydrate $2CaSO_4 \cdot H_2O$. They usually contain additives of various sorts. *Also known as Gypsum Cement.*

Plaster Primer *n* Primers with a degree of resistance to alkali which are used for priming plasters and cements of varying degrees of alkalinity. The primers must not only resist saponification but must insulate succeeding coats of paint from attack.

Plastic \ˈplas-tik\ *adj* [L *plasticus* of molding, from Gr *plastikos*, fr *plassein* to mold, form] (1632) (1) *adj* Originally, the term was used as an adjective to indicate a material was capabled of being "moleded or shaped." Indicating that the noun modified is made of or pertains to a plastic or plastics. The singular form is customarily used when the noun obviously refers to a particular, single plastic, as in "a plastic hose," and the plural form is often used when the noun could refer to several types of plastics, as in "the plastics industry." (Merriam-Webster's collegiate dictionary, 11th edn. Merriam-Webster, Springfield, MA, 2004) However, there has been a trend in Europe to use the plural form exclusively even when it results in ungrammatical phrases such as "a plastics hose." The intent of the ungrammatical pluralization is to distinguish between the synthetic polymers used in the plastics industry and other materials sometimes referred to as "plastic," such as hot glass, modeling wax, and clay in the wet, unfired state. The preference of most authors is to use the singular form when it is evident from context that the noun refers to a single material. (2) *adj* Capable of being deformed continuously and permanently without rupture at a stress above the yield value. (3) *n* A material that contains as an essential ingredient one or more high polymers, is solid in its finished state and, at some stage in its manufacture or processing into finished articles, can be shaped by flow. However, this definition is supplemented by notes explaining that materials such as rubbers, textiles, adhesives, and paints, which may in some cases meet this definition, are not considered to be plastics. The terms *plastic*, *resin*, and *polymer* are somewhat synonymous, but resin and polymer most often denote the basic materials as polymerized, while the term *plastic* or *plastics* encompasses compounds containing plasticizers, stabilizers, fillers, and other additives, (Whittington's dictionary of plastics. Carley James F (ed). Technomic, 1993). See also ▶ Elastomer.

Plasticate *v* (plastificate) To render a thermoplastic more flexible, even molten, by means of both heat and mechanical working. Sometimes used imprecisely for ▶ Plastify and incorrectly for ▶ Plasticize.

Plasticating Capacity *n* Of an extruder or injection molder, the maximum rate at which the machine can melt room-temperature feedstock and raise it to the temperature suitable for extrusion or molding. This rate is determined mainly by the quotient of the available screw power, divided by the means specific heat of the plastic of interest and the rise in temperature of the plastic from feed to die; and to lesser degrees by extruder length, screw design, and die design.

Plastication *n* In a screw extruder, plastication means that one starts with a solid feed (pellet form), which is brought to a melt state, so that plastication extrusion can proceed.

Plastic-Bonded Wallpapers *n* Papers made with a protein size with a plastic added. They are washable, but subject to staining by certain liquids.

Plastic-Coated Wallpapers *n* Papers that are more washable and stain-resistant. A thick plastic coating increases the resistance and permits vigorous washing.

Plastic Deformation *n* (1) A change in dimensions of an object under load that is not recovered when the load is removed. For example, squeezing a chunk of putty results in plastic deformation. The opposite of plastic deformation is *elastic deformation*, in which the dimensions return instantly to the original values when the load is removed, e.g., as when a rubber band is stretched and released. (2) In tough plastics, deformation beyond the yield point, appearing on the stress-strain diagram as a large extension with little or no rise in stress. A part of the plastic deformation may be recovered when the stress is released; the remainder is ▶ Plastic Flow.

Plastic Flow Irreversible flow above the yield point. The flow of molten or liquid plastics during processing. Deformation without change of stress. A liquid displays plastic flow when a yield stress must be overcome or exceeded before flow will take place. Plastic viscosity, U, is expressed as:

$$U = \frac{\text{shear stress} - \text{yield stress}}{\text{shear rate}}$$

Liquids which display plastic flow are called Bingham liquids. (Patton TC (1964) Paint flow and pigment dispersion. Interscience, New York)

Plastic Foam *n* (1943) Syn: ▶ Cellular Plastic.

Plasticity \pla-ˈsti-sə-tē\ *n* (ca. 1783) The complex property of a material involving a combination of the properties of mobility and of yield value, enabling it to be continuously deformed without rupture when acted on by a force sufficient to case flow and allowing it to retain its shape after the applied force has been removed.

Plasticize \ˈplas-tə-ˌsīz\ *vt* (1919) To render a material softer, more flexible and/or more moldable by the addition and intimate blending in of a plasticizer. Should not be confused with ▶ Plasticate and ▶ Plastify.

Plasticizer *n* A substance incorporated into a material such as plastic or rubber to increase its softness, processability and flexibility via solvent or lubricating action or by lowering its molecular weight. Plasticizers can lower melt viscosity, improve flow and increase low-temperature resilience of material. Most plasticizers are nonvolatile organic liquids or low-melting-point solids, such as dioctyl phthalate or stearic acid. They have to be non-bleeding, nontoxic and compatible with material. Sometimes plasticizers play a dual role as stabilizers or crosslinkers.

Plasticizer *n* (1) An additive in a paint formulation to soften the film, thus giving it better flexibility, chip resistance, and formability. (2) A substance of low or even negligible volatility incorporated into a material (usually a plastic or an elastomer) to increase its flexibility, workability, or extensibility, while reducing elastic moduli. A plasticizer may also reduce melt viscosity and lower the glass-transition temperature. Most plasticizers are nonvolatile organic liquids or low-melting solids that function by reducing the normal intermolecular forces in a resin, thus permitting the macromolecules to slip past one another more freely. Some are polymeric. Plasticizers are classified in several ways according to: their compatibility; their general structure (monomeric or polymeric); their functions [(flame-retardant, high-temperature, low-temperature, nontoxic (see ▶ Nontoxic Material), stabilizing, crosslinking, etc.]; and their chemical nature (see ▶ Adipate Plasticizer, ▶ Chlorinated Paraffin, ▶ Epoxy Plasticizer, ▶ Phosphate Plasticizer, and ▶ Phthalate Ester). Many thousands of compounds have been developed as plasticizers, or which perhaps less than 200 are in widespread use today. The main facts about over 3,509 plasticizers are tabulated in the "Plasticizers" data table of the *Modern Plastics Encyclopedia* for 1993 (and earlier years). About two-thirds of all plasticizers produced are used in vinyl compounds, in which field the three "workhorse" plasticizers are dioctyl phthalate, diisooctyl phthalate, and diisodecyl phthalate.

Plasticizer-Adhesive *n* An additive, partly replacing plasticizers, that improves the adhesion of plastics coatings to substrates. For example, polymersable monomers such as diallyl phthalate or triallyl cyanurate are added to PVC plastisols to improve their adhesion to metals, but these compounds also contribute to the plasticizing function.

Plasticizer Efficiency *n* (1) The parts by weight of plasticizer per hundred parts of resin (phr) required to produce a plasticized PVC resin of a particular hardness on the Durometer A scale. (2) Taking dioctyl phthalate as the industry standard of comparison, one may express the efficiency (in percent) of another plasticizer as 100 (n_o/n_1), where n_o is the phr of DOP required to achieve a particular Durometer value (or other desired property) and n_1 is the phr of the alternate plasticizer required to reach that same value.

Plasticizer Migration ▶ See Migration Of Plasticizer.

Plasticizer, Polymerizable *n* (reactive plasticizer) A special type of plasticizer, unique in that it functions as a plasticizer only before and during the processing step consisting of a monomer added to a plastisol to increase its fluidity, which monomer cures in the presence of catalysts to become rigid in the fused plastisol article. Among such monomers are polyglycol dimethacrylates, dimethacrylates of 1,3-butylene glycol and trimethylolpropane, and some trade named monomers whose compositions are proprietary. These polymerizable plasticizers enable one to liquid-cast very rigid articles that would otherwise have to be made, with very low plasticizer levels by injection molding. Monomeric styrene, not ordinarily thought of as a plasticizer, performs in much the same way in polyester laminating formulations, lowering viscosity during wetting-out and the initial moments of pressure molding, then polymerizing to form crosslinks of the strong, stiff finished product, (2) any of a new class of epoxy resins having the general structure.

Basic epoxy prepolymer structure

An example is Bisphenol-Epichlorohydrin Epoxy Prepolymer, in which the R groups may H, methyl, or ethyl, and n = 1 to 10. These are very miscible with epoxy resins, they provide non-migrating internal plasticization after curing, and they are useful in coatings, adhesives, and sealants.

Di-n-octyl phthalate

Plasticizer *n*, **Solid** A plasticizer that is solid at room temperature but melts during processing to improve processability of the polymer in which it is incorporated. Upon cooling it resolidifies and thus does not soften the finished article. Solid plasticizers are used in rigid PVC, one of the most common being diphenyl phthalate (mp = 75°).

Plastic Memory See Memory.

Plasticorder *n* (Plastograph) See ▶ Barbender Plastograph.

Plastic Paint *n* A heavy-bodied, thixotropic paint which can be worked after application, by stripping or by paint rollers having a textured pattern, to produce various textured or pattern surfaces. *Also called Textured Paint and Texture-Finished Paint.*

Plastic Paper *n* (synthetic paper) Paper-like products in which the skeletal structure is composed of synthetic resin. Three main types are *spunbonded sheet, film paper,* and *synthetic pump* (synpulp). Film papers are similar to thin films of oriented polystyrene or polyolefins but they are treated to obtain opacity and ink receptivity. Synpulps are papermaking pumps made usually from polyolefins by processes that produce fibrous pulps without the use of conventional spinning methods.

Plastic, Rigid See ▶ Rigid Plastic.

Plastics See ▶ Polymers.

Plastic, Semi-Rigid See ▶ Semirigid Plastic.

Plastic Solid *n* Solid that deforms continuously and permanently when subjected to a shearing stress in excess of its yield value.

Plastics, Recycling *n* A term embracing systems by which plastics materials that would otherwise immediately become solid wastes are collected, separated, or otherwise processed and returned to the economic mainstream in the form of useful raw materials or products.

Plastic Strain *n* Plastic flow above the yield stress expressed as a fraction or percent of the original dimension before applying stress.

Plastic Tooling *n* A term designating structures composed of plastics that are used as tools in the fabrication of metals or other materials including plastics. While they are usually made of reinforced and/or filled thermosets, flexible silicone or polyurethane tools are often used for casting plastics. Common applications of rigid plastics tooling are sheet-metal-forming dies, models for duplicators, drill fixtures, spotting racks molds for thermoforming thermoplastics, and injection molds for short runs. The tools are formed by the usual processes used for thermosetting resins, such as laminating, cating, and sprayup.

Plastic Viscosity *n* A term describing a flow property of a printing ink. The flow curve is a graph of shearing force versus rate of shear. See ▶ Viscosity, Plastic.

Plastic Viscosity *n* For a ▶ Bingham Plastic, the difference between the shear stress and the yield stress, divided by the shear rate.

Plastic Welding See ▶ Welding.

Plastic Wood *n* A paste of wood flour, plasticizer, resins and/or other materials dispersed in nitrocellulose or other binders and volatile solvents, used for repairing or filling holes in wood, etc.

Plastify *adj* To soften a thermoplastic resin or compound by means of heat alone, as in sheet thermoforming. Should not be confused with ▶ Plasticize or ▶ Plasticate.

Plastigel *n* A vinyl compound similar to a plastisol, but containing sufficient gelling agent and/or filler to provide a putty-like consistency. It may be molded to a shape-retaining form at room temperature, then heated and cooled to impart permanency.

Plastisol \ˈplas-tə-ˌsäl\ *n* [*plastic* + ^{4}*sol*] (1946) A suspension of a finely divided vinyl chloride polymer or copolymer in a liquid plasticizer which has little or no tendency to dissolve the resin at ambient temperature but which becomes a solvent for the resin when heated. At room temperature the suspension is very fluid and suitable for casting. At the proper temperature, the resin is completely dissolved in the plasticizer, forming a homogeneous plastic mass which upon cooling is a more or less flexible solid. Additives such as fillers, stabilizes, and colorants are also usually present. A plastisol modified with volatile solvents or diluents that evaporate upon heating is known as an ▶ Organosol. When gelling or thickening agents are added to produce a putty-like consistency at room temperature, the dispersion is called a ▶ Plastigel. The coined term *ridigsol* is used to denote a plastisol modified with polymerizable or crosslinking monomers so that the fused product is rigid rather than flexible. Products are made from plastisols by rotational casting, slush casting, dipping, spraying, film casting, and coating.

Plastograph See ▶ Barbender Plastograph.

Plastomer *n* (1) (Solprene®) Any of a family of thermoplastic-elastomeric, styrene-butadiene copolymers whose molecules have a radial or star structure in which several polybutadiene chains extend from a central hub, with a polystyrene block at the outward end of each segment. They are used in making footwear components, in adhesives and sealants, and are also blended with other resins to upgrade performance. (2) Late in 1992 this term was adopted as generic by Exxon Chemical and Dow Chemical for grades of ▶ Very-Low-Density Polyethylene, produced with so-called "exact" metallocene catalysts and offering the flexibility of rubber with the strength and processability of linear low-density polyethylene. Densities range from 0.880 to 0./905 g/cm^3.

Plastometer *n* An instrument for determining the flow properties of plastic materials. The term "plastometer" is not actually necessary, for a plastometer is essentially any viscometer capable of measuring the flow properties of a material exhibiting yield value. See ▶ Rheometer and ▶ Viscometer.

Plate Cylinder *n* That roller of an offset printing press which bears the printing plate.

Plated *vt* (14c) (1) A term to describe a fabric that is produced from two yarns of different colors, characters, or qualities, one of which appears on the face and the other on the back. (2) A term to describe a yarn covered by another yarn.

Plate Die *n* An inexpensive and easily modified die for extruding a plastic profile, into which a orifice of the desired shape has been machined typically by ▶ Electrical-Discharge Machining. The plate is bolted to the front of a universal die body.

Plate Dispersion Plug *n* Two small, perforated, parallel disks joined by a central connecting rod. Such assemblies were sometimes inserted in the nozzles of ram-type injection-molding machines to improve the distribution of colorants in the resin as it flows through the nozzle.

Plateless Engraving *n* A typographic printing process in which densely pigmented ink on printed sheets is dusted with a powdered resin of low melting point. After surplus powder has been blown from the paper, heat is applied to fuse the resin and yield characters which are raised above the paper.

Plate Mark *n* Any imperfection in a pressed plastic sheet resulting from the surface of the pressing plate (ASTM D 883).

Platen \\ˈpla-tᵊn\ *n* [MF *plateine*, fr. *plate*] (1541) (1) A flat plate in a printing press, which presses the paper against the inked type, thus securing an impression. (2) A plate of metal, especially one that exerts or receives pressure, as in a press used for gluing plywood. (3) Either of the sturdy mounting plates of a press, usually a pair, to which the entire mold assembly is bolted. Syn: ▶ Caul.

Platen Press *n* A press which prints a single sheet by pressing the latter against an entire frame or chase of type at the same moment.

Plate-Out *n* An objectionable coating gradually formed on metal surfaces of molds and calendering and embossing rolls during processing of plastics, and caused by extraction and deposition of some ingredients such as a pigment, lubricant, stabilizer, or plasticizer. In the case of vinyls, which are especially prone to this condition, plate-out can be reduced by using highly compatible stabilizers such as barium phenolates and cadmium ethylhexoate, or by incorporating silica in the formulation. Resins can play a role in the plate-out problem, although the degree and mechanisms of resin contributions to plate-out are controversial.

Platform Blowing *n* A special technique for blow molding large parts. To prevent excessive sag of the massive parison, the machine employs a table that after rising to meet the parison at the die, descends with the parison, but a little more slowly than the parison, so as to support its weight, yet not cause buckling.

Plating See ▶ Electroplating on Plastics.

Pleat \\ˈplēt\ *n* [ME *plete*] (15c) Three layers of fabric involving two folds or reversals of direction; the back fold may be replaced by a seam.

Pleochroism \plē-ˈä-krə-ˌwi-zəm\ *n* [ISV *pleochroic*, fr. *pleio-* + Gk *chrōs* skin, color] (1857) Property of crystals or minerals of giving different absorption colors for different directions of vibration of light within the crystal; in optical mineralogy this property is observed in crystals under a polarizing microscope with the lower nicol only and rotating the stage. See ▶ Dichroism (2) An optical phenomenon in which mineral grains within a rock appear to be different colors when observed at different angles under a polarizing petrographic microscope. Pleochroism is caused by the double refraction of light by a mineral. Light of different polarizations is bent different amounts by the crystal, and therefore follows different paths through the crystal. The components of a divided light beam follow different paths within the mineral and travel at different speeds, and each path will absorb different colors of light (Verma 2009). When the mineral is observed at some angle, light following some combination of paths and polarizations will be present, each of which will have had light of different colors absorbed. At another angle, the light passing through the crystal will be composed of another combination of light paths and polarizations, each with their own color. The light passing through the mineral will therefore have different colors when it is viewed from different angles, making the stone (Peckett 1992) seem to be of different colors. Tetragonal, trigonal and hexagonal minerals can only show two colors and are called dichroic. Orthorhombic, monoclinic and triclinic crystals show three and are trichroic. Isometric minerals cannot exhibit pleochroism. Tourmaline is notable for exhibiting strong pleochroism. Gems are sometimes cut and set either to display pleochroism or to hide it, depending on the colors and their attractiveness. Pleochroism is an extremely useful tool in mineralogy for mineral identification, since minerals that are otherwise very similar often have very different pleochroic color schemes. In such cases, a thin section of the mineral is used and examined under polarized transmitted light with a petrographic microscope. See ▶ Birefringence. References:'Verma PK (ed) (2009). Optical mineralogy. Taylor & Francis, New York; Peckett A (1992) Colours of opague minerals. Springer, New York.

Plessy's Green n $CrPO_4 \cdot nH_2O$. A bluish green pigment consisting essentially of hydrated chromium phosphate.

Plexidur n Poly(methyl methacrylate). Manufactured by Roehm and Haas, US.

Plexiglas® n Poly(methyl methacrylate) Famous trade name for acrylic resins and cast polymethyl methacrylate sheet. Manufactured by Röhm & Haas, Germany.

Plexol n Oil-soluble methacrylate copolymer (viscosity improver). Manufactured by Röhm & Haas, Germany.

Plied Yarn n A yarn formed by twisting together two or more singles yarns in one operation.

Plied Yarn Duck See ▶ Duck.

Pliofilm n Rubber hydrochloride, manufactured by Goodyear, US.

Pliolite NR Cyclo rubber. Manufactured by Goodyear, US.

Plissé \pil-ˈsā\ n [F plissé, fr. pp of plisser to pleat, fr. MF, fr. pli fold, fr. plier to fold] (1873) A cotton, rayon, or acetate fabric with a crinkled or pleated effect. The effect is produced by treating the fabric, in a striped or spotted motif, with a caustic-soda solution which shrinks parts of the goods.

Plochere Color System n A color order system based on substractive colorant mixture developed by mixing a limited number of pigments in systematically varied proportions.

Plucking n A condition found at the feed roll and lickerin section of the card when larger than normal clusters of fiber are pulled from the lap by the lickerin. This situation is normally caused by uneven laps or the inability of the feed rolls to hold the lap sheet while small clusters of fibers are being pulled from the lap by the lickerin. Plucking inevitably produces flaky webs.

Plug-and-Ring Forming n A technique of ▶ Sheet Thermoforming in which a plug, functioning as a male mold, is forced into a heat-softened sheet held in place by a clamping ring.

Plug-Assist Forming n (vacuum forming with plug assist) A sheet thermoforming process in which a convex mold half presses the softened sheet into the concave half, accomplishing most of the draw, after which vacuum is applied, drawing the sheet onto the concave surface. The method provides more nearly equal bottom and side thicknesses than straight vacuum forming and permits deeper draws.

Plug Flow n Movement of a material as a unit without shearing within the mass. This is an extreme seldom realized in practice, but can occur over the center of a Bingham-plastic stream or in a system where the fluid does not wet the bounding walls. As compared with Newtonian flow, the more pseudoplastic the plastic melt, the more nearly sluggish is its flow.

Plugging Value n In the manufacture of acetate fibers, a measure of filterability. It is the weight of solids in an acetate dope that can be passed through a fixed area of filter before the filter becomes plugged. It is expressed as weight of solids per square unit of filter area, e.g., g/cm^2.

Plumbago \ˌpləm-ˈbā-(ˌ)gō\ n [L plumbagin-, plumbago galena, fr. plumbum] (1747) Another name for graphite.

Plunger \ˈplən-jər\ n (1611) The part of a transfer-press or old-style injection machine that applies pressure on the unmelted plastic material to push it into the chamber, which in turn displaces the plastic melt in front of it, forcing it through the nozzle and into the mold. See also ▶ Ram, ▶ Force Plug, and ▶ Pot Plunger.

Plunger Molding n A variation of ▶ Transfer Molding in which an auxiliary hydraulic ram is employed to assist the main ram. The auxiliary ram rapidly forces the material through a small orifice, thereby generating high frictional heat. The higher temperature speeds the cure of the material, which when transferred into the mold by the main ram, cures very soon after the mold is filled.

Pluronics n Ethylene oxide/propylene oxide copolymer, manufactured by Wyandotte Chemical, U.S.

Plush \ˈpləsh\ n [MF peluche] (1594) A term describing a cut-pile carpet in which the pile yarns are only slightly twisted, dense, and very evenly sheared. A plush carpet has the look of a solid, flat velvet surface. Similar pile constructions are also used in upholstery fabric.

Plutonium \plü-ˈtō-nē-əm\ n [NL, fr. Pluton-, Pluto, the planet Pluto] (1942) A fissile element, artificially produced in the pile by neutron bombardment of U^{238}.

Ply \ˈplī\ n [ME plien to fold, fr. MF plier, from L plicare; akin to OHGr flehtan to braid, Latin plectere, Gk plekein] (ca. 1909) (1) The number of singles yarns twisted together to form a plied yarn, or the number of plied yarns twisted together to form cord. (2) An individual yarn in a plied yarn or cord. (3) One of a number of layers of fabric (ASTM). (4) The number of layers of fabric, as in a shirt collar, or of cord in a tire.

Plyfil® n A proprietary system of making twofold long- and-short staple yarns by using ultrahigh drafting. The slightly twisted ends produced are not useable yarns but are well suited for subsequent processing, i.e., twisting.

Plying *n* Twisting together two or more singles yarns or ply yarns to form, respectively, ply yarn or cord.

Ply Twisting See ▶ Plying.

Plywood \ ▮plī- ▮wúd\ *n* (1907) A cross-bonded assembly made of layers of veneer or veneer in combination with a lumber core or plies joined with an adhesive. Two types of plywood are recognized, namely *veneer plywood* and *lumber core plywood*. NOTE – Generally the grain of one or more plies is approximately at right angles to the other plies, and almost always an odd number of plies are used.

PMA *n* (1) An abbreviation for Phosphomolybdic Acid. (2) Applied to pigments which have been precipitated with phosphomolybdic acid to give it permanence and insolubility.

PMAC *n* Abbreviation for Polymethoxy Acetal.

PMAN *n* Abbreviation for Polymethacrylonitrile.

PMA Pigment See ▶ Precipitated Basic Dye Blues.

PMC *n* Abbreviation for Polymer-Matrix Composite.

PMCA *n* Abbreviation for poly(methyl-α-chloroacrylate), a member of the acrylic-resin family.

PMDA *n* Abbreviation for Pyromellitic Dianhydride.

PMMA *n* Abbreviation for Poly(Methyl Methacrylate).

PMP *n* Abbreviation for Poly(4-Methylpentene-1).

Pnicogen *n* A member of group VA in the periodic table.

PO *n* Phenoxy resin.

Pock Mark *n* An imperfection on the surface of a blow-molded article, an irregular indentation caused by inadequate contact of the blown parison with the mold surface. Contributory factors are insufficient blowing pressure, air entrapment, and condensation of moisture on the mold surface. See also ▶ PIT for an ASTM-approved definition for synonymous term that is not specific to blow molding.

Pockmarking *n* Film defect in the shape of irregular and unsightly depressions formed during the drying of a paint or varnish film. See ▶ Orange Peel. *Also called Pitting.*

Point Bonding See ▶ Bonding (2).

Pointillism \ ▮pwa^n(n)-tē- ▮yi-zəm\ *n* {*often capitalized*} [F *pointillisme*, fr. *pointiller* to stipple, fr. *point* spot] (1901) From the French, pointiller, meaning to dot, stipple. A system of late impressionist painting developed by Georges Surant (1859–1991) and Paul Signac (1863–1935). contrary to the practice of color mixing on the palette, pointillism consists of applying separated spots or dots of pre color, side by side on the canvas, e.g., red and yellow for orange, red and blue for mauve, etc. Theoretically, at the right distance, the spectator's eye automatically mixes the colors.

Pointing *n* Treatment of joins in masonry by filling with mortar to improve appearance or protect against weather.

Point Source *n* Any discernible, confined and discrete conveyance from which pollutants are or may be discharged.

Poise \ ▮póiz\ *n* [F, fr. Jean Louis Marie *Poiseuille* † 1869 French physician and anatomist] (1913) A fundamental and absolute unit of viscosity measurement. A substance is said to have a viscosity of 1 poise when a force of 1 dyne is required to move a surface of 1 cm^2 at a speed of 1 cm/s relative to another plane surface separated from it by a layer 1 cm thick.

$$1 \text{ poise} = 1 \text{ dyne s/cm}^2 = 1\text{g}/\text{cm s}$$

Poiseuille Equation See ▶ Hagen-Poiseuille Equation.

Poiseuille Flow *n* Laminar flow in a pipe or tube of circular cross section under a constant pressure gradient. If the flowing fluid is Newtonian, the flow rate will be given by the ▶ Hagen-Poiseuille Equation.

Poison \ ▮pói-z°n\ *n* [ME, fr. MF, drink, poisonous drink, poison, fr. L *potion-*, *potio* drink] (13c) (1) Any substance that is harmful to living tissues when applied in relatively small doses. (2) A substance that reduces or destroys the activity of a catalyst.

Poisson Distribution \pwä- ▮sō^n-\ *n* [Siméon D. *Poisson* † 1840 French, mathematician] (1922) A probability density function that is often used as a mathematical model of the number of outcomes obtained in a suitable interval of time and space, that has its mean equal to its variance, that is used as a approximation to the binomial distribution, and that has the form

$$f(x) = e^{\mu}\mu^x/x!$$

where μ is the mean and x takes on nonnegative interval values.

Poisson Ratio *n* The proportion of lateral strain to longitudinal strain under conditions of uniform longitudinal stress within the proportional or elastic limit. When the material's deformation is within the elastic range its results in lateral to longitudinal strains will also be constant.

Poisson Ratio Distribution *n* The constant relating the changes in dimensions which occur when a material is stretched. It is obtained by dividing the change in width per unit length by the change in length per unit length.

Poisson's Ration *n* (for S. *Poisson*; symbol μ or ν) In a material under tensile stress, the ratio of the transverse contraction to the longitudinal elongation. For metals, Poisson's ratio is bout 0.3, for concrete, 0.1. For many plastics, the ratio is in the range from 0.32 to 0.48 and for rubbers it is about 0.5. Mineral fillers in plastics reduce the ratio by 0.05–0.10. In nonisotropic structures such as uni- and bidirectional laminates, Poisson's ratio may be different in orthogonal directions; but the value perpendicular to the reinforcing fibers will be the one most likely to be relevant. ASTM Standards lists methods for determining Poisson's ratio for ceramics, concretes, and structural materials, but none for plastics or rubbers. For isotropic, elastic materials, the ratio is related to the three principal moduli by the equations: $\mu = (E/2G) - 1 = 0.5 - (E/6B)$, where E, G, and B are the tensile modulus (Young's modulus), shear modulus, and bulk modulus, respectively. These relationships may be used to estimate the ratio when the strains are small. However, because plastics are not truly elastic, and because moduli may vary with temperature they should be regarded as approximate when applied to plastics.

Polar \pō-lər\ *adj* [NL *polaris*, fr. L *polus* pole] (1551) Molecule or radical that has, or is capable of developing, electrical charges. Thus, polar molecules ionize in solution and impart electrical conductivity. Water, alcohol, and sulfuric acid are polar in nature; most hydrocarbon liquids are not. Carboxyl and hydroxyl often exhibit an electrical charge. The formation of emulsions and the action of detergents are dependent on this behavior.

Polar Covalent Bond *n* A covalent bond in which the bonding electron pair is not shared equally but is drawn closer to the more electronegative atom.

Polarizability \pō-lə-rī-zə-bi-lə-tē\ *n* [F *polariser*, fr. NL *polaris* polar] (1811) The ease by which the particles of a substance may be distorted or oriented by an electric field.

Polarization \pō-lə-rə-zā-shən\ *n* (1812) (1, dielectric) The slight shifting of molecular electric charges when a polymer is placed in a strong electric field, creating local electric dipoles. The shifting of charge takes finite time and generates friction. In a high-frequency field, the rapid shifting causes considerable dissipation of energy, which is the basis for ▶ Dielectric Heating of plastics. (2, light) Of the three types possible, the most useful and common is *plane polarization*. This occurs when ordinary, unpolarized light having wave motions in all directions perpendicular to the ray, passes through nicol prisms or polarizing filters that deliver an exit ray whose vibrations lie in one plane.

Polarized Light *n* A bundle of light rays with a single propagation direction and a single vibration direction. The vibration direction is always perpendicular to the propagation direction. It is produced from ordinary light by reflection, by double refraction in a suitable crystal or by absorption with a suitable pleochroic substance. Specific rotation is the power of liquids to rotate the plane of polarization, It is stated in terms of specific rotation or the rotation in degrees per decimeter per unit density.

Polarizer *n* A polarizing element which is placed below the preparation with its vibration direction preferable set in an E–W direction.

Polarizing Microscope *n* An optical microscope fitted above and below the specimen-holding stage with nicol prisms or polarizing filters. The lower filter (*polarizer*) imparts plane ▶ Polarization to the incoming light. The upper one (*analyzer*) is rotatable, but is usually set so that its plane of vibrations is at 90° to that of the lower one. With isotropic specimens, all light is blocked at the analyzer and the observer sees only darkness. With a birefringent specimen, if the original light source is white, the observer sees bands of colors related to the crystal structure and the specimen's refractive indices.

Polar Molecule *n* A molecule in which the centers of positive and negative charge do not coincide; a dipole.

Polars *n* Two identical polarizing elements in a polarizing microscope. The polar placed between the light source and substage condenser is called the polarizer; the polar placed between the objective and ocular is called the analyzer. The vibration directions of the two polars may be crossed 90° to achieve "crossed polars"; slightly uncrossing one polar gives "slightly uncrossed polars"; removing the analyzer results in "plane-polarized light."

Polar Solvents *n* Solvents with oxygen in their molecule. Water, alcohols, esters, and ketones are examples. All possess a degree of conductivity that inhibits static build-up characteristic of non-polar solvents such as tolvol, xylol, and naphthas.

Polar Winding *n* In filament winding, a winding in which the filament path passes tangent to the polar opening at one end of a chamber and tangent to the opposite side of the polar opening at the other end.

Polepiece *n* In reinforced plastics, the supporting part of the mandrel used in filament winding, usually on one of the axes of rotation.

Polimerisado Oil *n* A bodied liquid oiticica oil.

Polish \pä-lish\ [ME *polisshen*, fr. MF *poliss*-, stem of *polir*, fr. L *polire*] (14c) (1, *v*) See ▶ Burnish. (2, *n*) A solid powder or a liquid or semi-liquid mixture that imparts smoothness, surface protection or a decorative finish.

Polishing *n* (1) Smoothing and imparting luster to a surface by rubbing with successively finer abrasive-containing compounds or by filling the minute low areas of the surface with a wax or polymeric finish. (2) Smoothing rough edges by applying a jet of hot gas (to plastics) or a flame (to glasses). Flame polishing of plastics is generally not recommended because of the likelihood of degrading the surface and/or leaving residual stresses, either of which can cause crazing.

Polishing Roll *n* A roll, usually one of a set that has a highly polished, chrome-plated surface that is mirrored on all sheet or film extruded onto the roll or calendered through it (them).

Polishing Varnish *n* Very hard-drying, short oil varnish used for interior woodwork, furniture, etc., and capable of being rubbed with abrasive and mineral oil lubricants to a very smooth surface for a desired degree of gloss. Syn: ▶ Rubbing Varnish.

Pollopas *n* Urea-formaldehyde resin. Manufactured by Dynamit Nobel, Germany.

Pollution \pə-ˈlü-shən\ *n* (14c) The presence of matter or energy whose nature, location, or quantity produces undesired environmental effects.

Pollution Abatement *n* Ending pollution. Distinguished from pollution control, which may only reduce pollution, and penalties which principally punish violations.

Pollution Control *n* Reducing pollution. Distinguished from pollution abatement, which means ending pollution, and penalties, which principally punish violations.

Poly- A prefix meaning many. Thus, the term *polymer* literally means *many mer*, a mer being the repeating unit of any high polymer.

Polyacenaphthalene

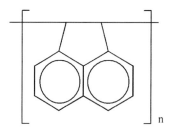

Polyacetal *n* A polymer in which the repeated structural unit in the chain is of the acetal type. See also ▶ Acetal Resin and Polyoxymethylene.

Polyacetylene *n* A polymer of acetylene, made with Ziegler-Natta catalysts and usually dark-colored, with the unusual property (for a polymer) of high electrical conductivity, achieved by doping the polymer with about 1% of ionic dopant such as iodine. It may become a useful solar-cell material because its absorption spectrum closely matches the solar spectrum, but mechanical properties and stability are poor. Also, practical processing methods have yet to be developed.

Polyacrylamide \ˌpä-lē-ə-ˈkri-lə-ˌmīd\ *n* (1944) Poly (2-Propenamide) A nonionic, water-soluble polymer prepared by the addition polymerization of acrylamide ($CH_2=CHCONH_2$). The white polymer is readily soluble in cold water but insoluble in most organic solvents. It is used as a thickener, suspending agent, and as an ingredient in adhesives.

Polyacrylate A polymer of an ester of acrylic acid or of esters of acrylic acid homologues or substituted derivatives. See ▶ Acrylic Resin. Polyacrylate n Sodium polyacrylate belongs to a family of water loving or hydrophilic polymers. It has the ability to absorb up to 800 times its weight in distilled water. Sodium polyacrylate is a powder takes the form of a coiled chain. There are two important groups that are found on the polymer chains, carbonyl (COOH) and sodium (Na). These two groups are important to the overall absorption potential of the polymer. When the polymer is in the presence of a liquid, the sodium dissociates from the carbonyl group creating two ions, carboxyl (COO−) and sodium cation (Na+). The carboxyl groups then begin to repel each other because they have the same negative charge. As a result of the repulsion between the like charges, the sodium polyacrylate chain uncoils or swells and forms a gel substance. The action of swelling allows more liquid to associate with the polymer chain. There are four major contributors to sodium polyacrylate's ability to absorb liquids or swell. These contributors are hydrophilic chains, charge repulsion, osmosis, and cross-linked between chains. Ions in the polymer chain such as carboxyl groups (COO−) and sodium (Na+) attract water molecules, thus making the polymer hydrophilic. Charge repulsion between carboxyl groups allow the polymer to uncoil and interact with more water molecules (see figure below).

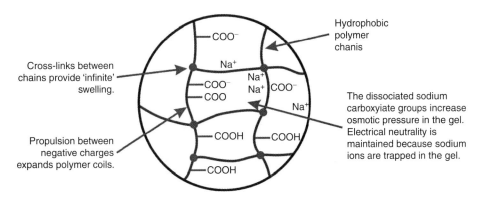

Charge repulsion between carboxyl groups in cross-linked SAP

Acrylate polymers – in a dry coiled state

When these dry coiled molecules are placed in water, hydrogen bonding with the HOH surrounding them causes them to unfold and extend their chains as shown in the figures below.

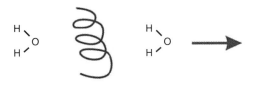

Interaction of SAP with water

When the molecules straighten out, they increase the viscosity of the surrounding liquid. That is why several types of acrylates are used as thickeners. Super absorbent chemistry requires two things: The addition of small cross-linking molecules between the polymer strands; and the partial neutralization of the carboxyl acid groups (−COOH) along the polymer backbone (−COO$^-$Na$^+$). Water molecules are drawn into the network across a diffusion gradient which is formed by the sodium neutralization of the polymer backbone. The polymer chains want to straighten but are constrained due to the cross-linking. Thus, the particles expand as water moves into the network.

Expansion between polymer chains with interaction with water

The water is tightly held in the network by hydrogen-bonding. Many soluble metals will also ion-exchange with the sodium along the polymer backbone and be bound.

Cross-linked polymer chains

A view of a crosslinks in super absorbent molecules is shown in the following figure, and the more practical polymer network in the following figure.

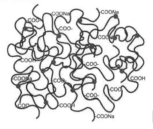

View of superabsorbent polymer network with cross-linked sites

Polyacrylates A thermoplastic resin made by the polymerization of acrylic ester such as methyl methacrylate. See ▶ Acrylic Resin.

Polyacrylic Acid (PAA) A polymer of acrylic acid, used as a textile size.

Poly(acrylic acid) A polymer of acrylic acid used as a sizing agent in the manufacture of nylon and other synthetic textiles.

Poly(acrylic-acid-co-styrene) Poly(2-propenoic acid-co-ethenyl benzene)

Polyacrylic Plastics Plastics based on polymers in which the repeated structural units in the chains are essentially all of the acrylic type. See ▶ Acrylic Plastics.

Polyacrylonitrile \ ˌpä-lē-ˌa-krə-lō-ˈnī-trəl\ n (1935) (PAN) Made by free-radical polymerization of acrylonitrile (CH_2=CHCN) in solution or suspension, this highly polar polymer is the basis of large-volume acrylic and modacrylic fibers.

Poly(acrylonitrile-co-butadiene) Common name for 2-Propenenitrile, polymer with 1,3 butadiene. It is also commonly known as Acrylonitrile-butadiene polymer.

Poly(acrylonitrile-co-styrene)

Polyaddition See ▶ Addition Polymerization.

Polyadipamide A polymer formed by the reaction of adipic acid with a diamine, NYLON 6/6 being the most important example.

Polyalcohol \ˌpä-lē-ˈal-kə-ˌhȯl\ n (1900) Alcoholic or hydroxy compound, containing more than one hydroxyl group. Typical polyalcohols are ethylene glycol with two hydroxyl groups, glycerol with three, and pentaerythritol with four hydroxyl groups. Polyalcohols are sometimes known as polyhydric alcohols. Syn: ▶ Polyol.

Polyalkenamer *n* A chlorine-containing elastomer developed by Goodyear, with properties similar to but somewhat better than those of neoprene rubber. It is a copolymer of the addition-reaction product of hexachlorocyclo-pentadiene or 1,5-cyclooctadiene and an olefin such as cyclopentane.

Polyalkylene Amide. See ▶ Amino Resin.

Polyalkylene Terephthalate *n* Any of a family of thermoplastic polyesters that are polycondensates derived from terephthalic acid, whose diol components may be any within a wide range. The principal members of the family are ▶ Polyethylene Terephthalate and ▶ Polybutylene Terephthalate.

Polyallomer *n* A crystalline block copolymer produced from two or more different monomers, usually ethylene and propylene, by alternately polymerizing the monomers in the presence of anionic, coordination catalysts, resulting in chains containing polymerized segments of both monomers. The polymer chains exhibit degrees of crystallinity normally found only in stereoregular homopolymers of propylene and ethylene, and the copolymers possess properties different from those of either blends of the homopolymers or copolymers prepared by conventional polymerization processes. Among such properties are high impact strength, low density, and flexural-fatigue resistance. The name "polyallomer" is derived from *allomerism*, meaning a similarity of crystalline form with a difference in chemical composition.

Polyalloys *n* Alloys that are combinations of two or more plastics which are mechanically blended; they do not depend on chemical bonds but often require special compatibilizers.

Polyallyl Diglycol Carbonate *n* A high-impact, transparent thermoplastic with excellent abrasion resistance, made from Pittsburg Plate Glass Industries' CR-39 monomer (and hence sometimes called CR-39). It is widely used in eyeglasses.

Polyamic Acid [(from *poly(amide-acid)*] A polymer containing both amide and acid groups. The aromatic varieties are precursors of ▶ Polyimides.

Polyamide \ˈpä-lē-ˌa-ˌmīd\ *n* [ISV] (1929) A synthetic polymer and the fibers made from it in which the simple chemical compounds used for its production are linked together by amide linkages (−NH−CO−). Also see ▶ Nylon Fiber.

Polyamide-Imide Resin *n* (PAI) Any of a family of polymers based on the combination of trimellite anhydride with aromatic diamines. In the uncured form (*ortho*-amic acid) the polymers are soluble in polar organic solvents. The imide linkage is formed by heating, producing an infusible resin with thermal stability up to 290°C. These resins are used for laminating, prepregs, and electrical components. Molding resins that behave as thermoplastics can be produced by thermally curing and modifying amide-imide polymers. These molding resins can be processed by compression molding, extrusion, and injection molding.

Polyamide-Imide Resins *n* Engineered thermoplastics characterized by excellent dimensional stability, high strength at high temperatures, and good impact resistance.

Polyamide Plastic *n* (polyamide) See ▶ Nylon and other listings following same.

Polyamide Thermoplastic Elastomers *n* Copolymers containing soft polyether and hard polyamide blocks having good chemical, abrasion, and heat resistance, impact strength, and tensile properties. Processed by extrusion and injection and blow molding. Used in sporting goods, auto parts, and electrical devices. Also called polyamide TPE.

Polyamide TPE See ▶ Polyamide Thermoplastic Elastomers.

Polyamide Resins *n* Condensation resins of an amine and an amine and an acid, the repeated structural unit in the chain being of the amide type.

Polyamides Polymers in which the monomer units are linked together by the amide group –CONH–.

Polyamine \ˈpä-lē-ə-ˌmēn\ *n* (1861) Any of a family of compounds containing multiple amines – primary, secondary, or tertiary – or mixtures of these, useful as hardening agents for epoxy resins.

Polyamine-Methylene Resin *n* A light-amber-colored resin derived from diphenylol and formaldehyde, used as an ion-exchange resin.

Polyamine Sulfone *n* A water-soluble copolymer of diallylamine monomer and sulfur dioxide, used as a paint additive, antistatic agent, synthetic-fiber modifier, and polishing agent for metal platings.

Polyaminobismaleimide Resin *n* (PABM) Any thermosetting resin of dark-brown color obtained by the addition reaction of an aromatic diamine and a bismaleimide. Typical prepolymers accept high percentages of fillers and can be cast and compression or transfer molded. PABMs have flow properties comparable to common thermosetting resins and thermomechanical properties exceeding those of some light alloys. They possess excellent dimensional stability and are flame- and radiation-resistant. They can be adapted to aircraft, electrical and electronic fabrication, ablation applications, and to chemical-process equipment where resistance to aromatic solvents, refrigerants, and acids is required.

Polyaminotriazole *n* (PAT) A family of fiber-forming polymers made from sebacic acid and hydrazine with small amounts of acetamide.

Polyary (–O–) in the chain. Sometimes used as generic for poly-2.6-dimethyl-1,4-phenylene oxide, several markers' brands are manufactured.

Polyarylamide Resins *n* Synthesized by melt polymerization of m-xylenediamoine with adipic acid. It is used as a high stiffness, high softening point engineering plastic. Polyarylamide resins have higher modules and yield stress values than nylon 66, however, its water sensitivity has precluded a wider usage.

Polyarylamides *n* Thermoplastic crystalline polymers of aromatic diamines and aromatic dicarboxylic anhydrides having good heat, fire, and chemical resistance, property retention at high temperatures, dielectric and mechanical properties, and stiffness but poor light resistance and processability. Processed by solution casting, molding, and extrusion. Used in films, fibers, and molded parts.

Polyarylate *n* A polyester made by the condensation of an aromatic diacid with a dihydroxy aromatic compound. Commercial resins are copolymers of iso- and terephthalic acid with bisphenol A, with properties comparable to those of polycarbonate and polyethersulfone.

Polyarylate Resins *n* Often defined as copolyesters involving bisphenol-A and a mixture of terephthalic acid and iosphthalic acid, resulting in two distinct repeat units.

Polyarylene Sulfide *n* Any of a polymer family prepared by polymerization and reaction of polyhalogenated aromatics with sodium sulfide in a polar solvent at high temperature. The best known (and only commercial) resin is ▶ Polyphenylene Sulfide.

Polyaryl Ether *n* A polymer having both aromatic rings and either links (–O–) in the chain. Sometimes used as generic for poly-2,6-dimethyl-1,4-phenylene oxide, of which several markers' brands are available.

Polyarylethersulfone Resin See ▶ Polyethersulfone.

Polyaryloxysilane *n* A family of polymers, resistant to high temperatures, made up of silicon atoms, oxygen atoms, and thermally stable aromatic rings, part organic and part inorganic in nature, like the ▶ Silicones.

Poly(aryl sulfone) *n* Also called Polyether-Sulfone. An aromatic polymer consisting of benzene rings linked by both sulfone ($-SO_2-$) groups and ether oxygen atoms. Poly(aryl sulfones) can be synthesized by solution polymerization in a polar solvent. Several commercial products of this type have been developed as reasonably high temperature resistant engineering plastics.

Polyarylsulfone Resins *n* Consists mainly of phenyl and biphenyl groups linked by thermally stable either and sulfone groups.

Polyazelaic Polyanhydride *n* (PAPA) A carboxyl-terminated, low polymer of approximately 2,300 molecular weight, used as a flexibilizing curing agent for epoxy resins.

Polybasic Acid *n* Acid containing more than one reactive hydrogen atom. With water-soluble ionizable acids which are polybasic, two or more hydrogen ions are made available, per molecule.

Polybenzamide (poly-*p*-benzamide) A fiber-forming polymer made from *p*-aminobenzoÿl chlorine by self-condensation polymerization. Fibers retain high modulus to high temperatures and have been used for composite reinforcement.

Polybenzimidazole *n* (PBI) A family of high-molecular-weight, strong, and stable polymers containing recurring aromatic units with alternating double bonds. PBIs are produced mainly by the condensation of 3,3',4,4'-tetraaminobiphenyl (3,3'-diaminobenzidine) and diphenyl isophthalate. The polymers are brown, amorphous powders, exhibiting a high degree of thermal and chemical stability. They are used to make fibers and films with excellent resistance to high temperatures and flame. Principal applications have been in the aerospace field, as protective coatings on missiles, radar antennas, and supersonic aircraft; and in reinforced laminates for critical applications. In 1992, Hoechst Celanese and KMI Inc., announced a joint venture to injection-mold bearings and other parts from PBI, heretofore considered not to be melt-processable using proprietary technology.

Polybenzimidazole Fiber *n* (PBI) A manufactured fiber in which the fiber-forming substance is a long chain aromatic polymer having recurrent imidazole groups as an integral part of the polymer chain (FTC definition). The polymer is made from tetraaminobiphenyl and diphenyl isophthalate and is dry spun from a dope with dimethylacetamide as a solvent. CHARACTERISTICS: A high-performance fiber with high chemical resistance that does not burn in air. It has no melting point and does not drip when exposed to flame. The fiber and fabrics from PBI retain their flexibility, dimensional stability, and significant strength without embrittlement even when exposed to flame or extreme heat. The fiber emits little smoke in extreme conditions. It processes well on conventional textile equipment, having processing characteristics similar to polyester. It can be used in 100% form or blended with other fibers. It has a high moisture regain and low modulus with comfort properties similar to cotton. The natural color of PBI is a gold–khaki shade, but it can be dyed to almost any medium to dark shade with conventional basic dyes. END USES: With excellent thermal, flame, and chemical resistance, combined with good comfort properties, PBI is a good fiber for many critical uses including: firefighter's protective apparel, aluminized proximity gear, industrial apparel such as pants, shirts and underwear, protective gloves, welder's apparel, aircraft fire-blocking layers, aircraft wall fabrics, rocket motor insulation, race car driver's apparel, and braided packings among others.

Polybenzothiazole *n* (PBT) A family of resins, one type of which has been made by cooking a toluidine with sulfur, or by reaction of, e.g., 3,3'-mercaptobenzidine with diphenyl phthalate. These polymers have outstanding thermal stability: glass-reinforced PBT composites have withstood 350°C for over 10 days. Though they are technically thermoplastics, they are not melt-processable. Applications are in coatings for high-temperature service and laminating.

Polybiphenylsulfone *n* An engineering thermoplastic (Union Carbide's Radel®) with a deflection temperature of 205°C and notched-Izod impact strength approaching 8 J/cm (15 ft-lb$_f$/in.), this resin is conventionally melt-processable.

Polyblend *n* A colloquial term – shortened from *poly*mer *blend* – used for physical mixtures of two or more polymers, for example, polystyrene and rubber or PVC and nitrile rubber. Such blends usually yield products with favorable properties of both components, sometimes opening markets not available to either of the neat resins. The term *alloy* is sometimes used for blends.

Polybutadiene \- ┃byü-tə- ┃dī- ┃ēn\ *n* (1939) A synthetic rubber made from 1,3-butadiene ($H_2C=CH-CH=CH_2$). The *cis* type has superior abrasion resistance and resilience, while the *trans* type is similar to natural rubber.

Polybutadiene-Carylic Acid Copolymer *n* A binder used in solid propellants.

Polybutadiene Resins *n* (PB, poly-1-butene) Any of a family of polymers consisting of isotactic, stereoregular, highly crystalline polymers based on 1-butene. Properties are similar to those of polypropylene and linear polyethylene, with superior toughness, creep resistance, and flexibility. FB has been used in pipe, wire coating, gaskets, and industrial packaging. PB pipe carries the highest design-stress rating of all flexible, thermoplastic piping materials, serving to temperatures of 90°C.

Polybutadiene-Type Resins *n* Unsaturated, thermosetting hydrocarbons cured by a peroxide-catalyzed, vinyl-type polymerization reaction, or by sodium-catalyzed polymerization of butadiene or blends of butadiene and styrene. Liquid systems, curable in the presence of monomers, are used for casting, encapsulation, and potting of electrical components, and in making laminates. Molding compounds, often containing fillers and modified with other resins or rubbers, may be compression or transfer molded. Syndiotactic 1,2-butadiene, introduced in 1974 in Japan, is thermoplastic, with semicrystalline nature, with good transparency and flexibility without plasticization. In the presence of a photosensitizer such as *p,p'*-tetramethyl diaminobenzophenone, this polymer can readily be cured by ultraviolet radiation. Transparent films of the nontoxic polymer are used for packaging, and cellular forms for shoe soles. It is biodegradable.

Polybutanediol Terephthalate *n* (PBTP, PBT) A crystalline polymer formed by the polycondensation of 1,4-butanediol and dimethylterephthalate.

Poly(1-butene) Produced commercially in isotactic for by Ziegler-Natta polymerization of butene-1.

Poly(1-butene co ethylene) *n* Alternative name for ethylene-butene-copolymer.

Poly(1-butene co ethylene co 1-hexene) *n* A copolymer produced commercially by the Phillips process and containing up to 5% butene-1. It is similar to high density polyethylene but has a slightly lower density and better resistance to creep.

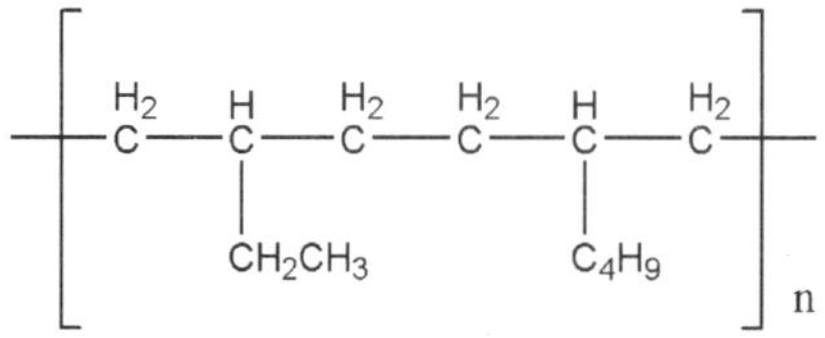

Polybutylene *n* Any of a family of low-molecular-weight polymers of mixed 1-butene, *cis*-2-butene, *trans*-2-butene, and isobutene. Depending on molecular weight, these polymers range from oils through tacky waxes, crystalline waxes, and rubbery solids. See also ▶ Butyl Rubber and Polybutylene Resin. Poly(1-butene) shown below.

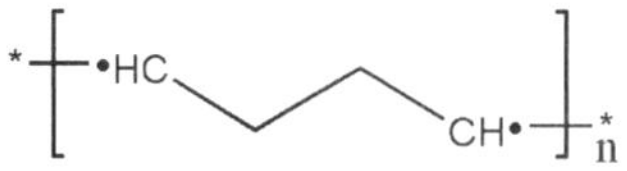

Polybutylene

Polybutylene Adipate Glycol *n* (PBAG) A polymeric diol used in the production of urethane elastomers.

Polybutylene Resin *n* (PB, poly-1-butene) Any of a family of polymers consisting of isotactic, stereoregular, highly crystalline polymers based on 1-butene. Properties are similar to those of polypropylene and linear polyethylene, with superior toughness, creep resistance, and flexibility. PB has been used in pipe, wire coating, gaskets, and industrial packaging. PB pipe carries the highest design-stress rating of all flexible, thermoplastic piping materials, serving to temperatures of 90 °C.

Polybutylene Terephthalate *n* (PBTP, PBT, polytetramethylene terephthalate) A member of the polyalkylene terephthalate family, similar to polyethylene terephthalate in that it is derived from a polycondensate of terephthalic acid, but with butanediol rather than glycol. PBTP can be modified easily to overcome its relatively low operating-temperature limit, making it equivalent to plastics used in construction and appliances. Grades are available for injection and blow molding, extrusion, and thermoforming. Properties include high tensile and impact strength, dimensional stability, low moisture absorption, and good electricals; also resistant to fire and chemicals when suitably modified.

Polycaprolactam *n* A cyclic amide type compound, containing six carbon atoms. See ▶ Nylon 6.

Polycaprolactone *n* (PCL) A low-melting (62°C) polyester resin with the linear structure $(-CH_2CH_2CH_2CH_2CH_2COO-)_n$, made by polymerizing ε-caprolactone. PCL is compatible with most thermosetting and thermoplastic resins and elastomers. It increases impact resistance and aids mold release of thermosets, and acts as a polymeric plasticizer with PVC. It is biodegradable, useful in containers for growing and transplanting trees and other plants. Unmodified PCL is completely consumed by soil microbes but the rate of degradation can be slowed by incorporating a non-biodegradable polymer. PCL is also useful in the production of polyurethane elastomers and foams, to which it imparts good-low-temperature properties and water resistance.

Polycarbodiimide *n* Polymers containing $-N=C=N-$ linkages in the main chain, typically formed by catalyzed polycondensation of polyisocyanates. They are used to prepare open celled foams with superior thermal stability. Sterically hindered polycarbodiimides are used as hydrolytic stabilizers for polyester-based urethane elastomers.

Polycarbonate \-ˈkär-bə-ˌnāt\ *n* (1930) A polymer that comprises the repeated structural unit in the chain is of the carbonate type, $[-O-CO-O-]$.

Polycarbonate Polyester Alloys High-performance thermoplastics processed by injection and blow molding. Used in auto parts.

Polycarbonate Resin *n* (PC) Any of a family of special polyesters in which groups of dihydric phenols are joined through carbonate linkages. They can be produced by a variety of methods, of which the most commercially important are (1) phosogeneration of dihydric alcohols, usually bisphenol A. (2) Ester interchange between diamyl carbonates and dihydric phenols, usually between diphenyl carbonate and bisphenol A. (3) Interfacial polycondensation of bisphenol A and phosgene. Bisphenol A polycarbonates with molecular weights close to 33,000 can be processed by injection molding, extrusion, thermoforming, and blow molding. Melt-casting and solution-casting processes are also employed. Such polycarbonates have high impact strength (to 8 J/cm of notched width), good heat resistance, low water absorption, good electrical properties, and no toxicity. They are vulnerable to some common organic solvents. Crystal-clear grades have been developed for safety glazing, including multilayer glass-and-PC, bulletproof structures. Other applications include dentures, food packages, electrical components, precision parts for instruments and household appliances and – a current large-volume use – compact disk (CD) records and data disks.

Polycarbonates See ▶ Polycarbonate.

Polycarboranesiloxane *n* (SiB) A polymer whose chain consists of alternating carbonane and siloxane groups. Commercial resins contain active end groups that may be vulcanized with peroxides to yield rubbers resistant to high temperatures (260°C in air).

Polycarboxane *n* Syn: ▶ Acetal Resin.

Polychlal Fiber *n* A manufactured, bicomponent fiber of polyvinyl alcohol and polyvinyl chloride. Some vinyl chloride is grafted to the polyvinyl alcohol (Japanese Chemical Fibers Association definition). The fiber is emulsion spun into tow and staple. CHARACTERISTICS: Polychlal fibers have a soft, lamb's wool-like hand and moderate moisture regain. The fibers are also characterized by high flame resistance and high abrasion resistance. END USES: Polychlal fibers are suitable for end uses such as children's sleepwear, blankets, carpets, curtains, bedding, upholstery, nonwovens, and papermaking.

Polychloroether *n* Syn: ▶ Chlorinated Polyether.

Polychloroprene *n* Polychloroprene is usually sold under the trade name Neoprene. It's especially resistant to oil. It was the first synthetic elastomer, or rubber to be a hit commercially. Polychloroprene is made from the monomer chloroprene. Syn: ▶ Neoprene.

Polychlorostyrenes *n* A polymer produced by the polymerization of one of the isomers of chlorostyrene or of a mixture of isomers. Polymerization conditions are similar to those use for styrene, but polymerization occurs more rapidly. These polymers have higher flame resistance than styrenes.

Polychlorotrifluoroethylene *n* (CTFE, PCTFE) A family of polymers made by polymerizing the gas $CIFC=CF_2$ by mass, emulsion, or suspension polymerization. The polymers range from oils, greases, and waxes of low molecular weight to the tough, rigid thermoplastics most commonly used in industry. Unlike polytetrafluoroethylene, PCTFE may be processed by conventional thermoplastic methods. It is also available as dispersions in xylene or ketones for application by dipping or spraying. The polymers are nontoxic, resistant to heat, chemically inert, and have outstanding electrical properties.

Poly(chlorotrifluoroethylene-co-vinyl fluoride) *n* Common name for Ethene, chlorotrifluoro-, polymer, with 1,1-difluoroethene.

Polychromatic Finish *n* (1) A multicolored paint finish. (2) A paint containing reflecting metallic flake and fine transparent pigments which appears as a variety of colors when viewed from different angles.

Poly(*cis*-1,4-isoprene) Poly(2-methyl-1,3-butadiene)

Polycondensate *n* Product of a condensation polymerization.

Polycondensation \-ˌkän-ˌden-ˈsā-shən\ *n* [ISV] (1936) See Condensation and Condensation Polymerization.

Polycoummarone Resins See ▶ Coumarone-Indene Resins.

Polycyclamide (1) A polyamide containing a cycloalkane ring. (2) Any linear, high-molecular-weight polyamide formed by condensing cyclohexane-1,4-bis (methylamine) with one or more dicarboxylic acids. These polymers have high melting points, but are sufficiently stable to permit melt processing at temperatures above 300°C without thermal decomposition. Their excellent physical and chemical properties indicate their usefulness as fibers, films, and moldings. See also ▶ Nylon.

Polycyclic *n* \ ▮pä-lē- ▮sī-klik\ *adj* [ISV] (1869) Of an organic compound, containing two or more rings, often with pairs of rings sharing two carbon atoms, as in the hydrocarbon anthracene ($C_{14}H_{10}$).

Polycyclization *n* Cyclization in which more than one internal ring formation occurs.

Polycyclohexylenedimethylene Ethylene Terephythalate *n* Thermoplastic polymer of cyclohexylenedimethylenediol, ethylene glycol, and terephthalic acid. Has good clarity, stiffness, hardness, and low-temperature toughness. Processed by injection and blow molding and extrusion. Used in containers for cosmetics and foods, packaging film, medical devices, machine guards, and toys. Also called PETG.

Polycyclohexylendeimethylene Terephthalate *n* [PCT, poly(cyclohexane-1,4-dimethylene terephthalate)] The newest member of the commercial thermoplastic-polyester family, PCT is produced by reacting 1,4-cyclohexane dimethanol with dimethyl terephthalate. It is superior to its siblings (PET and PBT) in that it can serve at higher temperatures, with 1.8 MPa deflection temperatures to 260°C with 30% glass-fiber content. Moisture absorption is lower, too, and it has excellent chemical resistance. It is being used in automotive parts and dual-ovenable cookware. It can be compounded for high flame resistance.

Polydentate Ligand *n* A polyatomic ligand with more than one lone pair of electrons which can simultaneously bond to the central ion in a complex.

Poly(Dichloro-p-Xylylene) *n* (Parylene D) See ▶ Parylene.

Polydicyclopentadiene Resins *n* Friable, thermoplastic, unsaturated aromatic resins derived from petroleum hydrocarbons whose principal component is dicyclopentadiene. Softening points from 10–140°C (50–285°F); sp gr, 0.99–1.11; iodine value, 170–200. Soluble in aliphatic, aromatic, chlorinated solvents; insoluble in water, alcohols, and glycols.

Poly(2,3-dimethyl butadiene) *n* A very elastic synthetic rubber, the monomer being made by conversion of acetone via pinacol to 2,3-dimethyl-butadiene.

Poly-2,6-Dimethyl-1,4-Phenylene Oxide See ▶ Polyphenylene Oxide.

Polydimethylsiloxane *n* (dimethyl silicone) Any of a family of silicones of the composition $[-(CH_3)_2SiO-]_n$. Those of low molecular weight – several hundred to 10,000 – are oils, some of which are widely used in aerosol mold releases for plastics that are not to be painted or printed. Polymers in the molecular-weight range near 105 are rubbers that are flexible at cryogenic temperatures but crystalline above –60°C. See also ▶ Silicone.

Poly(dimethyl siloxane) *n* Basis of most technical silicone oils, greases, rubbers and resins. The polymer is formed by the hydrolysis of dimethyldichlorosilane with water.

Polydisperse \-dis- ▮pərs\ *adj* [poly- + L *dispersus* dispersed, fr. pp of *dispergere* to disperse] (1915) Of a polymer, having a range of molecular weights as opposed to a single molecular weight (▶ Monodisperse) the usual state among commercial polymers. The broader the distribution relative to its center, the greater is its ▶ Polydispersity.

Polydispersity *n* The breadth of the molecular-weight distribution of a polymer. Two measures of polydispersity are in common use: (1) the ratio of the weight-average and number-average molecular weights M_w/M_n, and (2) the g-INDEX.

Polydispersity Index *n* Same value as polydispersity.

Poly(divinyl benzene co styrene)

Polyelectrolyte \ˌpä-lē-ə-ˈlek-trə-ˌlīt\ *n* (ca. 1947) (polyion, ionomer) (1) Any of several classes of polymers having fixed ionizable groups, such as polyacids (e.g., polyacrylic acid), polybases (e.g., polyvinyl trimethylammonium chloride), and sodium- or potassium-salt complexes of such polymers as polyethylene oxide. They are much more electrically conductive than ordinary plastics, their conductivities generally rising with temperature. Some are finding use in battery separators, photoelectrochemical cells, and humidity sensors. An allied class is described under ▶ Ionomer. (2) The presence of ionized groups on a polymer chain superimposes the effects of electrostatic interactions on the properties of the un-ionized polymer; e.g., poly(acrylic acid), sodium polyacrylate; useful for increasing viscosity in solvents including water, and water treatment. (Ku CC, Liepins R (1987) Electrical properties of polymers. Hanser, New York)

Polyester \ˈpä-lē-ˌes-tər\ *n* [ISV] (1929) (alkyd) A general term encompassing all polymers in which the main polymer backbones are formed by the esterification condensation of polyfunctional alcohols and acids. The coined term *alkyd* (see ▶ Alkyd Resin) is synonymous, chemically, with polyester. However, as more commonly used in the plastics industry, alkyd refers to polyesters modified with oils or fatty acids that are crosslinkable (see ▶ Alkyd Molding Compound). The term polyester is explained further under Polyester, Saturated and Polyester, Unsaturated.

Polyester, Aromatic See ▶ Poly-*p*-Hydroxybenzoic Acid.

Polyester Carbonates *n* A synthetic thermoplastic resin, a linear polymer of carbonic acid.

Polyester Fiber *n* Generic name for a manufactured fiber in which the fiber-forming substance is any long-chain synthetic polymer composed of at least 85% by weight of an ester of a dihydric alcohol and terephthalic acid (Federal Trade Commission). The polyester fiber in widest use throughout the world is derived from polyethylene terephthalate. Polyester filaments are produced by forcing the molten polymer at a temperature of about 290°C through spinneret holes about 0.23 mm (9 mils) in diameter, followed by air cooling, combining the single filaments into yarns, and drawing. The major end use of polyester fibers is in blends with cotton or wood to enhance crease retention and reduce wrinkling of garment fabrics. It is also used in carpeting and tire cords.

Polyester Imide *n* A polymer containing both ester and imide groups in the polymer chain.

Polyester Plasticizer *n* Any of a broad class of plasticizers characterized by having many ester groups in each molecule. They are synthesized from three components: (1) a dibasic acid such as adipic, azelaic, lauric, or sebacic acid, (2) a glycol (dihydric alcohol), and (3) a monofunctional chain terminator such as a monobasic acid. Molecular weights are low – from 500 to 5,000. Polyester plasticizers are noted for their permanence and resistance to extraction.

Polyester Resins *n* Group of synthetic resins which are polycondensation products of dicarboxylic acids with dihydroxyl alcohols. They are therefore a special type of alkyd resin. Oil-free alkyds are a class by themselves. Often these resins are dispersed in a suitable monomer.

Polyesters A polymer in which the monomer units are linked together by the group –COO– and are usually formed by polymerizing a polyhydric alcohol with a polybasic acid.

Polyester Saturated *n* Any polyester in which the polyester backbone has no double bonds. The class includes low-molecular-weight liquids used as plasticizers and as reactants in forming urethane polymers; and linear, high-molecular-weight thermoplastics such as polyethylene terephthalate. Usual reactants for the saturated polyesters are (1) a glycol such as ethylene-, propylene-, diethylene-, dipropylene-, or butylene glycol, and (2) an acid or anhydride such as adipic, azelaic, or terephthalic acid or phthalic anhydride. Some saturated, branched polyesters are used in high-temperature varnishes and adhesives.

Polyester Thermoplastic Elastomers *n* copolymers containing soft polyether and hard polyester blocks having good dielectric strength, chemical and creep resistance, dynamic performance, appearance, and retention of properties in a wide temperature range but poor light resistance. Processed by injection, blow, and rotational molding, extrusion casting, and film blowing. Used in electrical insulation, medical products, auto parts, and business equipment. Also called polyester TPE.

Polyester TPE See ▶ Polyester Thermoplastic Elastomers.

Polyester Unsaturated *n* A polyester family characterized by ethenic unsaturation in the polyester backbone that enables subsequent hardening or curing by copolymerization with a reactive monomer in which the polyester constituent has been dissolved. Unsaturated polyesters are made by agitating in a heated kettle a mixture of glycols, e.g., propylene- or diethylene glycol; unsaturated dibasic acids or anhydrides, e.g., fumaric acid or maleic anhydride; and, sometimes in order to control the reaction and modify properties, a saturated dibasic acid, or anhydride, e.g., iosphthalic acid or phthalic anhydride. After removal of water and cooling, the fluid polyester may be dissolved in a reactive monomer in the same kettle, or it may be shipped to users who add the monomer and catalyst in their plants. Styrene is most widely used as the reactive monomer. Others sometimes used as diallyl phthalate, diallyl isophthalate, and triallyl cyanurate. A peroxide catalyst is generally used for the final copolymerization. These unsaturated polyesters are thermosetting and are most widely used in reinforced plastics for making boat hulls, trays containers, and panels, and in potting of electrical assemblies.

Polyether n (1) Any polymer having the general structure $(-R-O-)_n$, where R may be simple or more elaborate. [Technically, polyoxymethylene, $(-CH_2-O-)_n$, is a polyether, though known as an ▶ Acetal Resin in the industry.] ▶ Polyphenylene Oxide is a well known polyether. (2) A low-molecular-weight polymer containing hydroxyl end groups, used as a reactant in the production of polyurethane foams. One type of polyether, widely used for rigid foams, is obtained by reacting propylene oxide with a polyol initiator such as a glycol glycoside in the presence of potassium hydroxide as a catalyst. rather than a polyester or other resin component. For rigid foams, polyethers often used as the propylene oxide adducts of materials such as sorbitol, sucrose, aromatics, diamines, pentaerythritol, and methyl glucoside. These range in hydroxyl numbers from 350 to 600. For flexible foams, polyethers with hydroxyl numbers ranging from 40 to 160 are used. Examples are condensates of polyhydric alcohols such as glycerine, sometimes containing small amounts of ethylene oxide to increase reactivity.

Polyether Glycol See ▶ Polyethylene Glycol.

Polyetherimide n One of the "advanced" thermoplastics, having both ether links and imide groups in its chain, as shown below.
Deflection temperature at 1.8 MPa is 199°C, tensile modulus is 3.0 GPa, strength is 96 MPa, and the resin has good fire resistance.

Polyether Imide n A polymer containing both ester and imide groups in the polymer chain. They are usually synthesized by polycondensation between a dianhydride containing aromatic ester links and a diamine. Their main use is as high temperature resistant wire.

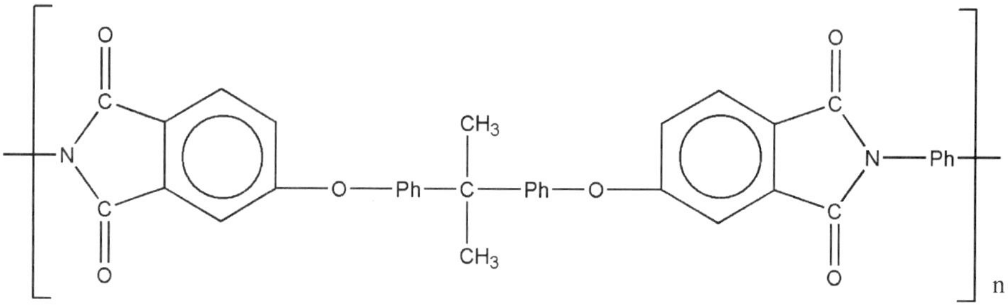

Polyether, Chlorinated See ▶ Chlorinated Polyether.

Polyetheretherketone (PEEK) n An "advanced" polymer chain It has excellent temperature resistance among processable thermoplastics, with a melting temperature of 334°C, deflection temperature at 1.8 MPa of 160°C, and tensile yield strength of 91 MPa. Reinforcement with 30% glass fiber elevates the deflection temperature to about 300°C and almost doubles the yield strength.

Polyethereketone Fiber (PEEK) n A manufactured fiber from polyetheretherketone polymer with high temperature and chemical resistance used in composites as a matrix material and in other industrial applications.

Polyether Foam n A type of ▶ Polyurethane Foam that has been made by reacting isocyanate with a polyether

Polyetherimide n One of the "advanced" thermoplastic, having both ether links and imide groups in its chain, as shown below.
Deflection temperature at 1.8 MPa is 199°C, tensile modulus is 3.0 GPA, strength is 96 MPa, and the resin has good fire resistance.

Polyetherimide Fiber (PEI) n A manufactured fiber spun from polyetherimide polymer having high temperature resistance, excellent processability, and toughness. Used for matrix materials in composites and in other industrial applications.

Polyetherimide Resins n An amorphous engineering TP characterized by high heat resistance, high strength and a high modulus, excellent electrical properties that remain stable over a wide range of temperatures and frequencies, and very good processability.

Polyetherketone *n* An "advanced" thermoplastic resin having both ether and ketone linkages in its chains, a close relative of ▶ Polyetheretherketone, above, and having the PEEK structure with the leftmost phenyl and ether oxygen deleted. This melt-processable polymer melts near 365°C, is fire-resistant, has good resistance to chemicals, and can be used at temperatures comparable to those for PEEK.

$$\left[-Ph-O-Ph-\overset{O}{\underset{\|}{C}}- \right]_n$$

Polyetherketoneketone *n* A polyetherketone which contains two ketone links between benzene rings to each link. It has the highest glass transition and melting temperatures of all the commercial aromatic polyether ketones.

Polyether Resins *n* Polymers in which the repeating unit contains a carbon–oxygen bond derived from aldehydes or epoxides or similar materials.

Polyethers *n* A polymer of general structure $[R-O]_n$, where R may be a simple alkene group.

Polyethers Resins *n* Polymers in which the repeating unit contains a carbon–oxygen bond derived from aldehydes or epoxides or similar materials.

Polyethersulfone *n* An "advanced" thermoplastic consisting of repeating phenyl groups (φ) linked by thermally stable either and sulfone ($-SO_2-$) groups, its structure being like that of PEEK, stated above, but with the right hand $-O-\phi-CO-$ section replaced by sulfone. The resin has good transparency and flame resistance, and has one of the lowest smoke-emission ratings among plastics. Both neat and reinforced grades are available in granule form for extrusion and molding. Unreinforced grades are used in high-temperature electrical applications, bakery-oven windows and medical components. Reinforced grades are used for radomes, structural aircraft and aerospace components, and corrosion-resisting applications in packaging and chemical-plant hardware.

Polyethersulfone Fiber *n* (PES) High molecular weight fibers from polymers containing sulfone ($-SO_2-$) groups and aromatic nuclei. They demonstrate high thermal stability and chemical inertness.

Polyethersulfone Resins *n* PES is a high temperature engineering TP in the polysulfone family.

Poly(ethyl acrylate co ethylene) *n*

$$\left[\underset{H}{\overset{COOC_2H_5}{\underset{|}{C}}}-\overset{H_2}{\underset{}{C}}-\overset{H_2}{\underset{}{C}}-\overset{H_2}{\underset{}{C}} \right]_n$$

Polyethylene \-ˈe-thə-ˌēn\ *n* (ca. 1862) (Pe, polyethene; *in Britain*, polythene) A huge family of resins obtained by polymerizing ethylene gas, $H_2C=CH_2$ and by far the largest-volume commercial polymer. Almost 10 Tg (11×10^6 t) was sold in the US in 1992, about one-third of all US resin sales. By varying the catalyst and methods of polymerization, properties such as density, melt-flow index, crystallinity, degree of branching and cross linking, molecular weight and polydispersity can be regulated over wide ranges. Further modifications are created by copolymerization, chlorination, and compounding additives. Low-molecular-weight polymers of ethylene are fluids used as lubricants; medium-weight polymers are waxes miscible with paraffin; and the polymers with molecular weights over 10,000 (to which the above sales figure applies) are the familiar tough and strong resins, flexible or stiff, to make a myriad of products, both consumer and industrial. Polymers with densities ranging from about 0.910 to 0.925 g/cm3 are called *low-density* polyethylene; those with densities from 0.926 to 0.940 are called *medium-density*; and those with densities from 0.941 to 0.965 and over are called *high-density* polyethylene. The low-density resins are polymerized at very high pressures and temperatures, and the high-density ones at lower pressures and temperatures, using special catalysts. Two newer types are *extra-high-molecular-weight* (EHMWPE) materials in the MW range from 150,000 to 1,500,000, and *ultra-high-molecular-weight* (UHMWPE) materials in the 1,500,000–3,000,000 range. Because UHMWPE does not melt and flow, it is processed by powder-molding and sintering techniques developed decades ago for polytetrafluoroethylene. Under carefully controlled conditions some EHMWPEs can be extruded, blow molded, and thermoformed on standard equipment. When fully crosslinked by irradiation or by the use of chemical additives polyethylene is no longer a thermoplastic, and has superior strength, impact resistance, and electrical properties. A still newer member of the family, much used in grocery bags, is ▶ Linear Low-Density Polyethylene. Another new subfamily are the ▶ Very-Low-Density Polyethylenes.

Poly(Ethylene-Chlorotrifluoroethylene) *n* (PE-CTFE, ECTFE copolymer) A high-molecular-weight, 1:1 alternating copolymer of ethylene and chlorotrifluoroethylene. Available in pellet and powder form, PE-CTFE can be extruded, injection, transfer, and compression molded, rotocast and powder coated. It is a strong, highly impact-resistant material that retains useful properties over a wide temperature range. Good electrical properties and chemical resistance make it useful in electrical and chemical ware and in packaging applications requiring corrosion resistance.

Polyethylene Copolymer *n* Thermoplastics polymers of ethylene with other olefins such as propylene. Processed by molding and extrusion. Also called PE copolymer.

Polyethylene Fiber *n* A manufactured fiber made of polyethylene, often in monofilament form as well as continuous filament yarns and staple. Ethylene is polymerized at high pressures and the resulting polymer is melt spun and cold drawn. It may also be dry-spun from xylene solution. CHARACTERISTICS: Polyethylene fibers have a low specific gravity, extremely low moisture regain, the same tensile strength wet and dry, and are resistant to attack by mildew and insects. These qualities have made polyethylene fiber suitable for industrial applications, geotextiles, outdoor furniture, and similar applications. Polyethylene fiber does not dye, and in most cases, it is colored by the addition of pigments and dyes to the material prior to spinning. It has a low melting point, a property that has restricted its use in apparel.

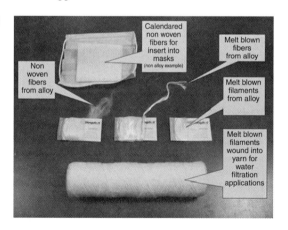

Polyethylene Foam *n* Low-density-PE foam, with foam densities as low as 0.03 g/cm^3, are made by thoroughly mixing a blowing agent with hot, molten polymer under pressure, then releasing the pressure and cooling. Foams are also made by extrusion, using pellets containing a heat-triggered foaming agent. Crosslinked PE foam is made by blending a peroxide crosslinking agent with the molten compound, then subsequently vulcanizing the molded shapes in a press. The denser foams have found application in packaging of electronic equipment.

Polyethylene Glycol *n* Any of several condensation polymers of ethylene glycol. These glycols general formula: $HOCH_2(CH_2OCH_2)_\eta CH_2OH$ or $H(OCH_2CH_2)_\eta OH$. Average molecular weights range from about 200 to 8,000, ranging from water-clear liquids to hard, waxy solids. They are used as plasticizers for polyvinyl alcohol, as intermediates, and in printing inks and mold releases. Syn: ▶ Polyether Glycol and ▶ Polyoxyethylene Glycol.

Polyethylene Glycol (200) Dibenzoate *n* $C_6H_5CO(OCH_2CH_2)_4OCO-C_6H_5$. A plasticizer compatible with cellulose acetate butyrate, ethyl cellulose, polymethyl methacrylate, polystyrene, and vinyl resins. Its major application is with phenol-formaldehyde resins in laminating applications, to improve flexibility without loss of electrical properties and high-temperature capability.

Polyethylene Glycol (600) Dibenzoate *n* A plasticizer similar to the preceding one but with 13 –OCH$_2$CH$_2$– groups, and with only partial compatibility with the resins listed for that one.

Polyethylene Glycol Di-2-Ethylhexoate *n* A plasticizer for most cellulosic plastics, polymethyl methacrylate, polystyrene, and vinyls.

Polyethylene Glycol (400) Dilaurate *n* A plasticizer for cellulose nitrate, PVC, and vinyl copolymers.

Polyethylene Glycol Terephthalate *n* A longer name for Poly Ethylene Terephthalate.

Polyethylene *n*, **High M.W.** Density = 0.97.

Polyethylene *n*, **Low M.W. (branched)** Density = 0.95.

Polyethylene Oxide *n* (PEO) Low-molecular-weight polymers of ethylene oxide are viscous liquids or waxes. Those of high molecular weight are tough, highly crystalline, ductile thermoplastics that can be processed by molding, extrusion, etc. All PEO resins are soluble in water, and thus are used in the form of packaging film for powdered detergents, insecticides, and other household, industrial and agricultural products that are dissolved in water prior to use. The film is heat-sealable and permeable to gases.

Polyethylene-Propylene Adipate Glycol *n* (PEPAG) A polymeric diol used in the production of urethane elastomers (Witco Corp., Formrez F 10–91).

Polyethylene Terephthalate *n* (PET, polyethylene glycol terephthalate) A saturated, thermoplastic polyester resin made by condensing ethylene glycol, and terephthalic acid, used for textile fibers, water-clear, biaxially oriented film (e.g., Mylar®) and, more recently, for extruded, thermoformable sheet (TV-dinner trays), injection-molded parts, and large, blow-molded, soft-drink bottles. It is extremely hard, wear- and chemical-resistant, dimensionally stable, and has good dielectric properties. See also ▶ Polyester, Saturated and ▶ Crystallized Polyethylene Terephthalate.

Poly(Ethylene-Tetrafluoroethylene) n (PE-TFE) A crystalline resin in which the proportion of ethylene to tetrafluoroethylene (E/TFE) may range, for the best combination of properties, between 2:3 and 3:2, modified with a vinyl copolymer for better toughness. It is stronger than either low-density polyethylene or polytetrafluoroethylene, has good electrical properties, high Izod-impact strength, and plastic memory that makes it useful for heat-shrinkable packaging.

Polyethylene, Ultra-high M.W. Density = 0.99.

Polyformaldehyde n See ▶ Acetal Resin, Paraformaldehyde and Polyoxymethylene.

Polyglycerols n Compounds of either-alcohol type derived from the interaction of two or more molecules of glycerol. Thus, for example, tow molecules of glycerol react to give diglycerol. The polyglycerols have some applications in alkyd and ester resin manufacture.

Polyglycidyl Polyepichlorohydrin Resin Any of a family of epoxy resins derived from epichlorohydrin and hydroxyl compounds, possessing flexibility and flame-retarding characteristics. They may be cured by themselves, or mixed with conventional epoxy resins to impart their favorable characteristics to laminates.

Polyglycol n A polyhydric alcohol of the monomeric glycol, of uncertain composition. Bp 230–250°C.

Polyglycol Distearate n (polyethylene glycol distearate) The di(stearic acid) ester of polyglycol, used as a plasticizer.

Polyhexafluoropropylene n A fully fluorinated polymer based on the gas $CF_3CF=CF_2$, not commercial. However, the copolymers of hexafluoropropylene and tetrafluoroethylene make up the family of ▶ Fluorinated Ethylene-Propylene Resins.

Poly(hexafluoropropylene-co-tetrafluoroethylene) n 1,1,2,3,3,3-hexafluoro 1-propene-co-1,1,2,2-tetrafluoroethene.

Poly(hexafluoropropylene-co-vinylidene fluoride) n 1,1,2,3,3,3-hexafluoro-1-propene-co-1,1-difluoroethene A fully fluorinated polymer based on gas $CF_3CF=CF_2$, not commercial. However, the copolymers of hexafluoropropylene and tetrafluoroethylene make up the family of ▶ Fluorinated Ethylene Propylene RESINs.

Poly(hexamethylene adipamide) n Explicit Syn: ▶ Nylon 6/6.

Polyhexamethylenesebacamide n Explicit name for ▶ Nylon 6/10.

Polyhexamethyleneterephthalamide n Explicit name for ▶ Nylon 6/T.

Polyhydric Alcohol Syn: ▶ Polyol.

Polyhydroxyether Resin See ▶ Phenoxy Resin.

Polyimidazopyrrolone n (ladder pyrrone, polypyrrolone) An aromatic, heterocyclic polymer that results from the reaction of an aromatic dianhydride with a tetramine. Due to the double-chain or ladder-like structure, these polymers have outstanding resistance to radiation, chemicals, and heat (no weight loss to 550°C). However, this structure also makes them difficult to process. To overcome this difficulty pyrrone prepolymers in the form of solutions and salt-like powders have been made available. The powders can be molded under conditions that complete the cyclization or conversion of the ladder-like molecular structure during the molding cycle. The cyclization reaction generates water, which must be removed from the part.

Polyimide *n* A polymer formed by the condensation of an organic anhydride or dianhydride with a diamine, in some cases followed by thermal dehydration (curing). The early polyimides from pyromellitic anhydride and aromatic diamines, when fully cured, had extremely high thermal stability but were unmeltable and required special processing. Later, addition-type polyimides based on reacting maleic anhydride and 4,4'-methylene-dianiline were developed. These are processable by conventional thermoset molding, film casting, and solution-fiber techniques. Molding compounds filled with lubricating fillers or fibers produce parts with self-lubricating wear surfaces. Thermoplastic polyimide reinforced with glass, boron, or graphic fibers can be molded into high-strength structural components. Polyimide solutions are used as laminating varnishes to produce radomes, printed-circuit boards, and other components requiring fire resistance, good electrical properties, and strength at high temperatures. Printed-circuit boards of polyimide-glass laminate handily endure high-temperature soldering. Recently, the introduction of thermoplastic polyimides containing aromatic rings in the polymer backbone and trifluoromethyl side group has opened these materials, to a wider field of applications because of improved processability. Film has been used as insulation in electric motors, magnet wire and missile wiring, and in dielectric applications.

Polyimide Fiber *n* A manufactured fiber formed from the condensation polymer of an aromatic dianhydride and an aromatic diisocyanate. The fiber is produced by dry spinning. It is a high-shrinkage fiber used in the formation of mechanically stable nonwoven fabrics. These fabrics are made without binders or resins; bonding apparently results from the local temperature and pressure that develop during shrinkage.

Polyimide Foam *n* A family of polyimide-precursor powders enables the production of flexible and rigid polyimide-foam structures. These powders are poured into molds and heated until sufficient integrity for removal is attained, then, they are subsequently cured at 300°C.

Polyimide Resin *n* Aromatic polyimides made by reacting pyromellitic dianhydride with aromatic diamines. Characterized by high resistance to thermal stress.

Polyimides *n* Thermoplastic aromatic cyclized polymers of trimellitic anhydride and aromatic diamine. Have good tensile strength, dimensional stability, dielectric and barrier properties, and creep, impact, heat, and fire resistance, but poor processibility. Processed by compression and injection molding, powder sintering, film casting, and solution coating. Thermoset uncyclized polymers are heat curable and have good processability. Processed by transfer and injection molding, lamination, and coating. Used in jet engines, compressors, sealing coatings, auto parts, and business machines. Also called PI.

Polyindene Resins See ▶ Coumarone-Indene Resins.

Polyisobutene See Polybutene.

Polyisobutylene *n* Poly(2-methyl-1-propene) Any of a family of polymers of isobutylene, $(CH_3)_2C=CH_2$, for which the IUPAC name is 2-*methylpropene*. Depending on molecular weight, they range from oily liquids to elastomeric solids. The higher-molecular-weight polymers are used as impact-resistance improvers in polyethylene and other plastics. The liquid polymers are used as tackifying agents in adhesives. A rubbery polymer of isobutylene which yields viscous solutions in aliphatic and aromatic hydrocarbons. Films of great elasticity can be obtained and the product is characterized by excellent chemical and light resistance, and absorbs very little. It is used chiefly in the manufacture of synthetic rubber.

Poly(isobutylene co isoprene) *n* Poly(2-methyl-1-propene-co-2-methyl-1,3-butadiene).

Polyisobutylvinyl Ether *n* (polyvinyliosbutyl ether) Any polymer of isobutylvinyl ether. Some are liquids, others are solid and crystalline. They are used as adhesives, surface coatings, laminating agents, and filling compounds in electrical cables.

Polyisocyanate See ▶ Isocyanate.

Polyisocyanurate *n* (PIR) A polymer containing isocyanurate rings, i.e., isocyanate trimer, and forming foams that have better fire resistance than rigid polyurethanes, but are more brittle, so are often used in mixtures with the latter.

Isocyanurate (1, 3, 5-triazinane-2, 4, 6-trione)
Also known as: 1, 3, 5-triazinetriol, s-triazinetriol, 1, 3,5-Triazine-2, 4, 6(1H, 3H, 5H)-trione, s-triazinetrione, tricarbimide, isocyanuric acid, and pseudocyanuric acid

Polyisoprene *n* A polymer of ▶ Isoprene. The *cis*-1,4- type of polyisoprene occurs naturally as the major polymer in natural rubber and is also produced synthetically. The *trans*-1,4- type resembles ▶ Gutta-Percha and has in the past been used in golf-ball covers and shoe soles. See ▶ Poly(*cis*-1,4-Isoprene).

Polyisoprene, Deutero *n* A polyisoprene in which heavy hydrogen (deuterium) atoms have replaced the ordinary hydrogen atoms. The *cis*-1,4-deuteropolyisoprene is more elastic than natural rubber.

Polyketone Resins *n* A new and unique family of aliphatic polymers composed of carbon monoxide, ethylene and minor amounts of other alpha olefins. This family of semicrystalline resins exhibits many of the properties of engineering resins while processing similarly to polyolefins.

Polylauryllactam See ▶ Nylon 12.

Poly(lauryl methacrylate) *n* Poly(propanoic acid, 2-methyl-, dodecal ester).

Polyliner *n* A perforated, longitudinally ribbed sleeve that fitted snugly inside the cylinder of a ram-type injection-molding machine, replacing the conventional torpedo. It improved the heat transfer, plastifying rate, and uniformity of melt temperature at the nozzle.

Poly(maleic anhydride co stilbene) *n* Poly (2-butenedioic acid-co-1,1'-(1,2-ethediyl)-bis-(E)-benzene.

Polymer \pä-lə-mər\ *n* [ISV, back-formation fr. *polymeric*, fr. Gk *polymerēs* having many parts, fr. *poly-+ meros* part] (1866) {*d* polymer *n*, *f* polymere, *s* polimero *m*.} (1) A chemical compound, or mixture of compounds, formed by a polymerization reaction and consisting essentially of repeating structural units. (2) A compound of high molecular weight derived either by the addition of many smaller molecules, as polyethylene, or by the condensation of many smaller molecules with the elimination of water, alcohol, or the like, as nylon. (3) A compound formed from two or more polymeric compounds. (4) A product of polymerization and (5) macromolecules.

Polymer Blend *n* A physical mixture of two or more polymers and possible additives, achieved by kneading or by high-intensity mixing of fine powders. Because hot working can cause chain scission in some polymers, some grafting of the component polymers is likely to occur during such operations. Blends are made to take advantage of synergistic gains in properties, some of which may be better than the same properties of the blend components alone. Often, costly resins of outstanding properties are blended with cheaper, compatible ones to achieve blends of intermediate properties at a cost lower than that of the more expensive member. Melt viscosities of blends may lie, at a given shear rate and temperature, between those of the separate components, or above or below both of them.

Polymer Chain Unsaturation See ▶ Chemical Unsaturation.

Polymer Concrete *n* A composite material consisting of graded aggregates with an organic binder or mixed organic and inorganic binders. Epoxy and other resins have been used, in contents between 8% and 20%. compressive and flexural strengths are several times those of Portland-cement concretes, they are impervious to liquids, and can be made to look like granite or marble. However, cost is about five times that of Portland concrete, so polymer concretes, so far, have been limited to special uses.

Polymer Gels *n* Similar to liquid emulsion polymer coatings, except furnished in paste form. Must be polished for luster. May contain some solvent.

Polymeric \ˌpä-lə-ˈmer-ik\ *adj*, **polymerism** \pä-ˈli-mə-ˌri-zəm\ *n* (1) Substance, the molecules of which consist of one or more structural units repeated any number of times; vinyl resins are examples of true polymers. The name is also frequently applied to large molecules produced by any chemical process, e.g., condensation in which water or other products are produced;

alkyd resins are examples of these. *Homopolymer* – Polymer of which the molecules consist of one kind of structural unit repeated any number of times; polyvinyl chloride and polyvinyl acetate are examples. *Copolymer* – Polymer of which the molecule consists of more than one kind of structural unit derived from more than one monomer; polyvinyl chloride-acetate, or polyvinyl acetate-acrylic copolymers are examples. See ▶ High Polymer. (2) A substance consisting of molecules characterized by the repetition (neglecting ends, branch junctions, and other minor irregularities) of one or more types of monomeric units. (IUPAC) *Also known as High Polymer and Macromolecule.*

Polymeric Modifier *n* A term applied to any polymer that is blended with the principal polymer to alter the latter's characteristics. See also ▶ Elasticizer, Impact Modifier, and Polymeric Plasticizer.

Polymeric Plasticizer *n* The term refers to plasticizers with molecules containing repeating mers and much larger than those of monomeric plasticizers that comprise virtually all other classifications of plasticizers. The two main types of polymeric plasticizers are the epoxidized oils of high molecular weight and ▶ Polyester Plasticizers. Polymeric plasticizers are noted for their permanence, which is due to the reduced tendency of the larger molecules to migrate and evaporate. However, the viscosity rises and the low-temperature properties of polymeric plasticizers decrease as their molecular weights increase. In cold weather, the high-molecular-weight polymerics may be difficult to handle and pump.

Polymeric Polyisocyanate *n* A generic term for a family of isocyanates derived from aniline-formaldehyde condensation products, used as reactants in the production of polyurethane foams.

Polymeric Sulfur Nitride See ▶ Sulfur Nitride Polymer.

Polymerizable Plasticizer See ▶ Plasticizer, Polymerizable.

1,2-Polymerization *n* Polymerization of a butadiene by 1,2 addition.

1,4-Polymerization *n* Polymerization of a butadiene by 1,4 addition.

Polymerization *n* Polymerization is the reaction in of two or more small molecules (*monomers*) that combine to form large molecules (polymers, macromolecules) that contain repeating structural units of the original molecules and have the same percentage composition as the small molecules if the small ones were of the same kind. There are two basic types of polymerizations, both with many variations: *addition* polymerization, which occurs when reactive, unsaturated monomers unit without forming any other products; and *condensation* polymerization (Odian GC (2004) Principles of polymerization. Wiley, New York), which occurs by combining of reactive end groups, accompanied by the elimination of a simple molecule such as water. Examples of condensation polymers are nylons and phenolic resins. Polymers polymerized via photoinitiators (free radical or chain polymerized) are photopolymerized. (Fouassier J-P (1995) Photoinitiation, photopolymerization and photocuring. Hanser-Gardner, New York) The majority of thermoplastics, aside from polyamides and polyesters, and a few thermosets, are made by addition polymerization, in which a pair of shared electrons in each monomer molecule is utilized to link the separate molecules into long chains. Polymerization processes and related terms are defined under the following heading. (Whittington's dictionary of plastics. Carley James F (ed). Technomic, 1993)

Polymerize *adj* The process of undergoing polymerization. The reaction of molecules that result in forming relatively long-chain molecules.

▶ Addition polymerization	▶ Precipitation polymerization
▶ Alternating copolymer	
▶ Block copolymer	▶ Radiation polymerization
▶ Branching	
▶ Bulk polymerization	▶ Random copolymer
▶ Chain-transfer agent	▶ Redox
▶ Condensation polymerization	▶ Solid-state polymerization
▶ Crosslinking	
▶ Emulsion polymerization	▶ Solution polymerization
▶ Free-radical polymerization	
▶ Gas-phase polymerization	▶ Stereoblock polymer
▶ Graft copolymer	▶ Stereograft polymer
▶ Interfacial polymerization	▶ Stereoregular polymer
▶ Ionic polymerization	
▶ Isotactic	▶ Stereospecific
▶ Network polymer	▶ Suspension polymerization
▶ Oxidative coupling	
▶ Autoacceleration	▶ Syndiotactic
▶ Bead polymerization	▶ Thermal polymerization
▶ Photopolymerization	

Polymerized Fatty Acid *n* Polycarboxylic acids produced by polymerizing acids from animal or vegetable fats and oils, in either an ester or free acid state, by means of heat alone or catalytically. See ▶ Dimer Acids.

Polymer Melts *n* Polymer that has been heated until it has reached a molten condition.

Polymer *n*, **Natural** A substance of high molecular weight occurring naturally, consisting of molecules that are, at least approximately, multiples of simple units. Natural polymers are often regarded as organic, but many inorganic minerals, such as quartz, feldspar, and asbestos are considered to be entirely or substantially polymeric. Examples of natural organic polymers are cellulose, natural rubber, lac, proteins such as collagen and keratin, and many natural fibers.

Polymerography *n* (resinography) The use of microscopic and metallographic techniques in the study of polymers.

Polymers, Mixing Behavior Of Mixing Behavior of Polymers image
Phase diagram of the typical mixing behavior of weakly interacting polymer solutions.

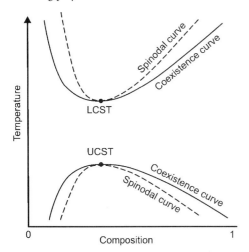

In general, polymeric mixtures are far less miscible than mixtures of small molecule materials. This effect results from the fact that the driving force for mixing is usually entropy, not interaction energy. In other words, miscible materials usually form a solution not because their interaction with each other is more favorable than their self-interaction, but because of an increase in entropy and hence free energy associated with increasing the amount of volume available to each component. This increase in entropy scales with the number of particles (or moles) being mixed. Since polymeric molecules are much larger and hence generally have much higher specific volumes than small molecules, the number of molecules involved in a polymeric mixture is far smaller than the number in a small molecule mixture of equal volume. The energetics of mixing, on the other hand, is comparable on a per volume basis for polymeric and small molecule mixtures. This tends to increase the free energy of mixing for polymer solutions and thus make solvation less favorable. Thus, concentrated solutions of polymers are far rarer than those of small molecules.

Furthermore, the phase behavior of polymer solutions and mixtures is more complex than that of small molecule mixtures. Whereas most small molecule solutions exhibit only an upper critical solution temperature phase transition, at which phase separation occurs with cooling, polymer mixtures commonly exhibit a lower critical solution temperature phase transition, at which phase separation occurs with heating.

In dilute solution, the properties of the polymer are characterized by the interaction between the solvent and the polymer. In a good solvent, the polymer appears swollen and occupies a large volume. In this scenario, intermolecular forces between the solvent and monomer subunits dominate over intramolecular interactions. In a bad solvent or poor solvent, intramolecular forces dominate and the chain contracts. In the theta solvent, or the state of the polymer solution where the value of the second virial coefficient becomes 0, the intermolecular polymer-solvent repulsion balances exactly the intramolecular monomer-monomer attraction. Under the theta condition (also called the Flory condition), the polymer behaves like an ideal random coil.

Polymer Structure *n* (1) A general term referring to the relative positions, arrangements in space, freedom of motion of atoms in a polymer molecule, and orientation of chains. Such structural details have important effects on polymer properties such as the second-order-transition temperature, flexibility, and tensile strength. (2) The microstructure of a polymer, as observed by light- or electro-microscopic techniques, and including crystalline structure, birefringence, distribution of sizes of filler particles and spherulites, and distribution of reinforcement directions. These, too, have important influences on macroscopic properties and behavior.

Polymer *n*, **Synthetic** The product of a polymerization reaction whose starting materials are one or more monomers. See ▶ Polymerization. When a single monomer is used, the product is called a *homopolymer*, *monopolymer*, or simply a polymer. When two monomers are polymerized simultaneously one obtains a *copolymer*. The term *terpolymer* is used for the polymerization product of three monomers. However, the term *heteropolymer* is also used for terpolymers as well as for products of more than three monomers. When no monomer is used, the product is known as a *nonomer*. The terms *polymer, resin, high polymer, macro-molecular material,* and *plastic* are often used interchangeably, although *plastic* also refers to compounds containing major additives. NOTE – The definition approved by IUPAC and ISO for polymer is "a substance composed of molecules characterized by the multiple repetition of one or more species of atoms or groups of atoms (constitutional units) linked to each other in amounts sufficient to provide a set of properties that do not vary markedly with the addition or removal of one or a few constitutional units." A polymer may be amorphous or may contain crystalline structures up to and exceeding half its specific volume, In a given polymer, the crystalline regions are always more dense than the amorphous ones; thus, percent crystallinity can be estimated from density.

Polymethacrylates *n* A polymer of a methacrylic ester, polymethyl methacrylate being the most important and useful member of the class.

$$\left[-CH_2-C(CH_3)(COOR)- \right]_n$$

Poly(methacrylic acid) *n* Poly(2-methyl-2-propenoic acid).

$$\left[-CH_2-C(CH_3)(COOH)- \right]_n$$

Polymethacrylonitrile *n* (PMAN) Poly(2-methyl-2-propenenitrile) A thermoplastic obtained by the polymerization of Methacrylonitrile, a vinyl monomer containing the nitrile group. The homopolymer has good mechanical strength and high resistance to solvents, acids, and alkalis, but discolors at molding temperatures.

$$\left[-CH_2-C(CH_3)(CN)- \right]_n$$

Polymethoxy Acetal *n* (PMAC) Any oligomer of methoxyl dimethyl acetal with degree of polymerization in the range 3–10. These oligomers are high-boiling, yellowish liquids used as modifiers for phenolic resins, and as solvents and plasticizers.

Polymethyl Acrylate *n* A polymer of methyl acrylate, having a glass-transition at 10°C, a leathery, tough material used in textile and leather finishing.

Poly(methyl acrylate) *n* Poly(2-methyl-2-propenoic acid).

$$\left[-CH_2-CH(COOCH_3)- \right]_n$$

Poly(methyl cyanoacrylate) *n* Poly(2-propenoic acid, 2-cyano-, methyl ester).

$$\left[-CH_2-C(CN)(COOCH_3)- \right]_n$$

Polymethylene *n* A polymer first made by polymerizing diazomethane (also called *azomethylene*) (CH$_2$N$_2$), with evolution of nitrogen gas. While this polymer has the same formula as polyethylene, it contains no side chains, so provides a standard with which branched ethylene polymers may be compared. Although long known in the laboratory, it is not a commercial resin.

Poly(5-methyl-hexene-1) *n*

$$\left[-CH_2-CH(CH_2CH_2CH(CH_3)_2)- \right]_n$$

Poly(methyl methacrylate) \\ˈpä-lē-ˌme-thəl-\ *n* (1936) (PMMA) Poly(2-propenoic acid, 2-cyano-, methyl ester). Belongs to the group of acrylic resins. It is a rigid amorphous polymer. A polymer of methyl methacrylate. The most important member of the family of acrylic resins, made by addition polymerization of the monomer, methyl methacrylate [CH_2=C(CH_3)$COOCH_3$]. Two outstanding characteristics of PMMA are its optical clarity (92% light transmission) and unsurpassed resistance to weathering. It also has good electrical properties, the ability to "pipe" light around bends, and is tasteless, odorless, and nontoxic. PMMA molding powders can be injection molded, extruded, and compression molded. The liquid monomer can be cast into rods, sheets, optical lenses, etc. Cast and extruded PMMA sheets are fabricated and thermoformed into many products such as aircraft canopies, skylights, lighting fixtures, and outdoor signs. See also ▶ Acrylic Resin.

Poly(methyl methacrylate co styrene) n Poly(2-propenoic acid, 2-cyano-, methyl ester-co-ethenyl benzene).

Poly(methyl methacrylate) Plastics n Plastics based on polymers made with methyl methacrylate as essentially the sole monomer.

Polymethylpentene n Thermoplastic stereoregular polyolefin obtained by polymerizing 4-methyl-1-pentene based on dimerization of propylene; having low density, good transparency, rigidity, dielectric and tensile properties, and heat and chemical resistance. Processed by injection and blow molding and extrusion. Used in laboratory ware, coated paper, light fixtures, auto parts, and electrical insulation. Also called PMP.

Poly(4-methylpentene-1) n (PMP) A polyolefin first introduced commercially in 1966 by Imperial Chemical Industries but now produced only by Mitsui Plastics ("TPX") and Phillips 66 C., ("Crystalor"). The monomer, 4-methylpentene-1, is produced by dimerization of propylene. Polymerization is conducted with a Ziegler-type catalyst. The polymers are supplied as free-flowing powders or as compounded granules, suitable for the usual thermoplastics processes. Properties of the resins are high light transmission (93%, better than many glasses), melting range near 240°C, rigidity and tensile properties similar to those of polypropylene, good electricals, high chemical resistance, and the lowest density of all commercial solid resins, 0.834 g/cm^3. It is approved for food contact. These properties account for its use in laboratory volumeware, slight glasses, high-frequency electrical components, coffee funnels, wire coatings, and microwave-safe cookware.

Poly(α-methyl styrene) n

Polymin n Poly(ethylene imine), manufactured by BASF, Germany.

Poly(Monochloro-p-Xylylene) n (Parylene C) See ▶ Parylene.

Polymorphism \ˌpä-lē-ˈmȯr-ˌfi-zəm\ n (1839) The ability of a substance to crystallize with different internal lattices thereby giving correspondingly different external crystal morphology and internal physical properties, Most, and probably all, elements and compounds exhibit polymorphism. The crystal systems of the two (or more) modifications (forms) of the compound are usually but not always, different.

Poly-m-Xylyleneadipamide n Syn: ▶ Nylon Mxd/6.

Poly-n-Butyl Methacrylate (PBMA) A rubbery polymer that enjoys some use as an adhesive and textile finish.

Polynosic Fiber n A high-wet-modulus rayon staple having a microfibular structure of fibers. The molecular chain length of the cellulose forming the fiber is about twice as long as in conventional rayon.

Polyoctanoamide n Syn: ▶ Nylon 8.

Polyol n (polyhydric alcohol, polyalcohol) An organic compound having more than one hydroxyl (−OH) group per molecule, In the cellular plastics industry, the term includes monomeric and polymeric compounds containing alcoholic hydroxyl groups such as polyethers, glycols, glycerol, and polyesters, used as reactants in polyurethane foam. See ▶ Polyalcohol.

Polyolefin \ ˌpä-lē-ˈō-lə-fən\ *n* (1930) Any of the largest genus of thermoplastics, polymers of simple olefins such as ethylene, propylene, butanes, isoprenes, and pentenes, and copolymers and modifications thereof. The two most important are polyethylene and polypropylene, which, together, accounted for just under half of all US resin sales in 1992. Polyolefin plastics are most usually processed into end products by extrusion, injection molding, blow molding, and rotational molding. Thermoforming, calendering, and compression molding are used to a lesser degree. An inherent characteristic common to all polyolefins is a nonpolar, nonporous, low-energy surface that is not receptive to inks, lacquers, etc., without special oxidative pretreatment. See the following:
- ▶ Ethylene-Propylene Rubber
- ▶ Ethylene-Vinyl Acetate Copolymer
- ▶ Ionomer Resin
- ▶ Poly(4-Methylpentene-1)
- ▶ Polypropylene

Polyolefin Fiber *n* A fiber produced from a polymerized olefin, such as polypropylene or polyethylene.

Polyolefin Plastics *n* Plastics based on polymers made with an olefin (or olefins) as essentially the sole monomer (or monomers).

Polyolefin Resins See ▶ Polyolefins.

Polyorganophosphazene *n* A polymer obtained by reaction of phosphorus pentachloride and ammonium chloride. A cyclic trimer, $(NPCl_2)_3$, or tetramer, $(NPCl_2)_4$, is formed which can be converted to polyorganophosphazenes, $(-N=PR_2-)_n$ where R represents an organic side group. The polymers have found some use in hose, gaskets, and seals in aviation-fuel-handling equipment. They have better solvent resistance and low-temperature elasticity than siloxane-carborane polymers, and are less costly.

Polyox *n* High-molecular-weight poly(ethylene oxide, manufactured by Union Carbide, US.

Polyoxamide Generic name for nylon-type materials made from oxalic acid and diamines. Their extremely high melting temperatures have kept them out of commerce.

Polyoxetane See ▶ Chlorinated Polyether.

Polyoxyamide Fiber *n* Copolymeric fiber with good comfort properties, particularly high moisture absorption and transfer, and intrinsic softness.

Polyoxyethylene Glycol See ▶ Polyethylene Glycol.

Polyoxymethylene *n* (POM) Linear polymers of formaldehyde or oxymethylene glycol with the formula $(-OCH_2-)_n$, in which n is above 100. Those in the range of $100 < n < 300$ are brittle solids used as intermediates. Those in the range $500 < n < 5000$ are ACETAL RESINs. NOTE – Polyoxymethylene is theoretically the simplest member of the generic class of polyacetals.

$$-\!\!\left[\!-CH_2O-\!\right]_n\!\!-$$

Polyoxypropylene Glycol *n* Any low-molecular-weight polymer with the structure $H[-OCH(CH_3)CH_2-]_nOH$, derived from propylene oxide and used in the production of polyurethane foams.

***Trans*-Polypentenamer** *n* An elastomer obtained by the polymerization of cyclopentane, using complex catalysts. Its structure is highly linear and the molecular weight has a wide range. Its properties are similar to those of natural rubber and *cis*-polybutadiene.

Polypeptide \ ˌpä-lē-ˈpep-ˌtīd\ *n* [ISV] (1903) Low molecular weight plastics of amino acids.

Polyphenone *n* A phenolic-like material developed in the early 1970s by Union Carbide, but still not commercial. Unlike phenolic, it was to be available in a range of light colors, with good moldability and electrical and physical properties equal to those of mineral-filled phenolics.

Polyphenylene Benzobisthiazole *n* (PBT, PBZ) A liquid-crystalline polymer from which very strong and heat-resistant fibers are made.

Poly-1,3-Phenylenediamine Isophthalate *n* A high-temperature fiber, trade named Nomex® by DuPont. This fiber resists common flame temperatures around 500°C for a short time and thus is suitable for fire-protective clothing and insulation of motors and transformers.

Polyphenylene Ether Nylon Alloys *n* Thermoplastics have imporved heat and chemical resistance and toughness. Processed by molding and extrusion. Used in auto body parts.

Polyphenylene Glycol See ▶ Polyethylene Glycol.

Polyphenylene Oxide (PPO) *n* A thermoplastic, linear, non-crystalline polyether obtained by the oxidative polycondensation on 2,6-dimethylphenol in the presence of a copper-amine complex catalyst. The resin has a wide useful temperature range, from below $-170°$ to $+190°C$, with intermittent use to 205°C possible. It has excellent electrical properties, unusual resistance to acids and bases, and is processable on conventional extrusion and injection-molding equipment. Because of its high coat PPO is also marketed in the form of polystyrene blends (see NORYL®) that are lower-softening (T$_g$ of PS is about 100°C vs. 208°C for PPO), and have working properties intermediate between those of the two resins.

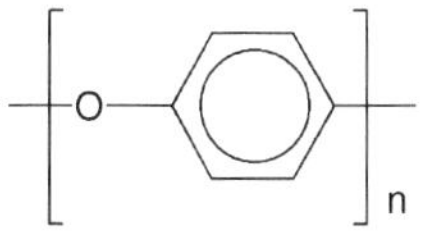

Poly(phenylene oxide) *n* Poly(phenylene oxide), or PPO, is a high-performance polymer. It's biggest strength is it's resistance to high temperatures. It has a very high glass transition temperature, 210°C. PPO is often made into blends with high-impact polystyrene (HIPS for short) due to its high resistance to heat. Blending PPO with HIPS makes the PPO easier to process, plus it gives PPO some resilience. Structurally, PPO is made of phenylene rings linked together by ether linkages in the 1,4 or para- positions, with a methyl group attached to carbon atoms in the 2 and 6 positions. PPO is made by what we call oxidative coupling polymerization of the monomer 2,6-dimethylphenol.

Poly(phenylene sulfide) *n* Poly(phenylene sulfide), or PPS, is a high-performance plastic that is very strong and can resist very high temperatures up to 300°C. To produce PPS you have to react paradi-chlorobenzene and sodium sulfide in a polar solvent like *N*-methyl pyrrolidone.

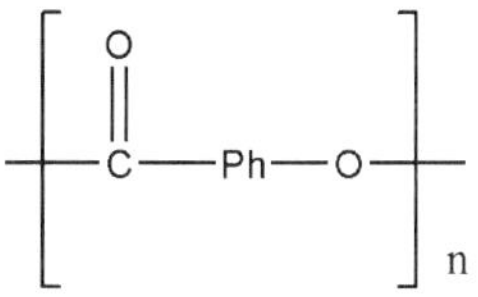

Polyphenylene Sulfide *n* (PPS) A crystalline polymer having a symmetrical, rigid backbone chain consisting of recurring *p*-substituted benzene rings and sulfur atoms. A variety of grades suitable for slurry coating, fluidized-bed coating, electrostatic spraying, as well as injection and compression molding are offered (Phillips 66 Co.'s Ryton® and others). The polymers exhibit outstanding chemical resistance, thermal stability, and fire resistance. Their extreme inertness toward organic solvents, and inorganic salts and bases make for outstanding performance as a corrosion-resistant coating suitable for contact with foods. Doping with arsenic pentrafluoride infuses the resin with usefully high electrical conductivity.

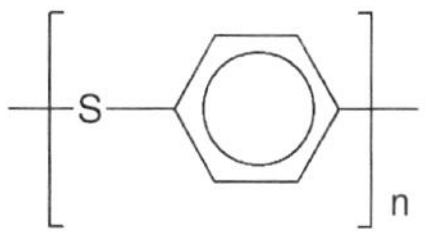

Polyphenylene Sulfide Sulfone *n* Thermoplastic having good heat, fire, creep, and chemical resistance and dielectric properties. Processed by injection molding. Used in electrical devices. Also called PPSS.

Polyphenylquinoxaline *n* (PPQ) Any of a family of high-performance thermoplastics that have potential for use as functional and structural resins in applications demanding high chemical and thermal stability. The most attractive synthesis is by copolycondensation of an aromatic bis(o-diamine) powder and a stirred solution or slurry of bis(1,2-dicarbonyl) monomer in an appropriate solvent such as a mixture of *m*-cresol and xylene. In solution form, the polymers can be used directly for prepreg and adhesive-tape formulations, film casting, etc. If desired, the polymer can be isolated from solution and compression molded. It is convertible to a thermoset form by rigidizing the linear polymer backbone with reactive latent groups and by crosslinking.

Polypheylsulfone *n*

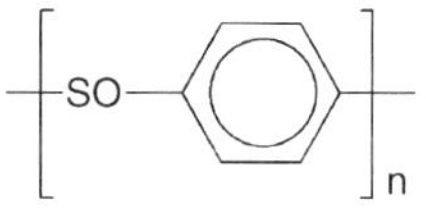

Polyphosphates *n*

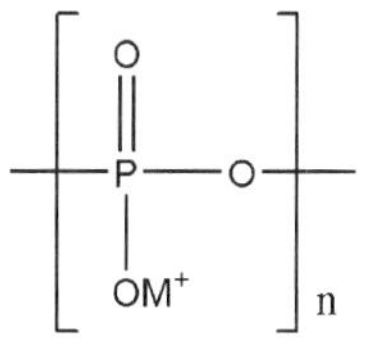

Polyphosphazene *n* A family of inorganic-base polymers having phosphorus-nitrogen backbones joined with fluorine or chlorine. Depending on which organic side groups are linked to the backbones, a wide variety of polymers can be made with properties ranging from rigid and flexible thermoplastics and elastomers to glass-like thermosets. Some grades outperform silicones in biomedical uses. See also ▶ Phosphazene Polymer.

Polyphosphazene Fluoroelastomer *n* Any of a family of elastomers developed primarily for fuel tanks to be used in the Arctic, having the typical chain-unit configuration $(CF_3CH_2O)_2PN(CH_2C_3F_6CH_2O)_2PN$. These elastomers are inert to aviation fuel and remain flexible to $-60°C$, lower than other elastomers previously used in this application.

Polyphthalamide n (PPA) Poly(1,2-Benzenedicarboxamide) A polyamide in which the residues of terephthalic or iosphthalic (or mixed) acid components are part of the mer unit of the chain. PPA is an advanced engineering polymer first commercially offered in 1991 by Amoco Chemical Co. under the trade name Amodel®.

$$\left[-NH-\overset{O}{\underset{\|}{C}}-Ph-\overset{O}{\underset{\|}{C}}-NH- \right]_n$$

Poly-p-Hydroxybenzoic Acid n A homopolyester of repeating p-oxyenzoyl units with a high degree of crystallinity. It does not melt below its decomposition temperature, 550°C, but can be fabricated at 300–360°C by compression sintering and plasma-spray processes. Copolymers with aromatic dicarboxylic acids and aromatic bisphenols are processable by normal means. Applications include electrical connectors, valve seats, high-performance-aircraft parts, and automotive parts.

Polypinvalotactone n A crystalline thermoplastic polymerized by ring opening from the cyclic monomer, $(CH_3)_2\underline{CCOOCH_2}$. It is useful for making high-strength fibers, also as a high-crystallinity (75%) matrix resin with carbon fibers.

Poly(p-oxybenzoate) n Poly(4-hydroxy-benzoic acid).

$$\left[-\overset{O}{\underset{\|}{C}}-Ph-O- \right]_n$$

Poly-p-Oxybenzoÿl See ▶ Poly-p-Hydroxybenzoic Acid.

Poly(p-phenylene)

$$\left[-\text{Ph}- \right]_n$$

Poly(p-Phenylene Sulfone) n (PPSU) Chemically similar to the polysulfones, this high-performance polymer has better impact resistance. It also has excellent heat resistance, low creep and good electrical properties.

It is difficult to process, however, which has limited its commercial acceptance.

Polypropylene \ˌpä-lē-ˈprō-pə-ˌlēn\ n (1935) (PP, polypropene) Any of several types of a large family of thermoplastics resins made by polymerizing propylene with suitable catalysts, generally solutions of aluminum alkyl and titanium tetrachloride. Its density (approximately 0.905 g cm^3) is among the lowest of all plastics. PP and copolymers enjoyed the third largest sales in the US in 1992, 3.8 Tg (4.2×10^6 t), about 13% of all US resins sales. As with the Polyethylenes, properties of the polymers vary widely according to molecular weight, method of production, and copolymers involved. The grades used for molding have molecular weights of 40,000 or more, are 90–95% isotactic with about 50% crystallinity. They have good resistance to heat, chemicals, and solvents, and good electrical properties. Properties can be improved by compounding with fillers, e.g., mica or glass fibers, by blending with synthetic elastomers, e.g., polyisobutylene; and by copolymerizing with small amounts of other monomers. Fibers are the single largest use of polypropylene. See ▶ Split-Film Fiber.

Polypropylene n Poly(1-propene) A plastic polymer of propylene, $(C_3H_5)_n$. Similar to polyethylene, but each unit of the chain has a methyl group attached. It is translucent, autoclavable, and has no known solvent at room temperature. A polymer of propylene (propene). *Also called Polypropene.* Abbreviation: PP.

$$\left[-\overset{H_2}{\underset{}{C}}-\overset{H}{\underset{CH_3}{C}}- \right]_n$$

Polypropylene Adipate n (polypropylene glycol adipate) A polymeric plasticizer for vinyl chloride polymers and copolymers formed by reacting propylene glycol and adipic acid.

Polypropylene Fiber n A manufactured, olefin fiber made from polymers or copolymers of propylene. Polypropylene fiber is produced by melt spinning the molten polymer, followed by stretching to orient the fiber molecules. CHARACTERISTICS: Polypropylene fibers have a number of advantages over polyethylene fibers in the field of textile applications. The degree of crystallinity, 72–75%, results in a fiber that is strong and resilient, and does not fibrillate like high-density polyethylene. Polypropylene has a high work of rupture, which indicates a tough fiber, and may be made with tenacities as high as 8.0–8.5 g per denier.

The melting point of polypropylene is 165°C, which is low by comparison with nylon or polyester, but is high enough to make it suitable for most textile applications. So light that it actually floats, polypropylene fiber provides greater coverage per pound than any other fiber. It is highly resistant to mechanical abuse and chemical attack. END USES: Polypropylene fibers are widely used in industrial, carpet, and geotextile applications. They have found important uses in fishing gear, in ropes, and for filter cloths, laundry bags and dye bags. The excellent chemical resistance of polypropylene fiber is of advantage in the filtration and protective clothing fields. Fibrillated polypropylene yarns are widely used in indoor-outdoor carpets. Staple fiber finds application in blankets, pile fabrics, underwear, and industrial fabrics; it is being developed for carpets, candlewicks, knitted outerwear, hand-knitting yarns, and upholstery.

Polypropylene Glycol *n* (PPG, polypropylene oxide) A family of nonvolatile liquids with the general formula $HOCH(CH_3)[-CH_2CH-(CH_3)O-]_nCH_2OH$. They are similar to the polyethylene glycols, but are more oil-soluble and less water-soluble. They are polyols used in producing polyurethane foams, adhesives, coatings, and elastomers.

Polyprotic Acid *n* An acid which has more than one available H+, or can donate more than one proton.

Poly(p-*tert*-butyl styrene) *n* Poly(1-(1,1-dimethylethyl)-4-ethenyl-benzene).

Poly-*p*-Xylylene *n* (Parylene N) See ▶ Parylene.

Polypyromellitimide *n* (PPM) The original polyimide family of polymers, having enhanced heat resistance and formed from polyamide carboxylic acids derived by reacting pyromellitic dianhydride with 4,4'-diaminophenyl ether. Grades of the polymer are used for forming films, paint components, and are processable as thermoplastics under special conditions. More easily processed copolymers have enjoyed greater commercial success.

Polypyrrole *n* A polymer of pyrrole, a five-membered heterocyclic substance with one nitrogen and four carbon atoms and with two double bonds. The polymer can be prepared via electrochemical polymerization. Polymers thus prepared are doped by electrolyte anion and are electrically conductive. Polypyrrole is used in lightweight secondary batteries, as electromagnetic interference shielding, anodic coatings, photoconductors, solar cells, and transistors.

Polysar Butyl *n* Isobutylene/isoprene copolymer, manufactured by Sarnia, Canada.

Polysilane *n* A polymer whose backbone is composed of covalently linked silicon atoms with organic side groups that may be aliphatic, aromatic, or mixed, and not to be confused with *polysiloxanes* (see ▶ Silicone). Polysilanes have been used as lithographic resists.

Polysiloxane *n* Polysiloxane is the proper name for silicones. See ▶ Polysiloxane where X and Y can be an alkyl groups.

Poly(siloxanes) *n* See ▶ Polysiloxanes and ▶ Poly (p-*Tert*-Butyl Styrene) above.

Polystyrene \ˈpä-lē-ˈstī-ˌrēn\ *n* (1927) An important family of workhorse plastics, the polymer of styrene (vinyl benzene), which has been commercially available for more than half of a century. In 1992 polystyrene, neat and modified, accounted for about 10% of US resins sales, i.e., 3.0 Tg (3.2×10^6 t). The homopolymer is water-white, has excellent clarity and sparkle, outstanding electrical properties, good thermal and dimensional stability, is hard, stiff, and resistant to staining, and is inexpensive ($1/kg in 12/92). However, it is somewhat brittle, and is often modified by blending or copolymerization to a desired mix of properties. High impact grades (HIPS) are produced by adding rubber or butadiene copolymers. Heat resistance is improved by including some α-methyl styrene as a comonomer. Copolymerizations with methyl methacrylate improve light stability, and copolymerization with acrylonitrile raises resistance to chemicals. Styrene polymers and copolymers possess good flow properties at temperatures safely below degradation ranges, and can easily be extruded, injection molded, or compression molded. Abbreviation: PS.

Poly(styrene-co-vinyl acetate) n Poly(ethenyl benzene -co- acetic acid ester).

Poly(styrene-co-vinyl alcohol) n Poly(ethenyl benzene–co- 1 hydroxyethylene).

Polystyrene Foam n (expanded polystyrene, EPS) A low-density, cellular plastic made from polystyrene by either of two methods. Extruded foam is made in tandem extruders, the first for plasticating the resin, the second to homogenize the blowing agent, which may be a gas or volatile liquid, such as nitrogen or pentane, and reduce the temperature of the melt before it reaches the die. As it emerges from the die the large drop in pressure frees the blowing agent and the mass expands to form a low-density "log" conveyed through a long cooling tunnel. The cooled slab is usually sliced into a large range of shapes marketed through building-materials dealers. In the other basic method, a volatile blowing agent, e.g., isopentane, is incorporated into the tiny PS beads as they are polymerized, or afterward. The beads are first pre-expanded, allowed to "rest" for about a day, then molded in a closed, steam-heated mold, and finally cooled with water in the mold members. This method, which generally produces closed-cell foams, is used to mold finished products such as coffee cups, packaging components, and life-preserver rings. Beads are also used to generate very large, thick slabs (6 × 1.2 × 0.6 m) by blowing live steam into an expanded, low-pressure mold charged with measured weight of beads. After cooling, these slabs are sliced with multiple hot-wire cutters to produce foam "lumber", as with the extruded foam.

Polystyrene Resins n Synthetic resins formed by polymerization of styrene.

Polystyrenes Polymers that contain the styrene monomer. Styrenes can be used as high impact grades, or heat resistance substances, or light stable substance or chemically stable polymer by simply blending the styrene homopolymer or creating a styrene copolymer.

Polystyrol n A rarely used term for polystyrene.

Polystyrylpyridine n (PSP) A thermosetting resin, resistant to high temperatures, formed by condensation of 2,4,6-trimethylpyridine and 1,4-benzyldialdehyde, and a useful matrix for carbon-fiber composites.

Polysulfide \-ˈsəl-ˌfīd\ n [ISV] (1849) A family of sulfur-containing polymers prepared by condensing organic polyhalides with sodium polysulfides in aqueous suspension. They range from liquids to solid elastomers.

Polysulfide Rubber n (T) A family of sulfur-containing polymers prepared by condensing organic polyhalides with sodium polysulfides in aqueous suspension. They range from liquids to solids elastomers. The first commercial polysulfide was Thiokol® A, polyethylene tetrasulfide, made from sodium tetrasulfide and ethyl dichloride. This elastomer had outstanding solvent resistance, but its poor mechanical properties and unpleasant odor limited its use to plasticizing acid-resistant cements. Modern T materials have the general structure $(-R-S_m-)_n$ where m is usually 2–4 and R is $(CH_2)_2$ or an ether group. These elastomers are used in hose, printing rolls, gaskets, and gas meter diaphragms. Polysulfide products have excellent resistance to oils, solvents, oxygen, ozone, light, and weathering, and low permeability to gases and vapors.

Polysulfonate Copolymer n (sulfonate-carboxylate copolymer) A family of transparent, thermoplastic polyesters, moldable at 250–300°C, and formed by reaction of a diphenol, generally bisphenol A, with an aromatic disulfonyl chloride and an aliphatic disulfonyl chloride or carboxylic acid chloride. These copolymers have good electrical and mechanical properties, and excellent resistance to hydrolysis and aminolysis.

Polysulfone *n* (PSU, PPSU) A family of sulfur-containing thermoplastics, closely akin to ▶ Polyethersulfone made by reacting bisphenol A and 4,4'-dichlorodiphenyl Sulfone with potassium hydroxide in dimethyl sulfoxide at 130–140°C. The structure of the polymer is benzene rings or phenylene units linked by one or more of three different chemical groups – a Sulfone group, an ether link, and an isopropylidene group. Each of these three linking components acts as an internal stabilizer. Polysulfones are characterized by high strength, very high service-temperature limits, low creep, good electrical characteristics, transparency, self-extinguishing ability, and resistance to greases, many solvents, and chemicals. They may be processed by extrusion, injection molding, and blow molding.

Polysulfone *n* Copolymer from bisphenol A + p,p'-dichlorodiphenyl sulfone. Manufactured by Shell, The Netherlands.

Polysulfones *n* (PSU, PPSU) A family of sulfur-containing thermoplastics, closely akin to ▶ Polyethersulfone made by reacting bisphenol A and 4,4'-dichlorodiphenyl Sulfone with potassium hydroxide in dimethyl sulfoxide at 130–140°C. The structure of the polymer is benzene rings or phenylene units linked by one or more of three different chemical groups – a Sulfone group, an ether link, and an isopropylidene group. Each of these three linking components acts as an internal stabilizer. Polysulfones are characterized by high strength, very high service-temperature limits, low creep, good electrical characteristics, transparency, self-extinguishing ability, and resistance to greases, many solvents, and chemicals. They may be processed by extrusion, injection molding, and blow molding.

Polyterephthalate See ▶ Polyester and ▶ Terephthalate Polyester.

Polyterpene Resins *n* Friable, thermoplastic saturated cyclic resins produced by the catalytic polymerization of monomeric alpha- beta- pinene which are principal constituents of turpentine, with softening points, 10–135°C, Gardner color, 3–6; sp gr, 0.93–0.97; iodine no., 48. Soluble in aliphatic, aromatic, chlorinated solvents, insoluble in alcohols, glycols, or water. The amber-colored resins, ranging from viscous liquids to solids, are used as tackifiers, wetting agents, and modifiers in the manufacture of adhesives, paints and varnishes, and caulking and sealing compounds. They are compatible with natural and synthetic rubbers, polyolefins, alkyd resins, other hydrocarbon resins, and waxes. See ▶ Hydrocarbon Plastics and ▶ Terpene Resins.

Polytetrafluoroethylene *n* (PTFE) The oldest of the fluorocarbon-resin family discovered in 1938 by R. J. Plunkett, developed by DuPont and marketed under the trade name Teflon®. It is made by polymerizing tetrafluoroethylene, $F_2C=CF_2$, and is available in powder and aqueous-dispersion forms. PTFE inert to virtually all chemicals, has a crystalline melting point of 327°C, though it does not truly liquefy. Molecular weighs of commercial PTFE powders are very high, on the order of 106. It has a very low coefficient of friction on most surfaces, resists adhesion to almost any material unless strenuously pretreated, and has excellent electrical properties. Its nonstick character has long been evidenced by its everyday use as an interior coating in cooking utensils. Its inability to form a true melt long ago forced the development of special extrusion, molding, and calendering processes in which the PTFE powder is pressed, then, sintered with heat. PTFE tape and film are made by skiving pressed (or extruded) and sintered rods. Continuous extrusion is accomplished by alternating strokes of two ram extruders feeding a single die block. PTFE's low modulus and tendency to creep under load can be substantially improved by addition in inorganic fillers or chopped glass fiber. See also ▶ Fluorocarbon Resins.

Polytetrafluoroethylene Fiber *n* (PTFE) Fluorine-containing manufactured fibers characterized by high chemical stability, relative inertness, and high melting point. Polytetrafluoroethylene Fiber is made my emulsion spinning, a process that essentially results in fusion of fibrils by passing an emulsion through a capillary, then drawing the resulting fiber. The fiber has a moderate tensile strength and is particularly resistant to the effect of high temperatures and corrosive chemicals. Having very low frictional coefficients, it has a slippery hand. Its principal uses are in packaging and filtration media.

Polytetrahydrofuran *n* (polytetramethylene ether, PTHF) A type of polycol made from tetrahydrofuran by ring opening, have the mer [–$CH_2(CH_2)_3$–O–] and –OH end groups, with low to moderate molecular weights. These polymers have long been used as prepolymers for polyurethane elastomers.

Polytetramethyleneadipamide *n* Syn: ▶ Nylon 4/6.

Poly(Tetramethylene Terephthalate) Syn: ▶ Polybutylene Tere-Phthalate.

Polythene \ˈpä-lə-ˌthēn\ *n* [by contraction] (1939) Poly(ethylene) (high pressure), manufactured by DuPont, US. The British name for ▶ Polyethylene.

Polythiazyl See ▶ Sulfur Nitride Polymer.

Poly(trans-1,4-isoprene) n

[chemical structure]

Polytrifluorostyrene n A clear, thermoplastic material introduced in 1965 and said to combine the oxidation resistance of polytetrafluoroethylene with the mechanical and electrical properties and ease of processing of polystyrene, but still not commercially available in 1992.

Polyureas n

[chemical structure]

Polyurethane \ˌpä-lē-ˈyúr-ə-ˌthān\ n [ISV] (1944) *Type I*, one-package, prereacted-urethane coatings characterized by the absence of any significant quantity of free isocyanate groups. They are usually the reaction product of a polyisocyanate and a polyhydric alcohol ester of vegetable oil acids and are hardened with the aid of metallic soap driers. *Type II*, one-package, moisture-cured urethane coatings characterized by the presence of free isocyanate groups and capable of conversion to useful films by the reaction of these isocyanate groups with ambient moisture. *Type III*, one-package, heat-cured urethane coatings that dry or cure by thermal release of blocking agents and regeneration of active isocyanate groups that subsequently react with substances containing active hydrogen groups. *Type IV*, two-package catalyst-urethane coatings that comprise systems wherein one package contains a prepolymers or adduct having free isocyanate groups capable of forming useful films by combining with a relatively small quantity of catalyst, accelerator or crosslinking agent such as a monomeric polyol or polyamine contained in a second package. This type has limited pot life after the two components are mixed. *Type V*, two-package poly-urethane coatings that comprise systems wherein one package contains a prepolymers or adduct or other polyisocyanate capable of forming useful films by combining with a substantial quantity of a second package containing a resin having active hydrogen groups, with or without the benefit of catalyst. This type has limited pot life after the two components are mixed. *Type VI*, one-package, nonreactive lacquer-urethane solution coatings characterized by the absence of any significant quantity of free isocyanate or other functional groups. Such coatings convert to solid films primarily by solvent evaporation.

[chemical structure]

Polyurethane Elastomer n (PU) Any condensation polymer made by reacting an aromatic diisocyanate with a polyol that has an average molecular weight greater than about 750. The aromatic diisocyanates usually employed are toluene diisocyanate (TDI) and diphenylmethane diisocyanate (MDI). The polyol component is either a polyester or polyether. In the "one-shot" system the elastomer is prepared directly in the mold that shapes the final product. The diisocyanate, polyol, and catalyst are rapidly and intimately mixed, then immediately poured or pumped into the mold. Prepolymers, in the form of liquids or low-melting solids, are used when a longer working time is desired. The prepolymers are mixed with catalysts, heated when necessary, degassed, and poured into the molds. Also available are millable guns and pellets containing all components, which have been reacted to a degree that permits further processing by methods used for rubber, including injection molding, compression molding, and transfer molding. The polyurethane elastomers most widely used are harder than natural rubber, and possess excellent resistance to flexural fatigue, abrasion, impact, oils and greases, oxygen, ozone, and radiation, but are susceptible to hydrolysis.

Polyurethane Fiber See Spandex Fiber.

Polyurethane Finish n An exceptionally hard and wear-resistant paint or varnish made by the reaction of polyols with a multifunctional isocyanate. See ▶ Polyurethanes.

Polyurethane Foam n (urethane foam, isocyanate foam) This family of foams differs from other cellular plastics in that the chemical reactions causing foaming occur simultaneously with the polymer-forming reactions. As in the case of ▶ Polyurethane Resins the polymeric constituent of urethane foams is made by reacting a polyol with an isocyanate. The polyol may be of the polyester or polyether type. When the isocyanate is in excess of the amount that will react with the polyol, and when water is present, the excess isocyanate will react with water to produce carbon dioxide which expands the mixture. The hardness of the cured foam is governed by the molecular weight of the polyol used. Low-molecular-weight polyols (approximately 700) produce rigid foams, and high-molecular-weight polyols (3,000–4,000) produce flexible foams. Polyols with molecular weights around 6,000 are used for the so-called "cold-cure," highly resilient foams. They are usually capped with ethylene oxide to provide terminal primary hydroxyl groups that increase the polyols' reactivity about threefold. Crosslinked foams are rigid or semirigid. Auxiliary blowing agents are often used, especially in rigid foams where they improve the insulation values. Other ingredients often incorporated in urethane foams are catalysts to control the speed of reaction, and a surfactant to stabilize the rising foam and control cell size. Three basic processes are used for making urethane foams: the prepolymers technique, the semi-prepolymer technique, and the one-shot process. In the prepolymers technique, a polyol and an isocyanate are reacted to produce a compound that may be stored and subsequently mixed with water, catalyst, and, in some cases, a foam stabilizer. In the semi-prepolymer process about 20% of the polyol is prereacted with all of the isocyanate, then, this product is later reacted with a masterbatch containing the remainder of the ingredients. See also ▶ One-Shot Molding, Isocyanate, Polyol, Polyether Foam, Reticulated Polyurethane Foam, and Integral-Skin Molding.

Polyurethane/Imide Modified Foam *n* A polyaryl polyisocyanate (PAPI) is reacted with a 3,3',4,'-benzophenone tetracarboxylic dianhydride (BTDA) to form an isocyanate prepolymer. This prepolymer can be compounded with a polyol, a blowing agent, a catalyst, and a cell stabilizer to form the modified foam. Such a foam containing 5% BTDA in the prepolymer has better thermal properties than conventional polyurethane foams.

Polyurethane Resin *n* (isocyanate resin) A family of resins produced by reacting diisocyanates with organic compounds containing two or more active hydrogen atoms to form polymers having free isocyanate groups. These groups, under the influence of heat or certain catalysts, will react with each other, or with water, glycols, etc., to form a thermosetting material.

Polyurethanes *n* Polyurethanes are the most well known polymers used to make foams. Polyurethanes can be elastomers, paints, fibers, or adhesives. Polyurethanes are called polyurethanes because in their backbones they have a urethane linkage. Polyurethane can be any polymer containing the urethane linkage in its backbone chain. Polyurethanes are made by reacting diisocyanates with di-alcohols.

Polyurethane *n*, **Thermoset Resins** A family of resins produced by reacting diisocyanates with organic compounds containing two or more active hydrogen atoms to form an polymers having free isocyanate groups. These groups, under the presence of heat or certain catalysts, will react with each other, or water, glycols, etc., to form a thermosetting material.

Polyvinyl \ ˈpä-lē- ˈvī-nᵊl\ *adj* [ISV] (1927) Of, relating to, or being a polymerized vinyl compound, resin, or plastic.

Polyvinyl Acetal *n* (1) Generically, a class of polymers derived from polyvinyl esters in which some or all of the acid groups have been replaced by hydroxyl groups and some or all of these hydroxyl groups have been reacted with aldehydes to form acetal groups. (2) Specifically, polyvinyl acetal made by the reaction of the hydroxyl group with acetaldehyde. (3) A vinyl plastic produced from the condensation of polyvinyl alcohol with an aldehyde. There are three main groups: polyvinyl acetal, polyvinyl butyral, and polyvinyl formal; used in lacquers and adhesives. Polyvinyl acetal resins are thermoplastics which can be processed by casting.

$$\left[\begin{array}{c} \text{H} \quad \text{H}_2 \quad \text{H} \\ \text{C}-\text{C}-\text{C} \\ | \qquad \quad | \\ \text{O} \qquad \text{O} \\ \diagdown \; \diagup \\ \text{CH} \\ | \\ \text{R} \end{array} \right]_n$$

Polyvinyl Acetate *n* A colorless, odorless, nontoxic, transparent, thermoplastic, water-insoluble, resinous high polymer derived from the polymerization of vinyl acetate with a catalyst; used as a latex binder in certain paints and as an intermediate in the synthesis of polyvinyl acetal and polyvinyl alcohol. The major use is in water-based latex paints, adhesives, fabric finishes, and lacquers. In the plastics industry, the *copolymers* of vinyl acetate, particularly with vinyl chloride, are of most interest. Abbreviation for PVA and PVAc.

$$\left[\begin{array}{c} \text{H} \quad \text{H}_2 \\ \text{C}-\text{C} \\ | \\ \text{OOCH}_3 \end{array} \right]_n$$

Poly(vinyl acetate co vinyl chloride) *n* Poly(1 acetoxy ethylene -co- chloro-ethene).

$$\left[\begin{array}{cccc} H & H_2 & Cl & H_2 \\ -C-&C-&C-&C- \\ | & & | & \\ OOCH_3 & & H & \end{array} \right]_n$$

Polyvinyl Alcohol *n* (PVA) Poly(vinyl alcohol), Poly (1-hydroxy-ethylene) (1) A colorless, water-soluble, thermoplastic polymer prepared by partial or complete hydrolysis of polyvinyl acetate with methanol or water. Although it can be extruded and molded, its principal uses are in packaging films, fabric sizes, adhesives, emulsifying agents, etc. The packaging films are impervious to oils, fats, and waxes, and have very low transmission rates of oxygen, nitrogen, and helium. Thus, they are often used as barrier coatings on other thermoplastics or coextruded with them. The water solubility of polyvinyl alcohol films can be regulated to some degree. The "standard" type, made from higher-molecular-weight polymers and plasticized with glycerine, is only weakly soluble in cold water. The other type, known as CWS (cold-water-soluble), is made from internally plasticized or lower-molecular-weight resins. It is used in synthetic resins. (2) Oil resistant plastic.

$$\left[-CH_2-CH(OH)- \right]_n$$

Polyvinyl Alcohol Resins *n* A water-soluble thermoplastic prepared by partial or complete hydrolysis of polyvinyl acetate with methanol or water. These resins are mainly used as packing films since they are impervious to oils, fats, and waxes, and have very low transmission rates of oxygen, nitrogen, and helium.

Polyvinyl Butyral *n* (PVB, polyvinyl butyral acetal) A member of the ▶ Polyvinyl Acetal family, made by reacting polyvinyl alcohol with Butyraldehyde, with some unreacted PVAL groups retained in the polymer. It is a tough, sticky, colorless, flexible solid, used primarily as the interlayer in automotive safety glass. Other applications include adhesive formulations; base resins for coatings, toners and inks, solutions of rendering fabrics resistant to water, staining and abrasion; and crosslinking with resins such as urea's, phenolics, epoxies, isocyanate, and melamine's to improve coating uniformity and adhesion, increase toughness, and minimize cratering.

Polyvinyl Butyral Resins *n* A member of the polyvinyl acetal family. Resins formed by reacting polyvinyl alcohol with Butyraldehyde. It is a rough, sticky, colorless, flexible solid, used primarily as the interlayer in automotive safety glass. Other applications include adhesive formulations, base resin for coatings, solutions for rendering fabrics resistant to water, staining, and abrasion.

Poly(N-Vinylcarbazole) *n* (PVK) A thermoplastic resin, brown, obtained by reacting acetylene with carbazole. It has excellent electrical properties and good heat and chemical resistance, and is used as an impregnate for paper capacitors. It is photoconductive, a property that has found use in xerography.

Polyvinyl Chloride *n* (PVC) (1933) Poly(vinyl chloride), Poly(1-chloro-ethylene) A vinyl polymer which is similar to polyethylene, but on every other carbon in the backbone chain, one of the hydrogen atoms is replaced with a chlorine atom. It is produced by the free radical polymerization of vinyl chloride. A white, water-insoluble, thermoplastic resin, derived by the polymerization of vinyl chloride. A hard and tough plastic solid. Stabilizers are necessary to prevent discoloration from exposure to light and heat. Used for plastics and coatings. Commonly known as vinyl. Abbreviation is PVC (PVC).

$$\left[-CH_2-CH(Cl)- \right]_n$$

Polyvinyl Chloride Acetate *n* (PVAc) An important copolymer family of vinyl chloride and vinyl acetate, usually containing 85–97% vinyl chloride. These copolymers are more flexible and more soluble in solvents than PVC, and are used in solution coatings as well as in most of the processes and applications employing PVC.

Polyvinyl Chloride-co-Vinyl Acetate *n* A copolymer of vinyl chloride and vinyl acetate. Abbreviation: PVC/VAC.

Polyvinyl Dichloride See ▶ Chlorinated Polyvinyl Chloride.

Poly(vinylmethyl Ether) *n* (PVME, PVM poly(methylvinyl ether)) A family of polymers polymerized from Vinylmethyl ether, $H_2C=CHO-CH_3$. The range from viscous liquids to stiff rubbers. The liquids, soluble in cold water but not in hot water, are used in pressure-sensitive and hot-melt adhesives for paper and polyethylene, and as a Tackifier in rubbers. PVM also designates copolymers of vinyl chloride and Vinylmethyl ether.

Polyvinyl Fiber *n* A manufactured textile fiber developed in Japan. It is made by dissolving polyvinyl alcohol in hot water and extruding this solution through a spinneret into a sodium sulfate coagulating bath. In Japan, the fiber is used in apparel, household, and industrial fabrics.

Polyvinyl Fluoride n $(-H_2CCHF-)_n$. The polymer of vinyl fluoride (Fluoroethylene). The fluorine atom forms a strong bond along the hydrocarbon chain, accounting for properties such as high melting point, chemical inertness, and resistance to ultraviolet light. In the form of film, PVF is used for packaging, glazing, and electrical applications. Laminates of PVF film with wood, metal, and polyester panels are being used in building construction. Although it cannot be dissolved in ordinary solvents at room temperature, coating solutions can be made by dissolving PVF in hot "latent solvents" such as dimethyl acetamide and the lower-boiling phthalate, Glycolate, and Isobutyrate esters. Such solutions are used to protectively coat the insides of rigid metal containers for chemicals and industrial compounds. Abbreviation is ▶ PVF.

Polyvinyl Formal n (PVFO, PVFM) A member of the ▶ Polyvinyl Acetal family, made by condensing formaldehyde in the presence of polyvinyl alcohol or by the simultaneous hydrolysis and acetylation of polyvinyl acetate. It is used mainly in combination with cresylic phenolics for wire coatings, and impregnating, but can also be molded, extruded, or cast. It is resistant to greases and oils and to moderately high temperatures.

Polyvinyl Halide n A term sometimes used (almost exclusively in patents) for polymers and copolymers of vinyl chloride. Aside from polyvinyl fluoride, which is more similar structurally to polyethylene, and brominated butyl rubber, which has enjoyed some use in the automobile-tire industry, no polymers containing the other halogens (bromine, iodine, and astatine) exist in commerce.

Polyvinylidene Chloride n Poly(1,1-dichloroethylene) Poly(vinylidene chloride) is a vinyl polymer and is made from the monomer vinylidene chloride, using free radical vinyl polymerization. Copolymers with vinyl chloride (15% or more) are widely used as packaging and food-wrapping films under the name SARAN. Abbreviation is PVDC.

Poly(vinylidene chloride) See ▶ Polyvinylidene Chloride.

Polyvinylidene Fluoride n $(-H_2CCF_2-)_n$. Thermoplastic fluorocarbon polymer derived from vinylidene fluoride. It is a fluoropolymer with alternating CH 2 and CF 2 groups. PVDF is an opaque white resin. Extremely pure, it is superior for non-contaminating applications. In film form it is characterized by superior weather and UV resistance. Abbreviation is PVDF.

Poly(vinylidene fluoride) n Poly(1,1-difluoroethylene) Poly(vinylidene fluoride) is made by free radical vinyl polymerization of the monomer vinylidene fluoride Poly(vinylidene fluoride) of PVDF has a has very high electrical resistance, PVDF resists ultraviolet and is often blended with poly(methyl methacrylate) (PMMA) to make it more resistant to UV light. It is a piezoelectric material and when placed in an electric field will change its shape.

Polyvinylisobutyl Ether See ▶ Polyisobutylvinyl Ether.

Poly(vinyl isobutyl ether) n 1-(ethoxy)-2 propane. Any polymer of isobutylvinyl ether. Some are liquids, others are solid and crystalline. They are used as adhesives, surface coatings, laminating agents, and filling compounds.

Poly(4-vinyl pyridine) n 4-ethenyl pyridine.

Polyvinyl Pyrollidone n $(C_6H_9NO)_n$. White free-flowing amorphous polymer. It is soluble in water and organic solvents, and is compatible with a wide range of hydrophilic and hydrophobic resins. Abbreviation: PVP.

Poly(1-Vinylpyrrolidone) *n* [PVP, poly(*N*-vinyl-2-pyrrolidone)] A highly water-soluble polymer prepared by the addition polymerization of ▶ 1-Vinyl-2-Pyrrolidone, (for structure). Molecular weights range from 10,000 to 360,000. Solutions of the polymer are used as protective colloids and emulsion stabilizers, and it has been used as a substitute for human blood plasma. PVP films are clear and hard, but can be plasticized.

Polyvinyl Stearate *n* A wax-like polymer of vinyl stearate, of limited use in the plastics industry. However, the monomer is copolymerized with vinyl chloride, acting as an internal lubricant.

Polywater \ˈpä-lē-ˌwó-tər\ *n* [*poly*meric *water*] (1969) In the late 1960s, it was reported that the Soviet physicist, Boris Derjaguin, had discovered a polymeric form of water, formed by condensing ordinary water on the inside quartz capillary tubing of very fine bore. Properties of the polymer, dubbed *polywater*, were said to be thermal stability up to 50°C, density equal to 1.4 g/cm^3, i.e., 40% greater than that of ordinary water, and solidification to a glass-like state at -40°C. There were subsequent reports that US scientists had confirmed the existence of polywater. Later, however, the discoverer of "polywater" admitted that the substance he had created was actually impurities dissolved from the quartz tubes used in the experiment.

POM *n* Abbreviation for Poly(Oxymethylene). Also see ▶ Acetal Resin.

Pompey Red *n* Another name for ferric oxide. See ▶ Iron Oxides, Natural.

Pongee \ˌpän-ˈjē\ *n* {*often attributive*} [Chinese (Beijing) *běnjī*, fr. *běn* own + *jī* loom] (1711) (1) A thin, naturally tan-colored silk fabric with a knotty, rough weave. (2) A cotton fabric made from yarns spun from fine-combed staple and finished with a high luster. This fabric is used for underwear. (3) Fabrics like cotton pongee made from manufactured fibers.

Pontianak \ˌpän-tē-ˈä-ˌnäk\ Manila type of semifossil copal obtained from Borneo. By reason of its alcohol solubility, it is used in spirit varnishes, but after running it becomes oil-soluble and is sometimes used in oleoresinous varnishes.

Pony Mixer See ▶ Change-Can Mixer.

Popcorn \ˈpäp-ˌkórn\ *n* A name for nonuseful, hard, tough, insoluble polymer, resembling popcorn, formed by polymerization in the manufacture of synthetic rubbers.

Popcorn Polymerization *n* Polymerization reaction in which the material's molecular matrix has been penetrated due to vapor pressure by uninhibited monomer.

Poplin \ˈpä-plən\ *n* [F *papeline*] (1710) A plain-weave fabric of various fibers characterized by a rib effect in the filling direction.

Popping (1) Eruptions in a film of paint or varnish after it has become partially set so that craters remain in the film. (2) Of plaster. A mild form of blowing. (3) In the coil coating industry, a film defect manifested as a pinhole completely through the film.

Poppyseed Oil *n* (14c) Oil with only fair drying properties, obtained from the seeds of *Papaver somniferum*, which grows in India, Russia, and France. Its main constituent acid is linoleic, which is present to approximately 62%. The oil is little used in the trade, its main applications being for artistic purposes. Sp gr, 0.925/15°C; iodine value, 134, saponification value, 192.

POR Elastomer from propylene oxide and allyl glycidyl ether.

Porcelain Enamel See ▶ Vitreous Enamel.

Porch Paint See ▶ Deck Paint.

Pores \ˈpōr, ˈpór\ *n* [ME, fr. MF, fr. L *porus*, fr. Gk *poros* passage, pore] (14c) Minute openings (holes) in surface of cured goods. May refer to minute bubbles within the article. See ▶ Pinholes.

Pore Size *n* The size of the openings of filters or screens, usually expressed in micrometres.

Poromeric *n* (from micro*porous* and poly*meric*) A material that has the ability to transmit moisture vapor to some degree while remaining essentially waterproof. The first plastic material of this type was DuPont's "Corfam," introduced in 1963 and vigorously marketed as a leather substitute in she uppers, at which task, it enjoyed only mediocre success. It was a composite of urethane polymers and polyester fibers. The most successful poromerics are the fabrics known as Gore-Tex®, developed by W L Gore Associates, and applied widely to raincoats sport garments, and camping gear.

Poromerics *n* The microporosity, air permeability and abrasion resistance of natural and synthetic leather.

Porosity \pə-ˈrä-sə-tē\ *n* (14c) The ratio of the volume of voids contained within a sample of material to the total volume, solid matter plus voids, expressed as a fraction, void fraction or percentage of voids. See ▶ Absorbency.

Porous Mold *n* A mold that is made up of bonded or sintered aggregates (powdered metal, pellets, etc) in such a manner that the resulting mass contains numerous connected interstices or regular or irregular shape and size through which air may escape as the mold is filled.

Portland Cement *n* A hydraulic cement produced by pulverizing clinker consisting essentially of hydraulic calcium silicates and usually containing one or more of the forms of calcium sulfate as an interground addition.

Portland Cement Paint See ▶ Cement Paint.

Positive Crystals *n* A unlaxial crystal is optically positive if $\varepsilon > \omega$. A bilaxial crystal is said to be optically positive if $\gamma - \beta > \beta - \alpha$.

Positive Mold *n* A compression mold in which the pressure is applied wholly on the material, and which is designed to prevent the escape of any molding material.

Positron \ ▎pä-zə- ▎trän\ *n* [*posi*tive + *-tron* (as in elec*tron*)] (1933) A particle with the same mass M_e, as an electron. It has a positive electrical charge of exactly the same amount as that of an ordinary electron (which is sometimes called negatron). Positrons are created either by the radioactive decay of certain unstable nuclei or, together with a negatron, in a collision between an energetic (more than 1 Mev) photon and an electrically charged particle (or another photon). A positron does not decay spontaneously but on passing through matter it sooner or later collides with an ordinary electron and in this collision the positron–negatron pair is annihilated. The rest energy of the two particles, which is given by Einstein's relation $E = mc^2$ and amounts to 1.0216 ▶ mev altogether, is converted into electromagnetic radiation in the form of one or more photons. (Freir GD (1965) University physics. Appleton-Century-Crofts, New York; Handbook of chemistry and physics, 52nd edn. Weast RC (ed). The Chemical Rubber, Boca Raton, FL)

Post Cure *n* A treatment (normally involving heat) applied to an adhesive assembly following the initial cure, to modify specific properties. *v* To expose an adhesive assembly to an additional cure, following the initial cure, for the purpose of modifying specific properties.

Post-Curing *n* Completing the cure of a thermosetting casting or molding after removal from the mold in which a partial cure has been accomplished. Post-curing usually involves heating, for example, in a circulating-air oven.

Poster Color *n* Opaque water color (gouache) obtainable in pots or tubes; often used by poster designers. *Also called Tempera (a misnomer) and Show Card Colors; these are Water Paints with a Gum binder.*

Postforming (1) The heating and reshaping of a fully or partially cured laminate, on cooling, the formed laminate retains the contours and shape to which it has been postfomed. (2) Operations applied to still warm extrudates, particularly some types of profile extrusions, in which limbs of the extrudate pass through fixtures that bend or curl them into their final shapes.

Pot (1) *n* A chamber to hold and heat molding material for a transfer mold. (2) *v* See ▶ Potting.

Potash \ ▎pät- ▎ash\ *n* [singular of *pot ashes*] (ca. 1648) A common name for potassium or potassium compounds. Potash is generally used to mean potassium carbonate.

Potash Blue See ▶ Iron Blue.

Potassium \pə- ▎ta-sē-əm\ *n* {*often attributive*} [NL, fr. *potassa* potash, fr. E *potash*] (ca. 1807) A silver-white soft light low-melting univalent metallic element of the alkali metal group that occurs abundantly in nature especially combined in minerals.

Potassium Bromide \- ▎brō- ▎mīd\ *n* (1873) A crystalline salt KBr with a saline taste that is used as a sedative and in photography.

$$K^+ \quad Br^-$$

Potassium Carbonate *n* (1885) K_2CO_3. A white salt that forms a strongly alkaline solution and is used in making glass and soap.

Potassium Chlorate *n* (1885) $KClO_3$. A crystalline salt that is used as an oxidizing agent in matches, fireworks, and explosives.

Potassium Dichromate \-(▎)dī- ▎krō- ▎māt\ *n* (1885) $K_2Cr_2O_7$. A soluble salt forming large orange-red crystals used ion dyeing, in photography, and as an oxidizing agent.

Potassium Hydroxide \-hī- ▎dräk- ▎sīd\ *n* (1885) A white deliquescent solid KOH that dissolves in water with much heat to form a strongly alkaline and caustic liquid and is used chiefly in making soap and as a reagent.

$$K^+ \quad OH^-$$

Potassium Permanganate \-(ˌ)pər- ˈmaŋ-gə- ˌnāt\ *n* (1869) KMnO₄. A dark purple salt used as an oxidizer and disinfectant.

$$O=Mn(=O)(-O^-) \quad K^+$$

Potassium Titanate \- ˈtī-tᵊn- ˌāt\ *n* $(K_2O)_{1/x}(TiO_2)_4$. The value of x in the formula is greater than 1.0 because the commercial pigment has the crystal structure of a leached tetratitanate. It is an acicular white hiding pigment that has high scattering power, and its ultraviolet reflection is significantly higher than that of commercial titanium dioxide. It has been used in the paper industry and in vinyl plastics. Density, 3.3 g/cm³ (27.5 lb/gal); O.A., approximately 80; particle size, diameter, 0.2 μm, length, 8–10 μm. Abbreviation: PKT.

$$K^+ \; ^-O-Ti(=O)(-O^-\,K^+)$$

Potassium Titanate Fiber *n* K₂O(TiO₂) in wherein n = 4 to 7. Highly refined, single crystals, approximately 6 μm long by 0.1 μm in thickness, used as reinforcing fibers in thermoplastic composites. The fibers melt at 1370°C; density is 3.2 g/cm³. They also act as white pigments.

Potential *n* (**Electric**) At any point is measured by the work necessary to bring unit positive charge from an infinite distance. Difference of potential between two points is measured by the work necessary to carry unit positive charge from one to the other. If the work involved is one erg then there is the electrostatic unit of potential. The potential at a point due to charge *q* at a distance *r* in a medium whose dielectric constant is *e* is,

$$V = \frac{q}{er}.$$

(Freir GD (1965) University physics. Appleton-Century-Crofts, New York)

Potential Energy *n* Energy associated with the position or configuration of an object.

Pot Life *n* (working life) (1) The period after mixing the ingredients of an active compound during which the viscosity remains low enough to permit normal processing. Dissatisfied with the vagueness of this definition, Breitigam and Ulrich (August, 1990) cautiously defend pot life or their epoxy compounds as the time, at room temperature, for the viscosity to reach double its initial value. (2) The length of time a paint material is useful after its original package is opened, or after catalyst or other ingredients are added. *Also called Usable Life, Spreadable Life.* See ▶ Working Time.

Pot Plunger *n* A plunger used to force softened molding material from the pot into the closed cavity of a transfer mold.

Pot Retainer *n* A plate channeled for passage of a heat-transfer medium (e.g., hot oil) and used to hold the pot of a transfer mold.

Pot Spinning *n* A method formerly used for making viscose rayon. The newly spun yarn was delivered into the center of a rapidly rotating, centrifugal pot, where it received twist and centrifugal force caused it to go to the wall of the pot. The yarn package so formed was called a cake.

Potting (1) *n* The process of encasing an article or assembly in a resinous mass, performed by placing the articles in a container that serves as a disposable mold, pouring a liquid resin into the mold to completely submerge the article, then curing the resin. The container remains attached to the potted article. The main difference between potting and ENCAPSULATION is that in the latter the mold is removed from the encapsulated article and reused. These processes are widely used in the electronics industry. (2) *v* The act of potting or "potting an article."

Potting Syrup See ▶ Casting Syrup.

Pounce \ ˈpaún(t)s\ (1) *n* Design on paper which has been formed by pricking it out with a sharp pointed instrument. The design is transferred to the surface on which it is to be painted, by laying the paper on the latter and shaking on it to a dry powdered pigment through a linen bag called a pounce bag. Some of the pigment passes through the pinholes in the paper and forms a replica of the design on the surface where it can then be rendered in paint. (2) *v* To transfer the outlines of a drawing by applying a dry powder through small perforations made in the outline.

Poundal \ ˈpaún-dᵊl\ *n* [¹*pound* + -*al* (as in *quintal*)] (1879) An obsolete, but still occasionally seen, unit of force (the force required to accelerate a 1-lb mass 1 ft/s²), analogous to the dyene in the cgs system, both created many years ago to perpetuate the illusion that Newton's law of momentum change needs no proportionality constant. See ▶ Force.

Pour Point *n* Temperature at which materials posses a defined degree of fluidity. One particular method of determining pour point involves recording the times of efflux of a specified volume of the molten product through an orifice of standard dimensions. From the figures thus obtained it is possible to arrive at a temperature which permits a standard volume of product to flow through the orifice in a specified time. This temperature is the pour point.

Powder Blend See ▶ Dry Blend.

Powder Bonded Nonvoven *n* A manufactured product in which a carded web is produced and treated with a thermoplastic powder that has a melting point less than that of the fiber in the web. The powder is heated to its melting point by through-air and infrared heating or by hot-calendering to effect bonding.

Powder Coatings *n* (1) A 100% solids coating applied as a dry powder and subsequently formed into a film with heat. (2) A coatings application method which utilizes a solid binder and pigment. The solid binder melts upon heating, binds the pigment and results in a pigment coating upon cooling. See also ▶ Fluidized Bed Coating.

Powder Compact *n* A molding material in the form of dry, friable pellets prepared by compacting dry-blended mixtures of resin (typically PVC) with plasticizers and other compounding ingredients. The powder compacts are about as easy to handle and process by extrusion as ▶ Pellets and offer the advantages of lower heat history and somewhat lower cost than equivalent materials in the form of fused pellets. See also ▶ Dry Blend.

Powder Density See ▶ Bulk Density.

Powdered Distemper See ▶ Calcimine.

Powdered Plastic *n* A resin or plastic compound in the form of extremely fine particles, for use in fluidized-bed coating, rotational molding, and various sintering techniques.

Powdered Quartz See ▶ Silica, Crystalline.

Powdering See Chalking powdering Of Paints See Chalking.

Powdering (of Polishes) The partial or total disintegration of the polish film resulting in a fine, light-colored material.

Powder Metallurgy *n* (1933) A branch of science or an art concerned with the production of a powdered metals or of metallic objects by compressing a powdered metal or alloy with or without other materials and heating without thoroughly melting to solidify and strengthen.

Powder Molding *n* A general term encompassing rotational molding, slouch molding, compression molding, and centrifugal molding of dry, sinterable powders such as polyethylene, nylon, PVC, polytetrafluoroethylene and ultra-high-molecular-weight polyethylene. The powders are charged into molds that are heated, and manipulated or pressed, according to the process being used. These actions cause the powders to sinter or fuse into a uniform layer or molding against the mold surfaces, which are then cooled. See also ▶ Fluidized-Bed Coating, Rotational Molding, Centri-Fugal Molding, Slush Molding, and Sinter Molding.

Powell-Eyring Model *n* (Eyring-Powell model) A complex rheological equation containing three parameters that must be evaluated by fitting experimental flow data. It has the form:

$$\tau_{xz} = C\left(\frac{dv_z}{dx}\right) + \frac{1}{B}\sinh^{-1}\left[\frac{1}{A}\left(\frac{dv_z}{dx}\right)\right]$$

where τ_{xz} is the z-directed shear stress perpendicular to x, v_z is the velocity in the z-direction, dv_z/dx is the shear rate at x, and A, B, and C are temperature-dependent constants characteristic of the flowing medium. The model calls for Newtonian-flow regions at both very low and very high shear rates, a type of behavior seen in a few polymer solutions but rarely in melts. The limiting viscosity at low heat rate is given by C + 1/AB, while the high-shear limiting viscosity is C. Applying this model to even simple flow geometries is cumbersome, especially when it is to be used to find shear rates, velocities, and flow rates from known stresses, so it has not seen much use. Compare ▶ Eyring Model. (Whittington's dictionary of plastics. Carley James F (ed). Technomic, 1993)

Power \ˈpaú(-ə)r\ *n* [ME, fr. OF *poeir*, fr. *poeir* to be able, fr, (assumed) VL *potēre*, alter. of L *posse*] (13c) The rate at which work is being done or energy expended. The SI unit is the watt (W), equal to 1 joule/second (J/s). Some conversions of other units are given in the Appendix. In purely resistive, direct-current electric circuits, power is given by the product of voltage drop times current ($\Delta e \times i$) and the watt is equal to 1 volt-ampere (V A). In sinusoidally alternating circuits, power is given by $\Delta e \, i \cos \phi$, in which ϕ is the phase angle between the current and the voltage.

Power *n* (in Watts for Alternating Current) $P = EI \cos \phi$ where E and I are the effective values of the electromotive force and current in volts and amperes respectively and φ the phase angle between the current and the impressed electromotive force. The ratio,

$$\frac{P}{EI} = \cos \phi$$

is called the power factor.

Power Cleaning See ▶ Blast Cleaning.

Power Developed by a Direct Current *n* The power in watts developed by an electric current flowing in a conductor, where E is the difference of potential at its terminals in volts, R its resistance in ohms, and I the current in amperes,

$$P = EI = RI^2$$

The work done in joules in a time t s is,

$$W = EIt : RI2t.$$

(Giambattista A, Richardson R, Richardson RC, Richardson B (2003) College physics. McGraw Hill Science/Engineering/Math, New York)

Power Factor *n* The ratio of actual power (wattage) being used in an alternating circuit to product of voltage drop (Δe) times current, (i), usually expressed as a percentage. When the load in the AC circuit is purely resistive, as with ovens and incandescent lamps, the wattage equal $\Delta e \cdot i$ and the power factor is 100%. When the load includes inductive elements such as motors and transformers, the current lags the voltage by the phase angle ϕ, which can range from 0 for a purely resistive load to 90° for a load that is wholly inductive, and the power factor is referred to as a *lagging* power factor. When the load in the AC circuit contains capacitive elements, the current *leads* the voltage by the phase angle and the term *leading* power factor is used. In a circuit containing all three types of elements, the net effect will generally be current lagging or leading, but with a relatively smaller phase angle. The difference between actual power and $\Delta e\, i$ is called *reactive power*, which increases as the power factor decreases. Reactive power does no useful work but costs the same as actual power used. Thus, low power factors increase power costs, cause overloading of motors and transformers, and reduce load-handling capacity of plant electrical systems. (2) In testing the behavior of plastics as dielectrics, power factor is the cosine of the phase angle when the voltage across the capacitor varies sinusoidally. In a perfect dielectric (pure capacitance), the current would lead the voltage and the phase angle (ϕ) would be 90°, it cosine 0. When loss occurs, the phase angle is 90 - δ, where δ = the *loss angle*, hence cos ϕ = sin δ. In the literature, tan δ is often called the power factor. In capacitor application, δ is usually very small, so the difference between sine and tangent is negligible. This might not be so in dielectric heating, say, or phenolics or vinyls, where power factors are higher. Dielectric loss depends on frequency. Because it is generated by oscillatory movement of molecular and atomic dipoles within the material, the loss spectrum over the frequency range of many decades will usually show one or more maxima and minima. (Ku CC, Liepins R (1987) Electrical properties of polymers. Hanser, New York)

Power Law *n* (Ostwald-deWaele model) The simplest representation of pseudoplastic flow, and characteristic of most polymer melts over several decades of shear rate. One versatile form of the model is

$$\tau_{xz} \infty - m \left| \frac{dv_z}{dx} \right|^{n-1} \left(\frac{dv_z}{dx} \right)$$

where τ_{xz} is the z-directed shear stress perpendicular to x, v_z = the velocity in the z-direction, dv_z/dx is the shear rate at x, and m and n are constants peculiar to the liquid. n is called the *flow-behavior index* and has a value between 0.25 and 0.9 for most polymer melts. The quantity m is analogous to viscosity and is temperature-dependent. For n = 1, the power law reduces to Newton's law of flow and $m = \mu$, the Newtonian viscosity. Over the limited range of shear rates occurring in a given process, the power law can often provide a sufficiently accurate approximation to the actual flow behavior. Chemical engineers often cast the power law into the simpler form:

$$\frac{\Delta P \cdot D}{4L} = K \left(\frac{8V}{D} \right)_n$$

in which the left side is the shear stress at the wall of a pipe of diameter D and length L, ΔP is the pressure drop over that length of pipe, K is a viscosity-like property (temperature-dependent), V = the average liquid velocity, and n is the flow-behavior index of the liquid. The quantity ($8V/D$) is the apparent Newtonian shear rate at the tube wall. (Patton TC (1964) Paint flow and pigment dispersion. Interscience, New York; Goodwin JW, Goodwin J, Hughes RW (2000) Rheology for chemists. Royal Society of Chemistry, August 2000)

Power Ratio *n* In telephone engineering are measured in *decibels*. The gain or loss of power expressed in decibels is ten times the logarithm of the power ratio. By reference to an arbitrarily chosen "power level" the actual power may be expressed in decibels. The numerical values thus used will not be proportional to the actual power level but roughly to the sensation on the ear produced when the electrical power supply to a telephone receiver produces approximately the smallest change in volume of sound which a normal ear can detect. (Serway RA, Faugh JS, Bennett CV (2005) College physics. Thomas, New York)

POY See ▶ Partially Oriented Yarns.

Po-yok Oil *n* An oil derived from *Afrolicania elaeosperma* and *Parinarium sherbroense*, which occur in West Africa. It is a drying oil since its main constituent acids are licanic and elaeostearic, but it has never become available in quantity. It bodies rapidly on heating and behaves generally something like a mixture of oiticica and tung oils. Sp gr, 0.9612/15°C; iodine value, 150; saponification value, 188; refractive index, 1.5082. *Also spelled "po-yoak" oil.*

PP *n* Abbreviation for Poly(Propylene).

PPA See ▶ Polyphthalamide.

ppb *n* Abbreviation for Parts Per Billion.

PPG *n* Abbreviation for Polyoxypropylene Glycol.

PPI *n* Abbreviation for Parts Per Million.

PPMI *n* Abbreviation for Polypyromellitimide.

PPO *n* Poly(2,6-dimethyl phenylene oxide. Manufactured by General Electric, US *n* Abbreviation for Poly (Phenylene Oxide).

PPOX *n* Abbreviation for Polypropylene Oxide. See ▶ Polypropylene Glycol.

PPS *n* Abbreviation for Polyphenylene Sulfide.

PPSS See ▶ Polyphenylene Sulfide Sulfone.

PPSU *n* Abbreviation for Poly(*p*-Phenylene Sulfone).

PRA *n* Abbreviation for Paint Research Association (British).

Prandtl Number *n* (Pr, Np$_r$) A dimensionless group important in the analysis of convection heat transfer, defined as (in consistent units) $C_p\mu/k$, where C_p = the specific heat of a fluid at constant pressures, μ = its viscosity, and k = its thermal conductivity. The Prandtl number is also the ratio of the kinematic viscosity to the thermal diffusivity (see both entries).

Prebond Treatment *n* See ▶ Surface Preparation.

Precipitate \pri-ˈsi-pə-ˌtət\ *n* NL *praecipitatum*, fr. L, neuter of *praecipitatus* (1594) Substances separated from a solution in solid form by application of cold or heat or by a chemical reaction.

Precipitated Barium Sulfate See ▶ Barium Sulfate.

Precipitated Basic Dye Blues *n* The more commonly used pigments of this group are the so-called PTA, PMA, and PTMA pigments. They are so designated because they comprise the precipitated products of the reaction between basic dyes (e.g., Victoria Blue B) and complex inorganic acids such as phosphotungstic and phosphomolybdic acids or their mixture. These pigments as a group are nearly black in masstone, accordingly, they are used chiefly for tinting purposes. In spite of certain deficiencies (e.g., poor bleed resistance and poor lightfastness on outdoor exposure), the brilliance and high tinting strength of these pigments make them suitable for various interior paints of more-or-less special character (e.g., foil lacquers, poster paints and some toy enamels). In general, PTA pigments are more lightfast than corresponding PMA pigments, although the latter are somewhat superior in tinting strength.

Precipitated Basic Dye Violets *n* The general properties of the phosphotungstic or molybdic acid violet pigments in paint are essentially the same as those discussed under Precipitated Basic Dye Blue in the section not adequate for exterior exposure. About its only application in paint would be lead-free finishes for interior use.

Precipitated Calcium Carbonate See ▶ Calcium Carbonate, Synthetic.

Precipitated Driers *n* Metallic soaps derived from the interaction of aqueous solutions of the alkali soaps of the drier acids, and of the metallic salts. The drying soaps are obtained as precipitates, and are usually characterized by much paler colors than those made by the fusion method.

Precipitation \pri-ˌsi-pə-ˈtā-shən\ *n* (1502) The formation of a condensed phase (solid or liquid) during a reaction.

Precipitation Number *n* Measure of the amount of solid matter precipitated from oil in a test. The number of millilitres of solid matter formed in a certain amount of mixture of oil and solvent.

Precipitation Polymerization *n* A polymerization reaction in which the polymer being formed is insoluble in its own monomer or in a particular monomer-solvent combination and thus precipitates as it is formed.

Precision \pri-ˈsi-zhən\ *n* (1740) (as distinguished from accuracy) The degree of mutual agreement between individual measurements, namely repeatability and reproducibility.

Preconditioning *n* Bringing a sample or specimen of textile material to a relatively low moisture content (approximate equilibrium in an atmosphere between 5% and 25% relative humidity) prior to conditioning in a controlled atmosphere of higher humidity for testing. (While preconditioning is frequently translated as predrying, specimens should not be brought to the overdry state.)

Precure *n* A partial or full state of cure existing in an elastomer or thermosetting resin prior to its use as an adhesive or in a forming operation.

Precursor \pri-ˈkər-sər\ *n* [ME *precursoure*, fr. L *praecursor*, fr. *praecurrere* to run before, fr. *prae-* pre- + *currere* to run] (15c) One who or that which precedes and suggests the course of future events. A compound or polymer that is later transformed into another material or polymer by chemical reaction.

Predrying *n* The drying of a resin or molding compound prior to its introduction into an extruder, a mold, or molding machine. Many resins and plastics compounds are hygroscopic and require this treatment, to prevent formation of bubbles in the product, particularly after exposure to a humid atmosphere. Predrying for extrusion or injection molding is usually accomplished by passing heated, bone-dry air up through the bed of pellets in an enclosed feed hopper. This has a bonus of reducing the heat input required from the extruder drive and boost extruder output. The exit, moistened air is recycled through a dryer packed with silica gel or other drying agent.

Prefabrication Primer *n* Quick-drying material applied as a thin film to a metal surface after cleaning. e.g., by a blast cleaning process, to give protection during the period before and during fabrication. Prefabrication primers should not interfere seriously with conventional welding operations or give off toxic fumes during such operations. Syn: ▶ Shop Primer.

Preform *n* (1) The "test tube" shape that is used to form the final blown product in injection blow molding, (2) a compressed tablet or biscuit of plastic composition used for efficiency in handling and accuracy in weighing materials, particularly thermosets, (3) foamed perform, (4) a preshaped fibrous reinforcement formed by the distribution of chopped fibers or cloth by air, water flotation, or vacuum over the surface of a perforated screen to the approximate contour and thickness desired in the finished part, and (5) a preshaped fibrous reinforcement of mat or cloth formed to the desired shape on a mandrel or mock-up before being placed in a mold press.

Preform \ˈprē-ˌfórm\ *vt* [L *praeformare*, fr. *prae-* + *formare* to form, fr. *forma* form] (1601) (1) The "test tube" shape that is used to form the final blown product in injection blow molding. (2) A compressed tablet or biscuit of plastic composition used for efficiency in handling and accuracy in weighting materials, particularly thermosets. (3) Formed perform. (4) A preshaped fibrous reinforcement formed by the distribution of chopped fibers or cloth by air, water flotation, or vacuum over the surface of a perforated screen to the approximate contour and thickness desired in the finished part, and (5) a preshaped fibrous reinforcement of mat or cloth formed to the desired shape on a mandrel or mock-up before being placed in a mold press.

Preform *n* (1) A compressed, shaped mass of plastic material or fibrous reinforcing material or a combination of both, prepared in advance of a molding operation for convenience in handling or for accuracy of loading by weighing the mass. The term also applies to tablets and biscuits of thermoplastic and thermosetting compounds. (2) In the reinforced-plastics industry, a preform is a mat of chopped strands boned together by a resin in approximately the shape of the end product, for use in processes such as matched-die molding, Or it may be a complex shape made by two- or three-dimensional weaving or braiding that, when wet out with resin and cured, will become the finished product. See ▶ Near-Net-Shape Configuration.

Preform *v* (1) To make plastic molding powders into pellets, tablets, or biscuits of known mass that facilitate accuracy in compression molding. (2) To prepare by hand cutting of reinforcing cloth or mat, or by blowing chopped fibers onto a contoured screen, the reinforcement for a fiber-reinforced molded object. The reinforcement, which has a shape close to that of the final molded object, is placed into or onto the mold along with the required amount of resin, then wet out and cured.

Preform Binder *n* A light application of resin applied to a mat or screened preform that provides enough shape stability to permit handling the preform into the mold without tearing the mat or shifting the fiber distribution.

Preform Molding See Matched-Metal-Die Molding.

Pregel *n* An unintentional, prematurely cured, or partially cured layer of resin on part of the surface of reinforced plastic prior to molding. Should not be confused with ▶ Gel Coat.

Pregl Method See ▶ Combustion Analysis.

Preheating *n* Heating of feedstock or material to be processed prior to the main processing step. In extrusion of coated wire, the wire is resistively heated just before entering the die to ease melt flow through the die and maintain coating quality at high wire speeds. (Compare ▶ Preheat Roll.) In compression molding with preforms, the preforms are commonly preheated electronically before being loaded into the mold, thus improving the flow, reducing curing time in the mold, and shortening the cycle, In some extrusion and injection-molding operations, pellet feedstocks are dried in the hopper with hot air, this predrying of the feed not only precludes splay marks, bubbles, etc., caused by moisture in the melt, but significantly reduces the amount of energy per unit mass of plastic that must be furnished by screw action, permitting higher throughputs with lower screw-energy input per unit of product delivered. (Strong AB (2000) Plastics materials and processing. Prentice Hall, Columbus, OH)

Preheating Hopper See ▶ Hopper Dryer.

Preheat Roll *n* In extrusion coating, a heated roll installed between the pressure roll and unwind roll, the purpose of which is to heat the substrate before it is coated, thereby providing better adhesion of the coating and permitting lower melt temperature and, possibly, a higher production rate.

Prehistoric Art *n* Painting and sculpture produced by artists of the Old, Middle, and New Stone ages. See ▶ Cave Painting. The earliest known piece of prehistoric sculpture is the famous "Venus of Willendorf" (Natural History Museum, Vienna), a small fertility image of Paleolithic origin dating around 11,000 BC.

Preimpregnation *n* A method of preparing fiber-reinforced molding material by forcing thermoplastic resin, or thermosetting resin advanced only to the B-stage, into mats or cloths of fiber reinforcement. The product, called a Prepreg is ready for molding, but storable for periods up to several months and is shippable.

Premature Vulcanization *n* Uncontrolled curing or setting up of material before final cure. See ▶ Bin Cure.

Premix *n* ("gunk") A term originally applied to mixtures of polyester resin with sisal or glass fiber reinforcement and fillers, usually prepared by molders shortly before use. The ASTM definition (D 883) specifies that the premix should not be in web or filamentous form. The term premix is now often used by molding compounds of any thermosetting resin mixed with fillers, reinforcements, and catalysts. (Strong AB (2000) Plastics materials and processing. Prentice Hall, Columbus, OH)

Premix Molding *n* ("gunk" molding) A variation of matched-die molding in which the ingredients, usually chopped roving, resin, pigment, filler and catalyst, are premixed and divided into accurately weighted charges for molding.

Preoxidized Fiber *n* In carbon fiber production, a fiber that results from a relatively low temperature (200–500°C) heat treatment in the presence of oxygen which converts the precursor fiber, PAN or rayon, to an infusible fiber that is stable to further processing.

Prepared Linseed Oil *n* In the printing ink industry, linseed oil which has been treated with litharge and other chemicals.

Prepasted *n* Adhesive applied to the back of wallcovering by the manufacture. Dipping in water before hanging activates the paste.

Preplasticization *n* In plunger-type injection molding, the technique of premelting molding powders in a separate chamber, then transferring the melt to the injection cylinder. The technique shortened molding cycles and provided a more homogeneous melt entering the mold. With the widespread adoption of screw-injection machines, the need for separate preplasticization has fallen sharply to a few special circumstances.

Prepolymer *n* A polymer of relatively low molecular weight, usually intermediate between those of the monomer or monomers and the final polymer or resin that may be mixed with compounding additives, and that is capable of being hardened by further polymerization during or after a forming process.

Prepolymer Molding *n* In the polyurethane-foam industry, a system whereby a portion of the polyol is prereacted with the isocyanate to form a liquid prepolymer with a viscosity suitable for pumping or metering. This component is supplied to end-uses with a second premixed blend of additional polyol catalyst, blowing agent, etc. When the two components are vigorously mixed, foaming and crosslinking occurs. For a contrasting method, see ▶ One-Shot Molding.

Prepreg \ˈprē-ˈpreg\ *n* [*pre-* + im*preg*nated] (1954) In the reinforced-plastics industry, a mat or shaped mass of reinforcing fibers, typically glass strands, impregnated with a thermosetting resin advanced in cure only through the B-stage. Such prepregs may be stored until needed for a molding or laminating operation. A prepreg containing a chemical thickening agent is called a Mold-Mat. The term "prepreg" also includes fabrics such as jute coconut fiber, or rayon yarn impregnated with a thermoplastic resin, e.g., vinyl, acrylonitrile-butadiene-styrene, or acrylic. For sheet forms, the term "prepreg" is being displaced by the more specifically descriptive "sheet molding compound (SMC)."

Prepreg Molding *n* A type of Matched-Metal-Die Molding in which the fibrous mat is preimpregnated with a partially cured, thermosetting resin.

Preprinting *n* In sheet thermoforming, the inversely distorted printing of sheets before they are formed. During forming, the stretching of the sheet brings the print into its proper size and spacing.

Preproduction Test *n* A test or series of tests conducted by (1) an adhesive manufacturer to determine conformity of an adhesive batch to established production standards, (2) a fabricator to determine the quality of an adhesive before parts are produced, or (3) an adhesive specification custodian to determine conformance of an adhesive to the requirements of a specification not requiring qualification tests.

Presensitized Plate *n* In photomechanics, a metal or paper plate that has been precoated with a light-sensitive coating, e.g. presensitized lithographic plate.

Preservative \pri-ˈzər-və-tiv\ *adj* (14c) A chemical incorporated in a material to prevent deterioration, mainly by living organisms, but more generally, also by heat, oxidation or weather. See also ▶ Antioxidant, ▶ Fungicide, and ▶ Stabilizer.

Press Cake *n* A pigment dispersed in water (obtained directly from a filter press) in which a water-insoluble resin is emulsified. A solvent-based dispersion is made by breaking this emulsion and removing the water.

Pressley Index A measure of the strength of fiber bundles determined under prescribed conditions and expressed in an arbitrary unit, pounds per milligram.

Press Polishing *n* (plainshine) A finishing process used to impart high gloss and improved clarity and mechanical properties to sheets of vinyl, cellulosic, and other thermoplastics. The sheets are hot-pressed against thin, highly polished metal plates.

Pressure \│pre-shər\ *n* [ME, fr. LL *pressura*, fr. L, action of pressing, pressure, fr. *pressus*, pp of *premere*] (14c) Force exerted over an area, expressed as force per unit area. The SI unit is the pascal (Pa), equal to 1 N/m^2, the same as the unit of stress. Since our pervading atmosphere keeps us all under a pressure of about 101 kPa, many pressure-sensing devices detect and indicate the "difference" between a process pressure and atmospheric, called "gauge" (or gage) pressure. Pressure referred to total vacuum is "absolute pressure." When exerted by solid contact, as by a ram on an elastic surface, pressure may vary over the contact area. Pressures of confined gases at rest are equal, everywhere within the container while liquid pressures can depend significantly on depth because of density and gravity.

Pressure Bag Molding See ▶ Bag Molding.

Pressure Break *n* As applied to a defect in a laminated plastic, a break apparent in one or more outer sheets of the paper, fabric, or other base visible through the surface layer of resin that covers it.

Pressure Drop *n* (1) A decrease in pressure that is caused by friction between a flowing liquid and a constricting container. The pressure drop is increased by a reduction in diameter of the container. (2) The change in pressure across a filter.

Pressure Dyeing See ▶ Dyeing.

Pressure Flow *n* (1) In general, any flow that is driven by a pressure gradient along a flow path, including any vertical component due to gravity's action on the fluid density. Flows through orifice-type rheometers and extrusion dies are pressure flows. (2) Specifically, in the metering section of an extruder screw, the rearward flow ("back flow") that would occur *if* the screw were not rotating and the pressure gradient were unaltered. In the rotating screw, pressure flow opposes the productive drag flow, reducing net output, but can never exceed it. In Newtonian flow equations for extruders the pressure flow is subtracted from the drag flow to obtain the net flow (throughput). In the actual plastics extruder, the nonNewtonian character of the melt invalidates the algebraic summing, yet because pressure flow is usually a third or less of the drag flow in a well designed system, the errors of this algebraic summing are seldom serious. If the simplified flow equation overstates the actual output, the difference is more likely to be due to insufficient feeding or poor plasticating action than the equation errors. On the other hand, there have been instances, with screws of high compression ratio, where the pressure of the melt entering the metering section was as high or higher than that at the die. In that case, there may be little or no pressure flow or even *positive* (forward) pressure flow. See also ▶ Drag Flow and ▶ Net Flow. (Strong AB (2000) Plastics materials and processing. Prentice Hall, Columbus, OH; Pittance JC (ed) (1990) Engineering plastics and composites. SAM International, Materials Park, OH)

Pressure Forming *n* A variant of sheet thermoforming in which pressure above atmospheric is used to push the heat-softened sheet against the mold surface, as opposed to using only a vacuum to suck the sheet against the mold. The cycle may be shortened. Pressure forming has been effectively used to form container lids form biaxially oriented polystyrene sheet without losing the orientation and good strength properties accruing therefrom. See ▶ Trapped-Sheet Forming.

Pressure Mark See ▶ Finger Mark.

Pressure Marking *n* Glossy or dull spots which become apparent as a strip is uncoiled. This is usually due to an undercured or soft film. Also can be caused by improper plasticizer balance.

Pressure Motting *n* The film distortion or uneven pattern that causes a change of gloss and a nonuniform appearance in the coated surface, as opposed to blocking. See ▶ Pressure Marking.

Pressure Pad *n* A reinforcement of hardened steel, several of which may be distributed around the dead area in the faces of a mold to help the land absorb the final pressure of closing without collapsing.

Pressure Roll In extrusion coating, a roll that presses the coating and substrate together to form a strong bond, continuous over the entire interface.

Pressure-Sensitive Adhesive *n* (PSA) An adhesive that develops a strong bond to most surfaces by applying only a moderate pressure. See ▶ Adhesive, Pressure-Sensitive.

Pressure Transducer *n* An instrument that converts a sensed fluid pressure into an electrical signal that in turn can be converted to a pressure reading and recorded. Transducers for extruders presented a difficult problem of temperature compensation and need for extreme ruggedness in service, and were pioneered by Dynisco in the 1950s. Several reliable makes are not available.

Pretension \prē-ten(t)-shən\ *vt* [*pre-* + *²tension*] (1937) The relatively low tension applied to remove kinks and crimp when mounting a specimen preparatory to making a test or to a textile processing operation, etc.

Pretreatment *n* Usually restricted to mean the chemical treatment of unpainted metal surfaces before painting.

Pretreatment Primer See ▶ Wash Primer.

Pretrimmed Papers Rolls of wallpaper form which selvage has been trimmed at factory.

PRI Abbreviation for Plastics and Rubber Institute (UK).

Primary \prī-mer-ē\ *adj* [ME, fr. LL *primarius* basic, primary, fr. LL, principal, fr. *primus*] (15c) In chemistry, a functional group at the end of a molecule's chain (or branch) in which only one of the hydrogen atoms has been replaced by some other link, as a primary alcohol, $-CH_2OH$, or primary amine, $-NH_2$. (Pittance JC (ed) (1990) Engineering plastics and composites. SAM International, Materials Park, OH)

Primary Amine Value *n* The number of milligrams of potassium hydroxide equivalent to the primary amine basicity in 1 g of sample.

Primary Backing *n* The material, usually woven or nonwoven polypropylene or jute, into which a carpet is tufted. The primary backing allows the positioning of each tuft and holds the tufts in position during processing, after which a secondary backing (q.v.) is applied to provide dimensional stability.

Primary Colors *n* (1612) In theory, those colors from which all other colors and white may be made. The primary colors in visible light are red, green, and blue. The so-called pigment primaries, each absorbing a light primary, would then be blue green cyan (minus red), magenta (minus green), and yellow (minus blue). Because of deficiencies in the available cyan and magenta colorants, confusion developed, so that red, yellow, and blue are now often referred to as the pigment primaries. See ▶ Primary Colors, Additive and ▶ Primary Colors, Subtractive. (Syszecki G, Stiles WS (1967) Color science: Concepts and methods, quantitative data and formulas. Wiley, New York; Billmeyer FW, Saltzman M (1966) Principles of color technology. Wiley, New York)

Primary Colors, Additive *n* Three colored lights from which all other colors can be matched by additive mixture. The three must be selected so that no one of them can be matched by mixture of the other two. Generally, a red, a green, and a blue are used. Additive primaries are the complements of the subtractive primaries.

Primary Colors, CIE *n* Red, green, and blue (violet) primaries defined by the CIE in terms of spectral distribution curves. They are imaginary primary lights so selected that all possible real colors can be matched by computation with positive amounts, to avoid the use of negative amounts, which are required to match all colors when using real colored lights. (Billmeyer FW, Saltzman M (1966) Principles of color technology. Wiley, New York)

Primary Colors, Subtractive *n* Colors of three colorants or colored materials which, when mixed together subtractively, result in black or a very dark neutral color. Subtractive primaries are generally cyan, magenta, and yellow, the three basic colorants used in printing, for example.

Primary Creep *n* The recoverable component of creep. Also see ▶ Delayed Deformation.

Primary High Polymer *n* One which is produced directly from small molecules, without chemical alteration subsequent to the polymerization. See ▶ Derived High Polymer.

Primary Plasticizer *n* A plasticizer that, within reasonable compatibility limits, may be used as the sole plasticizer, is completely compatible with the resin, and is sufficiently permanent to produce a composition that will retain its desired properties under normal service conditions throughout the expected life of the article.

Prime Pigments *n* Pigments which possess colorant value and hiding power. The refractive index of these pigments is 2.0 or higher in contrast to extended pigments.

Primer \prī-mər\ *n* (1819) First complete coat of paint of a painting system applied to a surface. Such paints are designed to provide adequate adhesion to new surfaces and are formulated to meet the special requirements of the surfaces. The type of primer varies with the surface, it condition, and the total painting system to be used. Thus, primers for new wood and certain other surfaces must provide for exceptional absorption of the medium. Primers for steelwork contain special anticorrosive pigments, such as red lead, zinc chromate, zinc powder, etc. See ▶ Metal Primer and ▶ Plaster Primer. A coating applied to a surface, prior to the application of an adhesive, to improve the performance of the bond. (Martens CR (1968) Technology of paints, varnishes and lacquers. Reinhold, New York; Paint / coatings dictionary. Compiled by Definitions Committee of the Federation of Societies for Coatings Technology, 1978)

Primer Surfacer See ▶ Surfacer.

Priming *n* The application of a primer.

Priming Paints See ▶ Primer.

Primitive Cell *n* A unit cell which has entities (atoms, molecules, ions) only at the corners of the cell.

Primrose Chrome *n* Complex, primrose-colored chromates to with the formula $PbCrO_4 \cdot 4PbSO_4 \cdot Al_2(OH)_6$ has been given. See ▶ Chrome-Yellow Pigment.

Primrose Yellow See ▶ Chrome-Yellow Pigment.

Principal Focus *n* In a lens or spherical mirror, the point of convergence of light coming from a source at an infinite distance.

Principal Quantum Number *n*, *n* A quantum number which specifies a shell for an electron in an atom.

Print *n* [ME *preinte*, fr. MF, fr. *preint*, pp of *preindre* to press, fr. L *premere*] (14c) (1) Name for etching, lithograph, woodcut, etc. (2) A fabric with designs applied by means of dyes or pigments used on engraved rollers, blocks, or screens. Also see ▶ Printing.

Printability *n* A collective term used to describe the properties required of all components in a printing process.

Print Bonding See ▶ Bonding (2).

Print Cloth *n* A medium weight, plain-weave fabric made of carded yarns, usually cotton or polyester/cotton blends, with counts from 28s to 42s. Millions of yards of print cloth are printed annually and other millions are finished as white goods. Large amounts of the goods are also used in the greige for bags, containers, and base fabric for coated materials.

Printing *n* (1) The process, art, or business of producing printed material by means of inked type and a printing press or similar means. (2) Forming a permanent impression in a semihardened paint film as a result of pressure from an object placed on it. (3) A process for producing a pattern on yarns, warp, fabric, or carpet by any of a large number of printing methods. The color or other treating material, usually in the form of a paste, is deposited onto the fabric which is then usually treated with steam, heat, or chemicals for fixation. Various types of printing follows. Also see ▶ Dyeing.

1. Methods of Producing Printed Fabrics:

Block Printing – The printing of fabric by hand, using carved wooden or linoleum blocks, as distinguished from printing by screens or roller.

Blotch Printing – A process wherein the background color of a design is printed rather than dyed.

Burn-Out Printing – A method of printing to obtain a raised design on a sheer ground. The design is applied with a special chemical onto a fabric woven of pairs of threads of different fibers. One of the fibers is then destroyed locally by chemical action. Burn-out printing is often used on velvet. The product of this operation is known as a burnt-out print.

Direct Printing – A process wherein the colors for the desired designs are applied directly to the white or dyed cloth, as distinguished from discharge printing and resist printing.

Discharge Printing – In "white" discharge printing, the fabric is piece dyed, then printed with a paste containing a chemical that reduces the dye and hence removes the color where the white designs are desired. In "colored" discharge printing, a color is added to the discharge paste in order to replace the discharged color with another shade.

Duplex Printing – A method of printing a pattern on the face and the back of a fabric with equal clarity.

Etching – See ▶ Burn-Out Printing.

Extract Printings – See ▶ Discharge Printing.

Heat Transfer Printing – A method of printing fabric of polyester or other thermoplastic fibers with disperse dyes. The design is transferred from preprinted paper onto the fabric by contact heat which causes the dye to sublime. Having no affinity for paper, the dyes are taken up by the fabric. The method is capable of producing well-defined, clear prints.

Ink-Jet Printing – Non-contact printing that uses electrostatic acceleration and deflection of ink particles released by small nozzles to form the pattern.

Photographic Printing – A method of printing from photoengraved rollers. The resultant design looks like a photograph. The designs may also be photographed on a silk screen which is used in screen printing.

Pigment Printing – Printing by the use of pigments instead of dyes. The pigments do not penetrate the fiber but are affixed to the surface of the fabric by means of synthetic resins which are cured after application to make them insoluble. The pigments are insoluble, and application is in the form of water-in-oil or oil-in-water emulsions of pigment pastes and resins. The colors produced are bright and generally fat except to crocking.

Resist Printing – A printing method in which the design can be produced: (1) by applying a resist agent in the desired design, then dyeing the fabric, in which case, the design remains white although the rest of the fabric is dyed; or (2) by including a resist agent and a dye in the paste which is applied for the design, in which case, the color of the design is not affected by subsequent dyeing of the fabric background.

Roller Printing – The application of designs to fabric, using a machine containing a series of engraved metal rollers positioned around a large padded cylinder. Print paste is fed to the rollers and a doctor blade scrapes the paste from the unengraved portion of the roller. Each roller supplies one color to the finished design, and as the fabric passes between the roller and the padded cylinder, each color in the design is applied. Most machines are equipped with eight rollers, although some have sixteen rollers.

Rotary Screen Printing — A combination of roller and screen printing in which a perforated cylindrical screen is used to apply color. Color is forced from the interior of the screen onto the cloth.

Screen Printing — A method of printing similar to using a stencil. The areas of the screen through which the coloring matter is not to pass are filled with a waterproof material. The printing paste which contains the dye is then forced through the untreated portions of the screen onto the fabric below.

Warp Printing — The printing of a design on the sheet of warp yarns before weaving. The filling is either white or a neutral color, and a grayed effect is produced in the areas of the design.

2. Methods of Producing Printed Carpets:

Millitron® Process — A computer-controlled, non-contact spray printing process that allows the production of intricate multicolored designs. Although this process was developed for carpets by Milliken & Co., it can also be used for upholstery, pile fabrics, and other textiles.

Mitter Printing Machine — A rotary carpet printing machine with up to eight stainless-steel mesh screens, and with cylindrical squeegees of moderately large diameter in each rotary screen. The unit has a streaming zone for dye fixation.

Stalwart Printing Machine — A carpet printing machine in which color is applied to the carpet with a neoprene sponge laminated to the pattern. The pattern is cut in a rubber base attached to a wooden roll. It is very similar to relief printing. Used primarily for overprinting random patterns on dyed carpets. Suitable for shags and plush carpets as well as level loop and needletuft types.

Zimmer Flatbed Printing Machine (Peter Zimmer) — A carpet printing machine that uses flat screens and dual, metal-roll squeegees. The squeegees are operated by electromagnets to control the pressure applied. The unit also has a steamer for dye fixation. The Zimmer flatbed machine is normally used for carpets of low to medium pile heights. Very precise designs are possible, but speeds are slower than with rotary screen printers.

Zimmer Rotary Printing Machine (Johannes Zimmer) — A three-step, rotary carpet printing machine consisting of: (a) rotary screens with small diameter steel-roll squeegees inside, with pressure adjusted electromagnetically for initial dyestuff application; (b) infrared heating units to fix dyes on the tips of the tufts; and (c) application of low-viscosity print paste, followed by steaming for complete penetration of dyes into tufts.

Zimmer Rotary Printing Machine (Peter Zimmer) — A rotary carpet printing machine in which each rotary screen has a slotted squeegee inside to feed print pastes through the screens to the carpet. Pressure of the print paste is adjusted by hydrostatic head adjustments. (Fairchild's dictionary of textiles. Tortora PG (ed). Fairchild Books, New York, 1997; Tortora PG, Merkel RS (2000) Fairchild's dictionary of textiles, 7th edn. Fairchild, New York; Complete textile glossary. Celanese Corporation, Three Park Avenue, NY, 2000; Elsevier's textile dictionary. Vincenti R (ed). Elsevier Science and Technology Books, New York, 1994)

Printing Ink *n* Any fluid or viscous composition of materials, used in printing, impressing, stamping, or transferring on paper or paper-like substances, wood, fabrics, plastics, films or metals, by the recognized mechanical reproductive processes employed in printing, publishing and related services. (Printing ink handbook. National Association of Printing Ink Manufacturers, 1976)

Printing on Plastics *n* Many methods commonly used on paper and other materials are also used for printing on plastics, with slight modifications such as the use of special links. Such processes are letterpress, offset, silk screen, electrostatic, and photographic methods. See also ▶ Electrostatic Printing, Flexographic Printing, Gravure Printing, Hot Stamping, Spanishing, and Valley Printing. Polyolefins are normally oxidatively treated before printing so as to make them receptive to inks. See ▶ Casing, Corona-Discharge Treatment, Flame Treating, and Ultraviolet Printing. (Printing ink handbook. National Association of Printing Ink Manufacturers, 1976)

Printing Plate *n* A surface carrying a design by which the ink is ultimately transferred to the material to be printed.

Printing Press *n* A mechanical device to apply ink to a surface, reproducing the pattern or design on the printing plate.

Printing Strength *n* A relative value indicating how much ink is required to given an equal depth of tint to a definite amount of white ink as compared with the same amount of a standard ink of the same consistency.

Print Paste *n* The mixture of gum or thickener, dye, and appropriate chemicals used in printing fabrics. Viscosity varies according to the types of printing equipment, the type of cloth, the degree of penetration desired, etc.

Print Resistance *n* The ability of a coating to resist taking on the imprint of another surface placed against it. See ▶ Printing (2).

Print Test See ▶ Print Resistance.

Prism (prismatic) \ˈpri-zəm\ *n* [LL *prismat-*, *prisma*, fr. Gk, literally, anything sawn, fr. *priein* to saw] (1570) Crystals made up of three, four, six, eight or 12 similar faces all parallel to a single axis.

Probability Density, ψ^2 *n* (1939) The probability of finding an electron in a small element of volume; the square of the wave function for an electron; the density of the electronic charge cloud.

Probability Density Function *n* (1957) (1) Probability Function. (2) A function of a continuous random variable whose integral over an interval gives the probability that its value will fall within the interval.

Probability Function *n* (1906) A function of a discrete random variable that gives the probability that a specified value will occur.

Probit \ˈprä-bət\ *n* [*prob*ability un*it*] (1934) A translation of origin of the scale of standard normal deviates (not a contradiction) to avoid the inconveniences of negative signs. For a given percentage point of the standard normal distribution, the probit = the corresponding standard normal deviate +5, i.e., e, $z + 5$. There is available normal probability paper that has a probit sale alongside the probability scale. The device is useful in plotting and discussing the results of testing by the Up-and-Down Method.

Processability \ˈprä-ˌse-sə-ˈbi-lə-tē\ *n* (1954) The ease with which a polymer, elastomer, or plastic compound can be converted to high-quality, useful products with standard melt-processing techniques and equipment. Some quantitative tests of processability have been devised; for example, see ▶ Molding Index and ▶ Thermoformability.

Process Characteristics See ▶ Processing Parameters.

Process Conditions See ▶ Processing Parameters.

Processing Additives See ▶ Processing Agents.

Processing Agents *n* Agents or media used in the manufacture, preparation and treatment of a material or article to improve its processing or properties. The agents often become a part of the material. Also called process media, processing aids, processing additives.

Processing Aid *n* A substance added to a compound to improve its behavior during processing. Many processing aids have been tried with rigid PVC because the neat resin is heat-sensitive, decomposing autocatalytically with evolution of toxic hydrogen chloride gas (CHl) at temperatures near 215°C. Processing aids may include heat stabilizers, lubricants, and other resins, even plasticizers.

Processing Aids See ▶ Processing Agents.

Processing Defects *n* Structural and other defects in material or article caused inadvertently during manufacturing, preparation and treatment processes by using wrong tooling, process parameters, ingredients, part design, etc. Usually preventable. Also called processing flaw, defects, flaw. See also ▶ Cracking.

Processing Flaw See ▶ Processing Defects.

Processing Methods *n* method names and designations for material or article manufacturing, preparation and treatment processes. Note – Both common and standardized names are used. Also called processing procedures.

Processing Parameters *n* Measurable parameters such as temperature prescribed or maintained during material or article manufacture, preparation and treatment processes. Also called process characteristics, process conditions, process parameters.

Processing Pressure *n* Pressure maintained in an apparatus during material or article manufacture, preparation and treatment processes. Also called process pressure. See also ▶ Pressure.

Processing Procedures See ▶ Processing Methods.

Processing Rate *n* Speed of the process in manufacture, preparation and treatment of a material or article. It usually denotes the change in a process parameter per unit of time or the throughout speed of material in a unit of weight, volume, etc. per unit of time. Also called process speed, process velocity, process rate.

Processing Time *n* Time required for the completion of a process in the manufacture, preparation and treatment of a material or article. Also called process time, cycle time. See also ▶ Time.

Process Inks *n* Used in reproducing illustrations by the halftone color separation process. The colors used base yellow, magenta (red) and cyan (blue); they are used with or without black.

Process Media See ▶ Processing Agents.

Process Parameters See ▶ Processing Parameters.

Process Pressure See ▶ Processing Pressure.

Process Rate See ▶ Processing Rate.

Process Speed See Processing Speed.

Process Time See ▶ Processing Time.

Process Variation *n* The degree to which measurements of the same process parameter, or characteristic or dimension of successive parts or products are different. See ▶ Standard Deviation and ▶ Range.

Process Velocity See ▶ Processing Rate.

Producer-Colored See ▶ Dyeing, Mass-Colored.

Producer's Risk *n* In quality control and acceptance sampling, the probability, under a given sampling plan, of making a Type I error, that is, of reflecting a lot whose true quality is at the desired acceptable level.

Producer-Textured Yarns *n* Continuous filament yarns that have been bulked during manufacturing by the fiber producer. Also see ▶ Texturing.

Producer Twist *n* Small amounts of twist, usually ½ turn per inch or less, applied to yarns by the manufacturer to provide cohesion of filaments for further processing.

Product \\'prä-(▪)dəkt\\ *n* [in sense 1, fr. ME, fr. ML *productum*, fr. L, something produced, fr. neuter of *productus*, pp of *producere*; in other senses, fr. L *productum*] (15c) A substance formed in a chemical reaction.

Profile \\'prō-▪fil\\ *n* [It *profilo*, fr. *profilare* to draw in outline, fr. *pro-* forward (fr. L) + *filare* to spin, fr. LL] (ca. 1656) (1) Any extruded product but those of the simplest cross sections, such as film, sheet, rod stock, pipe, and coated substrates. Examples of profiles are angle-stock and channels; square, triangular, and trapezoidal solids and annuli; house siding and refrigerator-door baskets. (2) The lineal variation of the smoothness/roughness of a finished surface. See ▶ Profilograph. (3) The pattern of variation of some process parameter over time, or more usually, distance. Examples are the channel-depth profile of an extruder screw and the temperature profile along an extruder cylinder.

Profile Angle *n* An angle, not necessarily an interfacial angle, used to describe a crystal. This angle is observed when the crystal is lying on a face. For example, a cube shows 90°; an octahedron 60° or 120°.

Profile Depth *n* Average distance between top of peaks and bottom of valleys on the surface of a coating.

Profile Die *n* A die used to form an extruded ▶ Profile. Two basic types are used: *plate* dies and *streamlined* dies. The former are cheaper to make and alter; the latter are essential when extruding rigid PVC and other heat-sensitive plastics, and are apt, with *any* compound, to permit higher extrusion rates of good product.

Profilograph *n* (profilometer) An instrument that measures the roughness of a surface, usually expressed as the local root-mean-square average in nanometer (or micrometer). The profile taken in any direction can be magnified and displayed graphically. (Reason RE (1970) The measurement of surface texture. Modern workshop technology, Part 2. The Macmillan, New York)

Progressive Aging *n* In a heat-aging test, stepwise raising of the temperature at preset time intervals.

Progressive Bonding *n* A method of curing thermosetting-resin adhesives in laminates or plywood slabs that are larger in area than the press platens in which they are being bonded. A partial area, say, a quarter of the laminate, is cured by application of heat and pressure. The press is then opened, and a different quarter of the laminate is moved between the platens and cured, and so on, until the entire laminate has been cured.

Progressive Proofs (Or Progs) *n* In color separation, a series of proofs of a color process reproduction pulled in each color, and in combinations of two, three, and four colors. Used to indicate color quality and as a guide for printing.

Projected Area *n* In molding, the area of a cavity, or all the cavities, or cavities and runners, perpendicular to the direction of mold closing force and parallel to the parting plane. In injection molding and blow molding, this area must be safely less than the quotient of the force applied to hold the mold closed divided by the maximum melt pressure or blowing pressure within the mold. In transfer molding, it must also be about 15% less than the cross-sectional area of the pot.

Projectile Loom *n* A shuttleless loom that uses small, bullet-like projectiles to carry the filling yarn through the shed. Fill is inserted from the same side of the loom for each pick. A tucked selvage is formed. Also see ▶ Weft Insertion

Projectiles *n* For bodies projected with velocity *v* at an angle *a* above the horizontal, the time to highest point of flight,

$$t = \frac{v \sin a}{g}$$

Total time of flight to reach the original horizontal plane,

$$= \frac{2v \sin a}{g}$$

Maximum height,

$$h = \frac{v^2 \sin^2 a}{2g}$$

Horizontal range,

$$R = \frac{v^2 \sin^2 2a}{g}$$

In the above equations the resistance of the air is neglected. *a* is the acceleration due to gravity.

(Handbook of chemistry and physics, 52nd edn. Weast RC (ed). The Chemical Rubber, Boca Raton, FL)

Promoter \-ˈmō-tər\ *n* (14c) (promotor) A chemical substance that, in very small concentrating, increases the activity of a ▶ Catalyst. The promotor may itself be a weak catalyst. Examples in the curing of polyester resins are cobalt octoate used as the promoter with methyl ethyl ketone peroxide, and *N*-alkyl anilines used with benzoÿl peroxide.

Proof \ˈprüf\ *n* [ME, alter. of *preove*, fr. OF *preuve*, fr. LL *proba*, fr. L *probare* to prove] (13c) A test photographic print or total impression in a printing process taken for correction or examination.

Proof *n*, **Apparent** The proof of a liquid as calculated from its specific gravity at 60°F. It is equivalent to the proof of a solution of pure alcohol and water having the same specific gravity at 60/60°F and the mixture in question. Since materials other than alcohol and water, such as denaturants or other soluble ingredients, affect the specific gravity of the solution, the apparent proof is not necessarily the true "alcohol proof" of the solution. (Russell JB (1980) General chemistry. McGraw-Hill, New York)

Proof Gallon *n*, **US** The amount of alcohol present in one wine gallon of 50% by volume of alcohol at 60°F. Proof gallons are calculated by multiplying the number of wine gallons at 60°F by the proof and dividing by 100. For example, one wine gallon of 190 proof alcohol contains 1.9 proof gallons. See ▶ Wine Gallon.

Proofing *n* The process of rubberizing fabrics, to render them impervious to water. It is an operation most commonly done by spreading a rubber cement of high viscosity or dough on the fabric, allowing the solvent to evaporate and curing in dry heat ovens or with sulfur chloride.

Proof Resilience *n* (energy to break) The work required to stretch an elastomeric test specimen from no elongation to its breaking point, expressed in joule per cubic centimeter of specimen volume.

Proof Spirit *n* (1790) (British) This corresponds with a definite mixture of absolute alcohol in water, and actually contains 49.24% by weight of alcohol. Sixty-four overproof (O.P.) industrial alcohol is commonly used for spirit varnish manufacture, containing approximately 90% of alcohol by weight.

Propagation \ˌprä-pə-ˈgā-shən\ *n* (15c) Chain propagation is the middle phase of any polymerization process during which monomers are extending polymer chain lengths by addition or condensation reactions. (Odian, G. C., *Principles of Polymerization*, John Wiley and Sons, Inc., New York, 2004)

Propanol See ▶ Propyl Alcohol.

2-Propanone See ▶ Acetone.

Propeller Mixer *n* A device comprising a rotating shaft with a propeller at its end, used for mixing relatively low-viscosity dispersions and holding contents of tanks in suspension. The propellers, of which there may be two or three on a single shaft, resemble boat propellers, having two to four broad, curved lobes. See also ▶ Paddle Agitator.

Propenal *n* Syn: ▶ Acrolein.

Propeneitrile *n* Syn: ▶ Acrylonitrile.

Propenoic Acid See ▶ Acrylic Acid.

Propiofan *n* Poly(vinyl propionate), manufactured by BASF, Germany.

Proportional Control *n* A method of controlling processes in which control action taken is proportional to the difference (process error) between the sensed state variable of a process and the desired target level of that variable. See ▶ On-Off Control.

Proportional Limit *n* The greatest stress a material is capable of sustaining without deviating from direct proportionality (linearity) between stress and strain (Hooke's law). See also ▶ Elastic Limit and ▶ Yield Point.

Proprietary Alcohol *n* Denatured ethyl alcohol.

Proprietary Solvents *n* Based on ethyl alcohol. Solvents containing more than 25% alcohol by volume which are manufactured from specially denatured alcohol, in accordance with authorized formulas. No permit is required to purchase proprietary solvents in the United States.

Propyl \ˈprō-pəl\ *n* {*often attributive*} [ISV *prop-* + *-yl*] (1850) C_3H_7. Either of two isomeric alkyl groups derived from propane.

Propyl Acetate *n* $C_3H_7COOCH_3$. Medium-boiling solvent used for nitrocellose. Bp, 102°C; sp gr, 0.897; flp, 12°C (53°F).

n-**Propyl Acetate** *n* (Propyl acetate) $C_3H_7OOCCH_3$. A clear, colorless liquid with a pleasant odor, used as a solvent for cellulosics, vinyls, acrylics, polystyrene, alkyds, and coumarone-indene resins.

Propyl Alcohol *n* $CH_3CH_2CH_2OH$. Used as a solvent and as a diluent for nitrocellulose lacquers. Bp, 97°C;

sp gr, 0.804/20°C; flp, 25°C (67°F). *Also known as n-Propanol.*

Propyl Benzoate *n* C$_6$H$_5$COOC$_3$H$_7$. Semipermanent plasticizer. Bp, 231°C.

Propyl Butyrate *n* CH$_3$CH$_2$CH$_2$COOC$_3$H$_7$. Medium-boiling solvent. Bp, 143°C.

Propyl Carbinol See ▶ *N-Butyl Alcohol.*
Propylene \ˈprō-pə-ˌlēn\ *n* (1850) H$_2$C=CHCH$_3$. 1-propene A flammable gas obtained from petroleum oils during the refining of baseline. Used in the polymerized form as polypropylene plastic.
Propylene Dichloride *n* C$_3$H$_6$Cl$_2$. Chlorinated hydrocarbon. Bp, 96°C; flp, 21°C (70°F); vp, 38 mmHg/20°C.

Propylene Glycol *n* (1885) CH$_3$CHOHCH$_2$OH. Dihydric alcohol used as an esterifying agent. Also used as a wet-edge additive. Bp, 187°C; sp gr, 1.038/20°C; flp 99°C (210°F); vp, 0.1 mmHg/20°C. *Also known as 1,2 Propanediol.*

1,2-Propylene Glycol Monolaurate *n* C$_{11}$H$_{23}$COOCH$_2$CH(OH)CH$_3$. A plasticizer for cellulosics, polystyrene, and vinyl resins.

1,2-Propylene Glycol Monoöleate *n* C$_{17}$H$_{33}$COOCH$_2$CH(OH)CH$_3$. A plasticizer for cellulose nitrate and ethyl cellulose.
Propylene Oxide *n* CH$_3$CH(O)CH$_2$ (1,2 –propylene oxide, 1,2 epoxypropane). A low boiling, liquid epoxide compound derived from the intermediate propylene chlorohydrin, which is itself produced by reacting propylene with chlorine and water. Propylene oxide is an important intermediate for the manufacture of polyglycols used for polyurethane foams and resins, and polyester resins.

Propylene Plastic See ▶ *Polypropylene.*
Propylene-Vinyl Chloride Copolymer *n* Any of a family of copolymers ranging from 2% to 10% by weight of propylene, that provides the application-properties advantages of PVC homopolymers plus the processing advantages attributable to the introduction of stable hydrocarbon structures as end groups, The copolymers are easy to mold and extrude, and have high thermal stability and low melt viscosity.
n-Propyl Oleate *n* C$_{17}$H$_{33}$COOC$_3$H$_7$. A monounsaturated fatty ester, and a plasticizer for ethyl cellulose, polystyrene, and, with limited compatibility, some vinyl and acrylic resins.

Propyl Propionate *n* CH$_3$CH$_2$COOC$_3$H$_7$. Medium-boiling ester solvent. Bp, 122°C; sp gr, 0.883; mp, −76°C.

Propyl Ricinoleate n $CH_3(CH_2)_5CHOHCH_2CH=CH(CH_2)_7COOC_3H_7$. Permanent plasticizer. Bp, 268°C/13 mm; sp gr, 9.908.

Protective Coatings A thin layer of metal or organic material, as paint applied to a surface, primarily to protect it from oxidation, weathering, and corrosion.

Protective Colloids n Materials such as gums, starches and proteins, polyacrylates and cellulose and cellulose derivatives, which are effective agents for protecting charged colloidal particles in aqueous media against flocculation.

Protective Factor (of an Antioxidant) n The proportion of millimoles of peroxide per kilogram in untreated oil to that in the oil containing an antioxidant.

Protein \ˈprō-ˌtēn *also* ˈprō-tē-ən\ n {*often attributive*} [F *protéine*, fr. LGk *prōteios* primary, fr. Gk *prōtos* first] (ca. 1844) Any of a group of complex nitrogenous organic compounds of high molecular weight that contain amino acids as their base structural units and that occur in all living matter and are essential for the growth of animal tissue. See ▶ Casein. (Merriam-Webster's collegiate dictionary, 11th edn. Merriam-Webster, Springfield, MA, 2004)

Protein-Aldehyde Resins n Plastic derived from casein and formaldehyde.

Protein Resin A generic term for resins derived from proteins, constituting ▶ Casein Plastics and ▶ Zein.

Protium \ˈprō-tē-əm\\n [NL, fr. Gk *prōtos* first] (1933) Ordinary hydrogen $^1_1 H$.

Proton \ˈprō-ˌtän\ n [Gk *prōton*, neuter of *prōtos* first] (1920) A nuclear particle that used to be considered elementary, having a positive charge equal to the negative charge of an electron but possessing a mass approximately 1,837 times that of an electron at rest, and slightly less than that of a neutron. The proton is in effect a hydrogen-atom nucleus.

Protonic Acids See ▶ Acids.

Prototype Mold n A temporary or experimental mold used to make a few samples to test product design or obtain market reactions. Such a mold is often made from a low-melting metal-casting alloy or from a filled and reinforced epoxy resin.

Protrusion \prō-ˈtrü-zhən\ n [L *protrudere*] (1646) Any raised area on a molded or painted surface, such as a blister, bump, or ridge. (Paint /coatings dictionary. Compiled by Definitions Committee of the Federation of Societies for Coatings Technology, 1978)

Prussian Blue \ˈpre-shən-\ n [*Prussia*, Germany] (1724) Brilliant deep blue pigment of excellent staining power, good lightfastness, but unstable in the presence of alkalis. It is usually obtained as a very fine powder. Its oil absorption is about 90. See ▶ Iron Blue. Syn: ▶ Erlangen Blue and ▶ Gas Blue.

Prussian Brown n Brown-colored, iron oxide pigment, obtained as a decomposition product of Prussian blue subjected to heat.

Prussiate of Iron See ▶ Iron Blue.

PS Abbreviation for Poly(Styrene).

PSB n Copolymer from styrene and butadiene.

Pseudoisohromatic Plate Test n General term applied to a type of test plates sued to determine defective or anomalous color vision. A PIC plate is a chromatic figure formed by dots on a background of different chromatic dots, varying sometimes in lightness and size. The figure is either an Arabic numeral or some other identifiable pattern. The simplest version is a dichotomous test which involves the perception of chromatic dot patterns on backgrounds of different chromatic dots, used to test for red–green confusion. More complex tests are double number types (color defectives see one number, normals a different number), special camouflage type (Ishihara), those designed to detect qualitatively type and degree of defect, and quantitative diagnostic types (Hardy-Rand-Rittler PIC Plate Test). Abbreviation: PIC Test.

Pseudomonas \-ˈmō-nəs\ n [NL, fr. *pseud-* + *monad-*, *monas* monad] (1903) A generic class of aerobic, mesophilic bacterium capable of releasing a variety of enzymes, including cellulose-decomposing "cellulose" enzymes; these enzymes are a factor in viscosity reduction of latex paints modified with cellulosic thickener, and may contribute to biodeterioration of paint films, enhancing their nutrient value for fungus (*mildew*) growth. (Black JG (2002) Microbiology, 5th edn. Wiley, New York)

Pseudoplastic Flow n Type of flow characterized by a consistency curve which shows no yield value (starts at the origin) and where the rate of flow increases faster than linearly with the shearing stress. See ▶ Viscosity.

Pseudoplastic Fluid *n* A solution or melt whose apparent viscosity decreases instantaneously and reversibly with increasing shear rate and stress without a yield stress (stress to initiate shearing). Most polymer solutions and melts are pseudoplastic. (Munson BR, Young DF, Okiishi TH (2005) Fundamentals of fluid mechanics. Wiley, New York) See also ▶ Power Law and ▶ Ellis Model. Pseudoplastic behavior is often confused with, and mistakenly labeled as ▶ Thixotropy. Most dispersions are pseudoplastic but with a yield stress or ▶ Yield Value, and this term is not to be confused with true pseudoplastic fluids. (Patton TC (1979) Paint flow and pigment dispersion: A rheological approach to coating and ink technology. Wiley, New York; Coussot P (2005) Rheometry of pastes, suspensions and grannular materials: Applications in industry and environment. Wiley, New York)

Pseudoplasticity *n* Time-independent shear thinning with no yield stress.

PSO SEE ▶ Polysulfones.

PSP *n* Abbreviation for Polystyrylpyridine.

PST *n* Poly(styrene) fiber.

PS-TSG *n* Injection-molding foam poly(styrene).

PSU *n* Abbreviation for Polysulfone.

Psychophysical \ˌsī-kō-ˈfi-zi-kəl\ *adj* (1847) Adjective used to describe the sector of color science which deals with the relationship between physical description or specification of stimuli and the sensory perception arising from them.

Psychrometer \sī-ˈkrä-mə-tər\ *n* [ISV] (1838) A wet-and-dry bulb type of hygrometer. Used for the determination of relative humidity.

PTA *n* (1) Abbreviation for Phosphotungstic Acid. (2) Applied also to toners and pigments which have been precipitated with phosphotungstic acid to give it permanence and insolubility.

PTA Pigment See ▶ Precipitated Basic Dye Blues.

PTB See Polybenzothiazole and Polybutylene Terephthalate.

p-t-Butyl Phenol *n* $(CH_3)_3CC_6H_4OH$. A white crystalline solid used as a plasticizer for cellulose acetate.

p-t-Butylphenyl Salicylate *n* A plasticizer approved by FDA for contact with foods, also used as a light-absorbing agent.

p-tert-Amyl Phenol *n* $(CH_3)_2C_2H_5CC_6H_4OH$. A white crystalline material made by alkylating phenol with amyl chlorides or amylenes, then separating by distillation. Resins made by reacting *p-tert*-amyl phenol with formaldehyde or paraformaldehyde are used in varnishes for wood, wire coating and coil insulation. They are also used as plasticizers and/or stabilizers in hot-melt adhesives based on ethyl cellulose.

PTF *n* Poly(tetrafluoroethylene) fiber.

PTFE Abbreviation for poly(tetrafluoroethylene).

PTFE Fluoroplastic *n* Polytetrafluoroethylene is prepared by free radical polymerization of tetrafluoroethylene in aqueous systems with persulphate or peroxide initiators to give granular or dispersion polymers. The polymers have exceptionally high thermal and thermo-oxidative stability and are completely solvent resistant. PTFEs are tough, relatively flexible materials which have outstandingly good electrical insulation properties as well as unusually low coefficients of friction.

PTHF *n* Abbreviation for Polytetrahydrofuran.

PTM *n* Abbreviation for Paint Testing Manual.

PTMA Pigment See ▶ Precipitated Basic Dye Blues.

PTMT See ▶ Poly(Tetramethylene Terephthalate).

p-type Semiconductor *n* A semiconductor in which the charged carriers are weakly localized holes (missing electrons).

PU *n* Polyurethane fiber. Abbreviation sometimes used in Europe for ▶ Polyurethane.

PUA Polyurea fiber.

Pucker \ˈpə-kər\ *n* [prob. irregular from ¹*poke*] (1750) Uneven surface caused by differential shrinkage of the yarns in a fabric or differential shrinkage of the fabric and sewing thread. A pucker may be desirable and planned, or undesirable. (Complete textile glossary. Celanese Corporation, Three Park Avenue, NY, 2000)

PUE *n* Segmented polyurethane fiber.

Pug Mill *n* Mill used for the preliminary mixing of pigments into oils or media to form stiff pastes prior to grinding. This process of mixing is sometimes known as *pugging*.

Pulforming *n* A modified pultrusion process developed to produce a changing volume/shape. See ▶ Pultrusion.

Pulldown See ▶ Drawdown.

Pulled-In Filling *n* An extra thread dragged into the shed with the regular pick and extending only a part of the way across the fabric.

Pulled Surface *n* Imperfections in the surface of a laminated plastic, ranging from a slight breaking or lifting of its surface in spots to pronounced separation of its surface from its body (ASTM D 883).

Puller *n* Any device used to pull an extrudate away from the extruder and through the cooling tank, playing a role in determining the dimensions of the product's cross section. See ▶ Caterpillar for a description of the kind most used.

Pulling *n* Resistance to the movement of a brush during the application of a material due to the viscous nature of the medium. Such a material is sometimes referred to as being sticky under the brush. See ▶ Drag.

Pulling Over *n* Process of leveling a cellulose lacquer film, usually on wood, by rubbing it with a soft cloth pad soaked in a mixture of organic solvents which is only a partial solvent for the lacquer film.

Pulling (under the Brush) See ▶ Drag.

Pulling Up *n* Action of a coat or paint or varnish which softens a previous coat to such an extent as to make brush application difficult and, in extreme cases, causes an objectionable intermingling of the two coats.

Pull-Out Strength *n* Of threaded inserts in plastics moldings, the force required to pull the insert out of the molding, It may be expressed as the force per unit area of the engaged outside surface.

Pull Strength *n* The bond strength of an adhesive joint, obtained by pulling in a direction perpendicular to the plane of the bond. This is an uncommon mode of test for adhesive bonds; the usual mode is to pull apart the ends of lap-joined specimens, thus testing the joint in shear. See ▶ Tensile-Shear Strength.

Pulp \\|pəlp\ *n* [ME *pulpe*, fr. MF *poulpe*, fr. L *pulpa* flesh, pulp] (14c) Press cake that has been further processed to yield a homogeneous paste of controlled solids content or controlled tinting strength for commercial sale.

Pulsed Positive/Negative-Ion Chemical Mass Spectrometry See ▶ Mass Spectrometry.

Pultrusion *n* A reinforced-plastics technique for continuously producing profiles of constant cross section, both solid and annular. Strands of reinforcing material are conveyed through a tank of resin – usually polyester but silicone and epoxy are also used – from which they are pulled through a long, heated steel die shaped to impart the desired profile. Both gelling and curing of the resin are sometimes accomplished entirely within the die length. Preheating of the resin-wet reinforcement is effected by dielectric energy prior to its entering the die, or heating may be continued in an oven after emergence from the die. In the past, the pultrusion process has mainly yielded continuous lengths of material with high unidirectional strengths, used for building siding, fishing rods, golf-club shafts, etc., but recent advancements in the technique permits multidirectional reinforcement and strengths. (Handbook of plastics, elastomers and composites, 4th edn. Harper CA (ed). McGraw-Hill, New York, 2002)

Pulverulent \\|pəl-\|ver-yə-lənt\ *adj* [L *pulverulentus* dusty, fr. *pulver-, pulvis*] (ca. 1656) Consisting of, or reducible to, a fine powder. Dusty, friable, crumbly.

Pumacite See ▶ Pumice.

Pumice \\|pə-məs\ *n* [ME *pomis*, fr. MF, fr. L *pumic-, pumex*] (15c) A highly vesicular (frothy), glassy, volcanic lava, usually rhyolitic (granitic) in composition; composed of complex aluminum, calcium, magnesium, iron, sodium and potassium silicates. Pumacite is the name for volcanic ash found in Kansas and Nebraska. Pumice is used as an abrasive, filler for plastics, polishing compounds and nonslip compounds. Density, 2.2 g/cm^3 (18.5 lb/gal). Syn: ▶ Pumacite.

Pumicing *n* A finishing method for molded plastics parts, consisting of the rubbing off of traces of tool marks and surfaces irregularities by means of wet pumice stones.

Pump Molding *n* A process by which a resin-impregnated pulp material is preformed by application of a vacuum and subsequently oven cured or molded. The pulp is first mixed with water and pumped into a tank wherein a mold, usually of wire mesh shaped like the finished article, is positioned. Air is evacuated from the mold to attract the pulp fibers, forming a preformed layer in contact with the screen. The mold is then removed from the vacuum tank, the pulp deposit is stripped off and dried, then, the preform is molded to final form by fluid pressure or conventional compression methods.

Pump Ratio *n* In single-screw extrusion with two-stage screws (as in vented operation), the ratio of the drag-flow capacity of the forward (final) pumping section to that of the rear (first) pumping section. This ratio is approximately equal to (but slightly less than) the ratio of the two pump depths, providing the lead angle is constant throughout, and is usually in the vicinity of 1.5.

Punching *n* A method of producing components, particularly electrical parts, from flat sheets of rigid or laminated plastics by cutting out shapes with a matched punch and die in a punch press.

Puncture Resistance *n* The ability of a plastic film or sheet to resist being penetrated by pointed objects. The most nearly relevant ASTM tests are two in which a specimen of films or sheets are punctured by not very pointy objects. One is D 1709, the free-falling dart method, in which a variably weighted dart having a hemispherical nose is dropped on a clamped specimen, a new specimen being used with each weight change and drop. By one of two testing techniques, the mean weight required for penetration is determined. In the other, more sophisticated test, D 3763, an instrumented plunger, also round-nosed, is forced at high speed through the clamped film or sheet specimen and load versus displacement trace is developed. (Handbook of plastics, elastomers and composites, 4th edn. Harper CA (ed). McGraw-Hill, New York, 2002)

PUR *n* The preferred (in US) abbreviation for ▶ Polyurethane.

Pure Black Iron Oxide See ▶ Black Iron Oxide.

Purging *n* In extrusion or injection molding, the cleaning of one color or type of material from the machine by forcing it out with the new color or material to be used in subsequent production, or with another compatible purging material. The operation goes faster when the purger is more viscous than the purgee. See also ▶ Dry Purge and Purging Compound.

Purging Compounds *n* A plastic compound especially designed to quickly purge most other plastics from an extruder or molder. It may contain organic fibers that help to scour the cylinder, and some purging compounds contain percentages of ultra-high-molecular-weight polyethylene which, because it does not actually melt in the extrusion, also tends to be an efficient purger.

Purified Stand Oil See ▶ Tekaol.

Purity, Colorimetric *n* Ratio of the luminance of the spectrum light, in mixture with the specified achromatic light required to match the light being described, to the luminance of the color itself. It is distinguished from excitation purity by the abbreviation P_c.

Purity, Excitation *n* Ratio of the straight line distance on a CIE chromaticity diagram between the chromaticity point of the sample and the achromatic or illuminant point on the diagram, to the linear distance between the point of intersection of this line with the spectrum locus and the illuminant point, It is properly designated as P_e, but is frequently abbreviation simply as P. The excitation purity, then, describes the relative distance from the neutral point and roughly corresponds in concept to the psychological description of saturation or chroma.

Purkinje Effect \(ˌ)pər-ˈkin-jē-\ A phenomenon associated with the human eye, making it more sensitive to blue light when the illumination is poor (less than about 0.1 lumen/ft^2) and to yellow light when the illumination is good.

Purl \ˈpər(-ə)l\ *n* [ME] (1526) (1) A knitting stitch that results in horizontal ridges across the fabric. It is made by drawing alternate courses through each side of the fabric. (2) A picot or small loop that edges needlework, lace, or ribbon. Sometimes spelled pearl. (Also see ▶ Picot.) (3) Coiled gold or silver thread used for embroidery. (Elsevier's textile dictionary. Vincenti R (ed). Elsevier Science and Technology Books, New York, 1994)

Purple Brown *n* Artificial red oxide pigment obtained by high-temperature treatment.

Purree See ▶ Indian Yellow.

Pushback Pin See ▶ Return Pin.

Pushing Flight *n* Syn: ▶ Leading Flight Face.

Push-Pull Molding *n* An injection-molding technique that uses twin injection units to fill a mold through well separated gates. By oscillating the advance and retraction of the injection screws or rams, the material in the mold is sheared and oriented, breaking up weld lines. It is particularly suited to molding of liquid-crystalline polymers.

Pushup *n* In the packaging industry, a container bottom with sufficient concavity to prevent rocking of the container when it is filled and placed on a flat surface.

Putty \ˈpə-tē\ *n* [F *potée* potter's glaze, literally, potful, fr. OF, fr. *pot* pot] (ca. 1706) (1) A heavy paste composed of pigment, such as whiting, mixed with linseed oil; used to fill holes and cracks in wood prior to painting, to secure and seal panes of glass in window frames. *Also called Painter's Putty, Glazing Compound*. (2) In plastering, a fine cement consisting of lump lime slaked with water; lime putty.

Putty Chaser See ▶ Edge Runner Mill.

PVA *n* Poly(vinyl ether). Abbreviation for either Polyvinyl Alcohol or Poly(Vinyl Acetate).

PVAC *n* Abbreviation for Poly(Vinyl Acetate).
PVAL *n* Abbreviation for Poly(Vinyl Alcohol).
PVB *n* Abbreviation for Poly(Vinyl Butyral).
PVC *n* (1) In the paint industry, abbreviation for pigment volume concentration. (2) Abbreviation for Poly(Vinyl Chloride).
PVCA *n* An abbreviation for copolymers of Vinyl Chloride and Vinyl Acetate.
PVD *n* A rarely used abbreviation for Polyvinyl Dichloride. See ▶ Chlorinated Polyvinyl Chloride.
PVDC *n* Poly(vinylidene chloride).
PVDF *n* Abbreviation for Poly(Vinylidene Fluoride).
PVF *n* Abbreviation for Poly(Vinyl Fluoride). The possibility of confusion exists because this abbreviation has been used in some literature for polyvinyl formal, for which the alternative abbreviations PVFM and PVFP have been employed.
PVFM, PVFO *n* Abbreviation for Poly(Vinyl Formal).
PVI *n* Abbreviation for Polyisobutylvinyl EtheR.
PVID *n* Poly(vinylidene cyanide).
PVK *n* Copolymer from vinyl ethers and vinyl chloride. Abbreviation for Poly(*N*-Vinylcarbazole).
PVM, PVME *n* Abbreviation for Poly(Vinylmethyl Ether).
PVOH *n* Abbreviation for Polyvinyl Alcohol. The abbreviation PVA is more commonly used.
PVP *n* Abbreviation for Polyvinyl Pyrollidone.
PVT Relationship *n* Pressure –(p) volume-(V) temperature-(T) relationship of boyle's law stating that the product of the volume of a gas times its pressure is a constant at a given temperature, $PV/T = R$, where R is Boltzmann constant.
PX *n* Abbreviation for *p*-Xylylene.
Pycnometer *n* \pik-ˈnä-mə-tər\ *n* [Gk *pyknos* + ISV – *meter*] (1858) A container whose volume is precisely known, used to determine the density of a liquid by filling the container with liquid and then weighing it. The same instrument may be used to measure the density of particular matter, such as plastic pellets, by immersing it in a liquid that is inert to, and significantly less dense than the solid matter. A dilatometer is a special pycnometer equipped with instruments to study SPECIFIC VOLUME as a function of temperature. See ▶ Weight-Per-Gallon Cup.
Pyramid (pyramidal) *n* \ˈpir-ə-ˌmid\ *n* [L *pyramid, pyramis*, fr. Gk] (1549) A group of three, four, six, eight, or twelve similar faces intersecting in a point or parallel to faces that would intersect in a point.
Pyranyl Foam *n* A type of rigid, pour-in-place, thermosetting foam similar to a polyurethane foam, but with superior resistance to high temperatures, It is formed in the same manner as polyurethane foams, using as the monomer a pyranyl (radical) derived from polypropylene by heating and oxidation to form an Acrolein dimmer, which ultimately forms the pyranyl.
Pyrazolone Red *n* $C_{36}H_{28}N_8O_6Cl_2$. Pigment Red 38 (21120). A metal-free diazo pigment based on a pyrazolone. A red powder used in the rubber, plastics, and flour products industries. Density, 1.35–1.58 g/cm^3 (11.3–13.2 lb/gal); O.A., 41–70.
Pyroabietic Acids *n* Acidic products obtained from the abietic acid of rosin by heating.
Pyrogenic Silica *n* See ▶ Fumed Silica.
Pyrogram *n* A chromatogram (see ▶ Chromatography) obtained from the pyrolysis products of a sample.
Pyrolysis \pī-ˈrä-lə-səs\ *n* [NL] (ca. 1890) Heating of a plastic or other material to temperature which cause decomposition and production of by-products; the process is temperature dependent.
Pyromellitic Dianhydride *n* (PMDA, 1,2,4,5-benzenetetracarboxlic anhydride) A triple-ring heterocyclic with the structure shown below,
PMDA is a curing agent for epoxies giving cured products of high deflection temperatures. It is the least costly starting material, reacted with diamines, for producing polyimides and related high-temperature-resistant polymers.

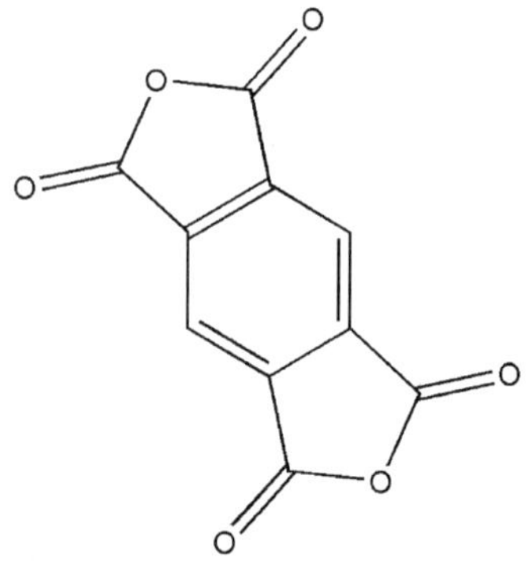

Pyrometer \pī-ˈrä-mə-tər\ *n* [ISV] (1796) (1) Instrument for measuring temperatures beyond the upper limit of the usual liquid thermometer. They may operate on the differential expansion of two metallic strips joined together, the measurement of changes of resistance, and the measurement of current flowing through two joined pieces of metal. In addition, radiation pyrometers are based on the measurement of heat radiated from a hot body, and optical pyrometers on the measurement of the intensity of light emitted from a hot body. (2) An ▶ Infrared Pyrometer.

Pyrophoric \ˌpī-rə-ˈfór-ik\ *adj* [NL *pyrophorus*, fr. Gk *pyrophoros* fire-bearing, fr. *pyr-* + *-phoros* carrying] (1836) Igniting spontaneously in air.

Pyroxylin \pī-ˈräk-sə-lən\ *n* [ISV *pyr-* + Gk *xylon* wood] (ca. 1847) Name given to the more soluble types of cellulose nitrate, and confined roughly to those containing less than 12.4% nitrogen. See ▶ Nitrocellulose.

Pyrrone See ▶ Polyimidazopyrrolone.

Q

Q \kyü\ *n* {*often capitalized, often attributive*} Symbol, in electronics, for the ratio of the reactance to the resistance of an oscillatory circuit, and then often called the *quality factor* of the circuit. $Q/2\pi$ = the ratio of energy stored to energy dissipated per cycle. A closely analogous measure applies to mechanical oscillating systems and, when the system is oscillating at or near its resonant frequency, Q is proportional to that frequency. (*College Physics*, McGraw Hill Science/Engineering/Math, New York, 2003)

QA Abbreviation for ▶ Quality Assurance.

QC Abbreviation for ▶ Quality Control.

QCT See ▶ Cleveland Condensing Humidity Cabinet.

Q-E Scheme The Q-E scheme is used for quantitatively correlating relative monomer reactivities in copolymerization reactions, introduced by Alfrey and Price in 1947 for the purpose of defining an equation for each cross-propagation rate constant (k_{12} or k_{21}), in a copolymerization reaction in terms of three constants characteristic of P is considered to be a function of the structures of the monomer: P, Q and e.

$$k = P_1 Q_2 exp(-e_1 e_2)$$

The constant P is considered to be a function of the reactivity of the radical only, and the constant Q is considered to be a function of the reactivity of the monomer only (both are determined by resonance effects), and the constant e is considered to be a reflection of the polar characteristics of both the radical and monomer (Odian GC, *Principles of Polymerization*, John Wiley and Sons, Inc., New York, 2004; Elias HG, *Macromolecules Vol. 1–2*, Plenum Press, New York, 1977; Tanford, C., *Physical Chemistry of Macromolecules*, John Wiley and Sons, Inc., New York, 1961).

Qiana /kē-än-ə/ (Trade name for silk like fiber material) Fiber from *trans*-diamino dicyclohexyl methane + dodecane dicarboxylic acid, manufactured by DuPont, U.S. New fibers should help the synthetics to capture an even bigger piece of the total. One of them is Qiana, introduced, after 20 years of experimentation, by E. I. du Pont de Nemours and Company, the largest U.S. producer of synthetic fibers. The new fiber has the appearance and feel of silk but has wrinkle-resistant and wash-and-wear properties that are as good or better than other synthetics. Qiana has a polyamide structure and thus is related to nylon fibers, but it is produced from different chemical ingredients and by different processes than previous nylons, according to du Pont. The use of Qiana is expected to be limited to high-fashion women's apparel at first. (Hounshell DA, Smith Jr. JK, *Science and Corporate Strategy: DuPont R&D, 1902–1980*, New York, 1988; DuPont Heritage: Innovation and Technological Development, www.heritage.dupont.com).

Q2 Polyamide from 1,4-*bis*(aminomethyl) cyclohexane + suberic acid. Manufactured from Eastman, U.S.

Quadripolymer (tetrapolymer) A rarely used term for the product of simultaneous polymerization of four monomers (*Complete Textile Glossary*, Celanese Corporation, Three Park Avenue, New York, NY).

Quadrupole Spectrometer \kwä-drə-pōl spek-trä-mə-tər\ *n* A type of mass spectrometer with two dipoles that provide better separation of ionic masses than can a single dipole. Carrying the idea even further are modern triple-quadrupole spectrometers. See ▶ Mass Spectrometry.

Qualification Test *n* A series of tests conducted by the procuring activity, or an agent thereof, to determine conformance of materials, or materials system, to the requirements of a specification which normally results in a qualified products list under the specification. NOTE – Generally, qualification under a specification requires conformance to all tests in the specification, or it may be limited to the conformance to a specific type or class, or both, under the specification.

Quality \kwä-lə-tē\ *n* [ME *qualite* fr. OF *qualité*, fr. L *qualitat-*, *qualitas*, fr. *qualis* of what kind; akin to L *qui* who] (14c) For acoustical purposes, the quality or timbre of a sound depends on the coexistence with the fundamental of other vibrations of various frequencies and amplitudes. See ▶ Seconds and ▶ Yarn Quality.

Quality Assurance *n* (1982) (QA) A system of activities whose purpose is to provide assurance, with documentation, that the overall quality-control function for any product, operation, service, or entire organization is in fact being accomplished.

Quality Characteristic Any dimension, property, aspect of appearance, surface finish, or performance specification that helps to determine the acceptability of a product or its ability to perform particular design functions. Most quality characteristics are measurable and therefore objective, but some, such as odor or texture, may be subjective and may be determined by the judgment of an expert or a panel of potential consumers of the product.

Quality Control *n* (1935) (QC) The techniques, measurements, and other activities that monitor and maintain product-quality characteristics within stated limits. These means require sampling of the product, measurement of important quality characteristics, statistical analysis, continuous presentation of the data (typically by means of *control charts*), and taking of decisions on whether lots are to be accepted, reworked, or scrapped.

Quantitative Analysis \ ˈkwän-tə- ˌtā-tiv-\ *n* (ca. 1847) A branch of chemistry, encompassing very many methods and techniques, whose scope is to determine the amounts of the different elements in substances or the percentages of molecular entities in a mixture of gases, liquids, or solids. (Harris DC, *Quantitative Chemical Analysis*, W. H. Freeman Co., New York, 2002; DeLevie R, *Principles of Quantitative Analysis*, Mc-Graw-Hill Higher Education, New York, 1996)

Quantitative Differential Thermal Analysis Differential Thermal Analysis in which the equipment used is designed to produce quantitative results in terms of energy and/or other physical parameters (ISO). (*Handbook of Thermal Analysis and Calorimetry*, Kemp RB, Elsevier Science and Technology Books, New York, 1999)

Quantity of Electricity or Charge The electrostatic unit of charge, the quantity which when concentrated at a point and placed at unit distance from an equal and similarly concentrated quantity, is repelled with unit force. If the distance is one centimeter and force of repulsion one dyne and the surrounding medium a vacuum, we have the electrostatic unity of quantity. The electrostatic unit of quantity may be defined as that transferred by electrostatic unit current in unit time. The quantity transferred by one ampere in one second is the coulomb, the practical unit. The faraday is the electrical charge carried by one gram equivalent. The coulomb = 3 × 10⁹ electrostatic units. Dimensions –

$$[e^{\frac{1}{2}} m^{\frac{1}{2}} l^{\frac{3}{2}} t^{-1}]; [\mu^{-\frac{1}{2}} m^{\frac{1}{2}} l^{\frac{1}{2}}]$$

(*CRC Handbook of Chemistry and Physics*, Lide DR (ed) CRC Press, Boca Raton, FL, 2004 Version).

Quantization of Energy The restriction of the energy of a system to certain specific, discrete amounts.

Quantum \ ˈkwän-təm\ *n* [L, neuter of *quantus*] (1567) Unit quantity of energy postulated in the quantum theory. The *photon* is a quantum of the electromagnetic field, and in nuclear field theories, the *meson* is considered to be the quantum of the nuclear field.

Quantum Mechanics *n* {plural but singular or plural in construction} (1922) The branch of physics which describes the behavior of small particles by assigning wavelike properties to them. *Also known as wave mechanics.*

Quantum Number *n* (1902) A number used to describe the state of an electron (*CRC Handbook of Chemistry and Physics*, Lide DR (ed), CRC Press, Boca Raton, FL, 2004 Version).

Quarry Tile Unglazed tile, usually six square inch or more in surface area and 0.5 to 0.75 in. (1.3–1.0 cm) in thickness made by the extrusion process from natural clay or shales.

Quartz \ ˈkwórts\ *n* [Gr *Quarz*] (ca 1631) The most common of minerals, of the rhombohedric crystal habit, occurring in a myriad of minerals and colors, but in its purest form, silicon dioxide (SiO_2), colorless, clear, very hard and transparent to both visible and ultraviolet light. The crystallized silica, when reduced to powder, is used as an extender. The term is also used for synthetically produced, amorphous fused quartz or vitrified Silica. See ▶ Silica and ▶ Crystalline. (*McGraw-Hill Dictionary of Geology and Mineralogy*, McGraw-Hill, New York, 2002; Nesse WD, *Introduction to Optical Mineralogy*, Oxford University Press, New York, 2003; Callister WD, *Materials Science and Engineering*, John Wiley and Sons, New York, 2002)

Quartz Fiber Fiber produced from natural quartz crystals of high purity (99.95% SiO_2). Quartz melts at 1610°C and is immune to thermal shock. Quartz- and silica-fiber reinforced composites are used in jet aircraft, rocket nozzles, and reentry nose cones. Quartz whiskers are also in use where their high cost is justified. They are among the strongest and stiffest of all fibers, comparable with graphite whiskers, with strength of 21 GPa and modulus of 700 GPa. Density is 2.65 g/cm³. (Callister WD, *Materials Science and Engineering*, John Wiley and Sons, New York, 2002; *McGraw-Hill Dictionary of Geology and Mineralogy*, McGraw-Hill, New York, 2002)

Quaternary Ammonium Compound *n* (ca 1934) Any of numerous strong bases and their salts derived from ammonium by replacement of the hydrogen atoms with organic radicals and important especially a surface-active agents.

Quaterpolymer The IUPAC term for a ▶ Copolymer derived from four species of monomers.

Quench \ ˈkwench\ *v* [ME, fr. OE –*cwencan*; akin to OE – *cwincan* to vanish, Old Frisian *quinka*] (12c) (1) A box filled with water into which fabric is run after singeing to prevent sparks or fires. (2) See ▶ Cabinet. Also see ▶ Quenching.

Quench Bath The cooling medium used in ▶ Quenching.

Quenching A process of shock cooling thermoplastic materials from the molten state, usually done experimentally with thin films of crystal-forming polymers in order to minimize the crystalline content and to study the nearly amorphous material.

Quench Spacer The "quiet" zone below the spinneret in which there is no quench airflow. Quench spacer distance is important in controlling fiber orientation and birefringence.

Quench-Tank Extrusion An extrusion process wherein the extrudate is conducted through a water bath for rapid cooling.

Quercitron \ ˈkwər-ˌsi-trən, ˌkwər-ˈ\ n [blend of NL *Quercus* and ISV *citron*] (1794) $C_{15}H_{10}O_7C_{21}H_{20}O_{12}$. (1) The inner part of a North American black oak tree (*Quercus velutina*), containing tannin and used in tanning and dyeing. (2) A yellow dye made from this bark. (3) Natural yellow matter obtained from the bark of *Quercus nigra* (oak) or *Quercus tinctorial*, and it is used primarily for the production of yellow lake pigments. The principal constituents are quercitrin and quercetin (*Merriam-Webster's Collegiate Dictionary*, 11th Ed., Merriam-Webster, Inc., Springfield, Massachusetts, U.S.A., 2004).

Quetsch The nip rollers of a padding machine (*Complete Textile Glossary*, Celanese Corporation, Three Park Avenue, New York, NY).

Quick-Burst Pressure (of a pipe, tube, or pressure vessel) See ▶ Hydrostatic Strength.

Quick-Hardening Lime A hydraulic lime.

Quicklime \ ˈkwik-ˌlīm\ n (14c) See ▶ Calcium Oxide.

Quickpeek A simple piece of equipment by means of which reproducible offset or letterpress ink films can be obtained (*Printing Ink Handbook*, National Association of Printing Ink Manufacturers, Inc., 1976).

Quill \ ˈkwil\ n [ME *quil*; akin to MHGr *kil*] (15c) A light, tapered tube of wood, metal, paper, or plastic on which the filling yarn is wound for use in the shuttle during weaving. (*Elsevier's Textile Dictionary*, Vincenti R (ed), Elsevier Science and Technology Books, New York; Joseph ML *Textile Science*, 5th Ed., CBS College Publishing, New York, 1986)

Quilling vt (1783) The process of winding filling yarns onto filling bobbins, or quills, in preparation for use in the shuttle for weaving (*Complete Textile Glossary*, Celanese Corporation, Three Park Avenue, New York, NY).

Quilting n (1609) (1) A fabric construction consisting of a layer of padding, frequently down or fiberfill, sandwiched between two layers of material and held in place by stitching or sealing in a regular pattern across the body of the composite. (Also see ▶ PINSONIC® Thermal Joining Machine.) (2) The process of stitch bonding a batting or composite, (*Complete Textile Glossary*, Celanese Corporation, Three Park Avenue, New York, NY).

Quinacridone Reds and Magentas Pigments of outstanding lightfastness and other resistance properties.

Quinacridone Golds These are very transparent, red-shade yellows of particular interest in formulating the best quality metallic finishes; often in blends with transparent iron oxides and or other, less transparent, organic pigments. See ▶ Quinacridone Pigments. (*Paint/Coatings Dictionary*, Compiled by Definitions Committee of the Federation of Societies for Coatings Technology, 1978)

Quinacridone Pigments A family of organic pigments based on substituted and unsubstituted forms of linear *trans*-quinacridones. Colors available include several shades of red, violet, gold, orange, magenta, and maroon. These pigments have good light-fastness, intensity of hue, resistance to bleeding and chemical attack, good transparency, and heat resistance. Quinacridones are unique pigments that are utilized in many diverse applications. Their outstanding light fastness, excellent bleed and heat resistance in combination with bright tones, good tinting value and working properties permit them as a right candidate in color formulation in the coatings, plastic, textile and ink industries to achieve the highest degree of quality performance. These pigments most closely parallel the phthalocyanines in properties. Quinacridone pigments are nonbleeding, heat and chemical resistant, and give outstanding exterior durability, even in light shades. However, their cost is high. Depending on substitution and crystal form, a variety of orange, maroon, scarlet, magenta and violet colors are available. Large particle size grades are used when opaque pigments are needed, and fine particle size grades for use in metallic automotive top coats (*Printing Ink Manual*, 5th Ed., Leach RH, Pierce RJ, Hickman EP, Mackenzie MJ, Smith HG, Blueprint, New York, 1993; *Organic Coatings Science and Technology*, 2nd Ed., Wicks ZN, Jones FN, Pappas SP, Wiley-Interscience, New York, 1999; Herbst W, Hunger K; *Industrial Organic Pigments*, John Wiley and Sons, Inc.).

Quinacridone Violet A versatile member of the quinacridone pigment family. It is used as a toning pigment; used to neutralize yellow tones in whites. Various shades from yellowish red to bright violet, resistant to most environmental conditions, very lightfast and used where superb properties are required, but very expensive in comparison to other pigments of the same hue (*Printing Ink Manual,* 5th Ed., Leach RH, Pierce RJ, Hickman EP, Mackenzie MJ, Smith HG, Blueprint, New York, 1993). It is also used in blends with molybdate orange to obtain relatively low-coat, durable, nonbleeding, bright reds. Its high level of transparency is also very desirable for styling automotive metallized finishes. For maximum saturation when toning phthalocyanine blues to redder hues, the quinacridone reds or magentas may be preferred to the violet. *(Paint/Coatings Dictionary,* Compiled by Definitions Committee of the Federation of Societies for Coatings Technology, 1978)

Quinol \kwi-nᵊl\ See ▶ Hydroquinone.

Quinone \kwi-ˈnōn, ˈkwi-\ *n* [ISV *quinine* + *-one*] (1853) See ▶ p-Benzoquinone.